Technical
Report
Writing
Today

FOURTH EDITION

Technical Report Writing Today

Steven E. Pauley

Daniel G. Riordan

HOUGHTON MIFFLIN COMPANY Boston

Dallas   Geneva, Illinois   Palo Alto   Princeton, New Jersey

Cover photograph © 1983 by Lou Jones.

Library of Congress Catalog Card Number: 89-80953

ISBN: 0-395-43355-X
ABCDEFGHIJ-RM-96543210-89

Acknowledgments

10 and 48–49 St. Croix Valley Memorial Hospital From the "Dr. Heart Plan." Reprinted by permission of St. Croix Valley Memorial Hospital.

74–75 *Management Technology* From Paul Finney, "A Layman's Guide to Network Bafflegab (A Glossary)." Copyright © 1983 *Management Technology*. Used by permission of Management Technology/Teleconnect.

99 Dialog Information Services, Inc. © Dialog Information Services, Inc. 1984, 1985, 1986. All rights reserved. Reproduced with permission.

110 *Packaging* "New Generation of Anti-Static Foams." Copyright © 1986 *Packaging*. Used by permission of *Packaging*.

148 and 150 Belinda Collins, et al. From *A New Look at Windows* NBSIR 77-1388. Used by permission of U.S. Department of Commerce, National Bureau of Standards, Center for Building Technology.

177 Honeywell, Inc. Reprinted by permission of Honeywell, Inc.

236–237 Used by permission.

243–244 Used by permission.

384 William P. Coldwell Letter reprinted by permission of the author.

ABCDEFGHIJ-RM 9543210/89

Contents

3   Defining Audience   40

4   Technical Writing Style   55

SECTION TWO   Technical Writing Techniques

5   Researching   81

8   Definition   160

9   Description   175

10   Sets of Instructions   201

SECTION THREE Technical Writing Applications

11 Memorandums and Informal Format 223

12   Formal Format   247

13   Recommendation and Feasibility Reports   272

16 Operator's Manuals 329

17 Oral Reports 354

SECTION FOUR   Professional Communications

18   Letters   369

19   Job Application Materials   387

To the Instructor

Technical Report Writing Today, Fourth Edition, continues to be a complete yet concise text. Thoroughly updated, this edition is appropriate for today's technical writing classes, where students' majors may represent a wide range of fields — engineering, business, industrial technology, and service professions, to name a few. The real-world examples and models, both student and professional, from many different fields; the accessible style of the text; and the many exercises and writing assignments help students focus on the writing demands they will face both in college and on the job.

NEW TO THE FOURTH EDITION

The aim of this book has always been to introduce students to professional and technical writing by helping them to internalize the skills and standards necessary to produce good, clear writing. Concise yet topically complete, the text emphasizes writing and producing technical documents. Streamlined treatment of conceptual material on audience, purpose, and the writing process moves students quickly into mastering technical writing skills such as definition and description and guides students in applying skills in such writing tasks as memos, informal reports, and proposals. The thoroughly revised fourth edition includes the following:

- **Added and improved coverage** of topics of growing importance in the field of technical writing, such as formatting documents, word processing, creating and using visual aids, and group writing
- **Streamlined presentation** that uses lists and visuals to help students find, use, and retain information and guidelines
- **New chapter** (Chapter 6), "Summarizing and Outlining," that teaches two skills important to technical writers
- **Many new examples** from a variety of fields, with marginal annotations that show how the writer successfully handles important issues

- **New full-length models** of technical reports that students can emulate
- **Worksheets** in each chapter that prompt students to plan and revise their reports with attention to audience, organization, format, and visual aids

Exercises and writing assignments throughout *Technical Report Writing Today* stress the issues presented in the chapter while allowing students to use their own interests and expertise in writing.

ANCILLARY MATERIALS

The new **Instructor's Manual** provides the following for each chapter in the text: an abstract, teaching suggestions, and comments on each exercise and writing assignment.

A package of twenty overhead **transparencies** consists of exercises and projects appropriate for in-class use. A section of the Instructor's Manual gives suggestions for using the transparencies. The transparencies are available upon adoption of the text by contacting your local Houghton Mifflin representative.

ACKNOWLEDGMENTS

Technical writing teachers from around the country offered valuable and insightful comments about the manuscript:

Susan Ahern   University of Houston — Downtown
Faye Angelo   Hinds Junior College, Mississippi
Virginia Book   University of Nebraska — Lincoln
Rita Bova   Columbus State Community College
Alma G. Bryant   University of South Florida
Stephen Flaherty   DeVry Institute of Technology, Ohio
Crystal Gromer   Vermont Technical College
Marshall Kremers   New York Institute of Technology
Mark Larson   Humboldt State University, California
Sherry Burgus Little   San Diego State University
Peter J. McGuire   Georgia Institute of Technology
Patricia McKinney   Oregon State University
William S. Pfeiffer   Southern College of Technology, Georgia
Scott P. Sanders   University of New Mexico
Pearl I. Saunders   Saint Louis Community College at Florissant Valley

Henrietta Nickels Shirk   Northeastern University — Haverhill, Massachusetts
Brenda Sims   University of North Texas
Sherry Southard   Oklahoma State University
Mel A. Topf   Roger Williams College, Rhode Island
Mark Weadon   U.S. Air Force Academy

In addition, we thank our students, who over the years have demanded
clear answers and clear presentations. For allowing us to reprint their
work, we thank these students:

Jill Adkins	Karen Johnson
Diana Albers	Philip Kuhns
Keith C. Balke, Jr.	Holly B. Maas
Jeffrey D. Barsness	Laurie L. Meyer
Jon Bauch	Michelle Migas
Gerry R. Bentzler	David Mitchell
Rich Biehl	Michael B. Mundy
Thomas Blodgett	Russell Nicol
Danette Boezio	Debra L. Peer
Richard A. Brueckner	Karen Pernsteiner
Timothy C. Buege	Pegeen S. Planton
Brenda L. Buske	Beth Reid
Julie Carmody	Frank Ries
Charles C. Clark	Edward R. Salmon
Ann M. Cleereman	Tom Schendel
Marcia L. Cody	Brian Schmitz
Jeff Davis	Steven Schneeburger
William Diesch	Connie Seilheimer
Jeff Edwards	David A. Siskoff
Louise Esaian	Julie Smith
Craig Ethier	Rick Stephen
Curt Everson	Jerry Swita
Frank L. Evertz	Ted J. Umentum
Joseph A. Frank	Steven P. Vande Walle
Brian Frederickson	Matthew R. Vander Wegen
John F. Furlano	David Weissinger
Scott Gerondale	Jeanne Wendlandt
Todd Hainer	Liz Wessley
Kevin L. Harris	Janet L. Wilson
Cynthia Hauswirth	Pat Wingo
Karen M. Hoff	John K. Wise
Karl S. Jerde	Ted J. Umentum
Daniel W. Johnson	

The following people of the University of Wisconsin-Stout deserve thanks for their gracious help: Brooke Anson, Ray Barlow, Gerane Dougherty, Noel Falkofske, Denise Madland, Dr. Art Muller, and Janet Polansky. And finally, thanks to family — Mary, Nathan, Clare, Simon, and Jane Riordan — who offered encouragement and support throughout the project.

S.E.P.
D.G.R.

Technical Report Writing Today

Technical Writing Basics

1

Technical Writing

WRITING IN THE WORKPLACE
PRELIMINARY DEFINITIONS
DEFINITION OF TECHNICAL WRITING

In industry and business today, technical writing is an important part of everyone's career. People write to propose projects, to document their own actions, to help others understand the results of research, to analyze and solve problems, to describe procedures and objects. Done well, technical writing is an exciting, fulfilling experience. Done poorly, it is frustrating, even harmful to career advancement.

This chapter will introduce you to this rewarding and challenging dimension of your professional life. This chapter reviews the importance of writing in professional careers, defines basic concepts that this book uses to discuss writing, and defines technical writing.

WRITING IN THE WORKPLACE

In study after study, researchers have discovered that college graduates report similar trends about the writing they do on the job. The graduates surveyed consistently stress that they spend much time writing and that writing affects job performance and advancement. They also consistently mention as important certain skills and the ability to write certain kinds of documents.

The Importance of Time Spent Writing

People in careers do a lot of writing. College graduates spend the equivalent of one day per week — and often more — writing. Carol Barnum and Robert Fischer, who interviewed engineering technologists, found that 70 percent write a minimum of one day a week (9). Gilbert Storms, who surveyed business school graduates, found that 24 percent write one-third or more of their time and 57 percent write one-half to one full day of every week (14).

Writing is extremely important for moving ahead in any profession. As Figure 1.1 shows, almost all college graduates feel that writing is important, and three-quarters feel that writing is either *very* or *critically important* in their jobs.

Graduates and employers also feel that writing ability helps individuals gain promotions. Barnum and Fisher found that nine out of ten graduates feel that "the ability to communicate has helped in their advancement" (10). In another survey, one employer was very blunt: "Good writing skills are critical to career advancement. Without such skills ad-

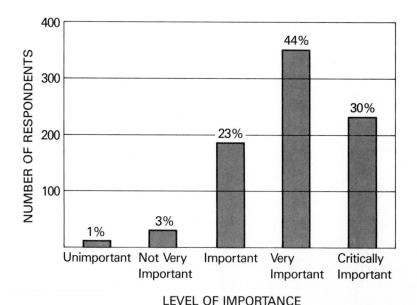

FIGURE 1.1
Importance of Writing in the Workplace
Source: Gilbert Storms, "What Business School Graduates Say about the Writing They Do at Work," *ABCA Bulletin* 46.4 (1983):15. Used by permission.

vancement opportunities are limited" (Minors 6). Another study reported that employers feel this way: "Communication skills are as important or — in the minds of many top executives — more important for their [entry level employees'] advancement than their engineering skills" (Cronin 82).

Types of Documents/Types of Skills

Whatever the field, college graduates report that they write the same types of documents and need the same writing skills. The most common types of documents are

> memos
> letters
> short reports

Graduates also write progress reports, proposals to customers, technical descriptions, instructions, in-house proposals, long reports, and scripts for speeches or presentations (Barnum and Fischer 11, Storms 16).

The most important writing skills are

> Knowing how to organize
> Writing clearly

The other skills graduates frequently mention are structuring sentences, formatting reports, stating purpose to reader, and writing concisely (Barnum and Fischer 11).

The rest of this book explains the writing types and skills that you need to know, and provides many opportunities to practice them. As you practice your writing, you will grow more confident in your ability to generate clear, effective technical writing, and you will be investing in your own career.

PRELIMINARY DEFINITIONS

Before you begin to practice technical writing, you must understand several basic concepts. These concepts are *communication, medium, document,* and *generate.*

Communication is the act of transmitting an idea from one person to another. Communication always requires at least two people — the sender and the receiver of the message. In this book, the sender is the writer, and the receiver is the reader or audience. Communication succeeds when the audience, or reader, understands the message exactly (or almost exactly) as the receiver intended. If I want you to pick up your hand, or attend a meeting, or know the reasons why I want to attend a convention, I need to transmit that idea to you. When you pick up your hand, or attend the meeting, or agree or disagree with my reasons, you indicate that you have received my transmission, or that I have communicated with you.

A *medium*, literally "a thing in the middle," is the means by which I convey ideas from me to you (see Figure 1.2). Since you cannot read my mind, I send you a message through a medium. I record or "encode" my message in the medium and then transmit it to you. You complete the transmission by interacting with the medium to understand or "decode" the message. If I say, "Be careful, that's hot!" I transmit the message by the medium of sound waves, which you interact with and decode. If I merely think, "Be careful, that's hot!" but do not use a medium, you will not receive the message. In addition to sound waves, other media include writing, visual representations ranging from photographs to abstract art, and even dance and body position. While this book focuses on writing , it also explains how to use visual art and speech to communicate. You will learn how to control these media in order to express your ideas clearly.

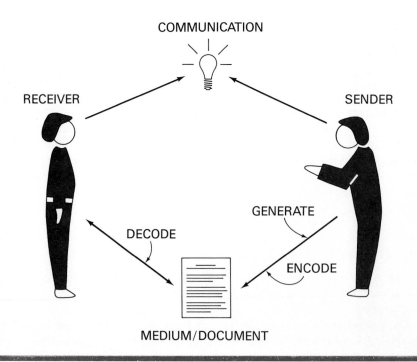

FIGURE 1.2
The Communication Situation

Document is the generic term for a written object. Other terms, such as *essay* or *paper* or *work*, do not clearly reflect the many types of technical writing. For instance, calling a set of instructions an *essay* or a *paper* is not really accurate, and calling them a *work* sounds too artistic. Throughout this book, the term *document* will designate any individual piece of technical writing.

Generate means to perform all the activities that result in a final document. Since these activities often include choosing or creating visual aids and designing pages for visual effect, *write* is not always an accurate term, yet *create* sounds more appropriate for poetry. *Generate* can mean just writing, and the term will sometimes be used that way in this book, but it often means a process that includes more than decisions about language.

DEFINITION OF TECHNICAL WRITING

Technical writing is difficult to define. Researchers in the field simply have not agreed on a definition. To help you, though, this section proposes an operational definition and then explains technical writing's purposes and characteristics.

An Operational Definition of Technical Writing

Technical writing is the practical writing that people do as part of their jobs. Whatever their position — from executive to middle manager, from specialized research scientist to secretary — people generate documents as an expected part of their responsibilities. These documents enable businesses; corporations; and public agencies, including governmental units, schools, and hospitals, to achieve their goals and maintain their operations. Some documents may be brief — a memo on the progress of a project could be one page long. Other documents are lengthy — a report on the feasibility of a certain site for a new plant could be hundreds of pages long; some military manuals are thousands of pages long. Regardless of the type of document produced or its length, technical writing is a medium that transmits the knowledge people need to fulfill their roles in organizations.

Purposes of Technical Writing

Although technical writing occurs in many places and takes many forms, it has two basic purposes: to inform and to persuade. Most technical writing informs. To carry out their functions in the workplace, people must supply or receive information constantly. They need to know or explain the scheduled time for a meeting, the division's projected profits, the physical description of a new machine, the steps in a process, the results of an experiment.

Technical writing also persuades. On the job, people must persuade others to follow certain courses of action. A writer not only describes two sites for a factory but also persuades readers to accept one of them as the best. Another writer describes a problem in a certain situation — perhaps a bottleneck in a production process — then persuades readers to implement the solution he or she proposes. In a slightly different vein, persuading is teaching or instructing. Someone must tell consumers how to use their new purchase, whether it is a clock radio or a mainframe computer. Someone must tell medical personnel exactly how to react when a patient has a heart attack in a hospital. The documents that achieve these purposes are called technical writing.

Characteristics of Technical Writing

Technical writing has four common characteristics (Cunningham). It engages a specific audience; uses plain, objective language; stresses clear organization; and uses visual aids.

Specific Audience   Technical writing engages a specific audience about a practical matter. The workers who must attend a meeting receive the memo that announces it. The consumer who must program the VCR receives the manual that explains the process (see Figure 1.3). The executive who must choose between two alternatives receives the feasibility report that explains them. The reader receives the document because of his or her role in the situation. If the reader has no role in the situation, he or she neither receives nor searches out the document. Knowing this, good technical writers always generate documents whose goal is to address the needs of specific readers.

Objective Language   Technical writing is written in plain, objective language. Since its purpose is to inform or persuade a reader about a specific practical matter, technical writing focuses the reader's attention on the relevant facts. The reader should respond only to the subject, whatever it is. As much as possible, the words should not cause the readers to add their own personal interpretations to the subject.

Contrast this kind of writing with writing designed to engage emotions — poems, novels, plays, and reflective essays. Here is a description of a death from the novel, *A Thief of Time,* by Tony Hillerman. Notice that the wording is designed to engage the reader's feelings, to help the reader feel the stunning finality of death.

"There's no good way to tell you this, Mr. Leaphorn," the voice had said. "We lost her. Just now. It was a blood clot. Too much infection.

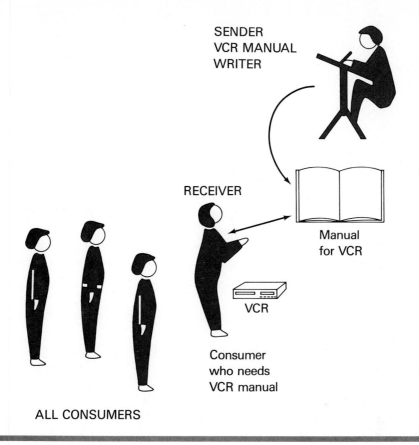

SENDER
VCR MANUAL
WRITER

RECEIVER

Manual
for VCR

VCR

Consumer
who needs
VCR manual

ALL CONSUMERS

FIGURE 1.3
A Specific Audience

Too much strain. But if it's any consolation, it must have been almost instantaneous."

He could see the man's face — pink-white skin, bushy blond eyebrows, blue eyes reflecting the cold light of the surgical waiting room through the lenses of horn-rimmed glasses, the small, prim mouth speaking to him. He could still hear the words, loud over the hum of the hospital air conditioner. It was like a remembered nightmare. Vivid. But he could not remember getting into his car in the parking lot, or driving through Gallup to Shiprock, or any of the rest of that day. He could remember only reviving his thoughts of the days before the operation. Emma's tumor would be removed. His joy that she was not being destroyed, as he had dreaded for so long, by the terrible, incurable, inevitable Alzheimer's disease. It was just a tumor. Probably not malignant. Easily curable. Emma would soon be herself again, memory restored. Happy. Healthy. Beautiful.

"The chances?" the surgeon had said. "Very good. Better than ninety percent complete recovery. Unless something goes wrong, an excellent prognosis."

But something had gone wrong. The tumor and its placement were worse than expected. The operation had taken much longer than expected. Then infection, and the fatal clot. (16–17)

Now consider a set of instructions prescribing courses of action in a life-or-death situation — a heart attack in a hospital. When a head nurse describes the procedures to follow in this situation, she chooses words that enable the reader to focus on the actions that will save the victim. She neither wishes nor tries to engage the subjective emotional dimension of the situation. To accomplish her goal, she states the desired actions plainly and objectively. Here is such a set of instructions ("Dr. Heart" 2). Notice the precise wording that is designed to produce a professional, unemotional response:

1. The ICU/CCU-trained RN, or supervisor, shall take charge of arrest situation and designate responsibilities.
2. Record arrest events and treatment on Critical Care Flow Sheet. (Person to be designated by RN in charge.) If extra RN is available, she will critique code by using the code rating scale (appointment by charge RN).
3. Place the cardiac board under patient when crash cart arrives.
4. Use ambu bag and Elder valve to replace initial mouth-to-mouth ventilation.
5. Connect patient to monitor.
6. Plug in defibrillator and turn on.
7. Start IV with 5% dextrose using largest gauge needle possible and an"addit" IV tubing set.
8. Prepare suction apparatus for use.
9. Prepare intubation equipment for use when qualified person arrives.
10. Administer IV medications if the physician orders. Administer NaHCO 1 amp every 5 minutes x 2.
11. When the physician arrives, explain time elapsed and patient condition.

Of course, human beings are not machines, and emotions are part of all human activity. A nurse working on a heart attack victim might experience a traumatic emotional reaction, caused perhaps by the victim's resemblance to someone close to her. The point, however, is that the set of instructions does not encourage an emotional, subjective response. The instructions, written in plain, objective language, focus the reader's attention only on the act of saving a life.

Clear Organization   Technical writing is clearly organized, which makes it easier to read and comprehend. To indicate organization clearly, good technical writers employ and emphasize words and phrases that point out structure. Since strategies for organizing are discussed fully in Chapter 4, only three will be briefly mentioned here. Technical writers "set up" a document, they use obvious repetition, and they emphasize transitions at the beginnings of paragraphs and sentences.

To "set up" a document means to follow the old rule, "Tell them what you're going to say, then say it." At the beginning of the document writers often name the topic idea and list the topic's subdivisions. To use obvious repetition means to repeat key words. To emphasize transitions means to use words that clearly indicate the start of a new section. Careful writers place words like *first, second,* or *another* at the beginning of sentences. Readers respond positively to such devices because the material is easier to comprehend. The following brief paragraphs use these devices.

SPECIFICATIONS

A specification is a detailed description of the requirements for the design and construction of parts, assemblies, or a complete product. Specifications can also be the requirements for a process in manufacturing. A product developer will write three kinds of specifications: testing, processing, and initial fabricating.

    Testing specifications are taken from the operating environment in which the part will function. The engineer examines such elements as temperature, pressure, moisture, time of use, and number of cycles in the service life. Then he or she specifies tests to duplicate the worst possible scenario that the part will be expected to endure.

    Processing specifications are the guidelines necessary for making the part. Processing conditions such as heating, rate of cooling, mold temperature, injection pressure, and rate of flow are all examples of controls to be specified.

    Initial fabricating specifications usually come from the blueprints and contain tolerances for all dimensions. Tolerances are the range from the minimum acceptable size to the maximum acceptable size for all dimensions. Fabricating specifications also contain instructions for creating unusual shapes or angles.

Margin notes:
- Title names subject
- Document begins with a definition of the subject
- List of three kinds of specifications sets up further descriptions
- Key word repeated at start of paragraph
- Key word repeated

Visual Aids   Technical writing communicates visually with graphs, tables, and drawings, and with carefully formatted pages. Other types of writing — essays, novels, poems — seldom use visual aids to communicate; newspapers sometimes use visual aids in their analytical articles. But visual aids appear commonly in technical writing. Documents that explain

Working With Windows

A window frames and displays its contents. There are six windows in MacWrite: the document window, the Header window, the Footer window, the Find window, the Change window, and the Clipboard window. There are also windows for desk accessories.

A window always has a title bar, and may have a close box, a scroll bar, or a size box. You can move a window, change its size, or close it.

To close a window, work in it, or change its size, you must activate the window first by clicking anywhere inside it.

To Move a Window

■ Position the pointer anywhere on the title bar, except on the close box.

■ Drag an outline of the window to the new location.

The window itself moves when you release the mouse button.

After you release the mouse button, the window is activated if it wasn't already.

To move the window without activating it, hold down the Command key while dragging the window.

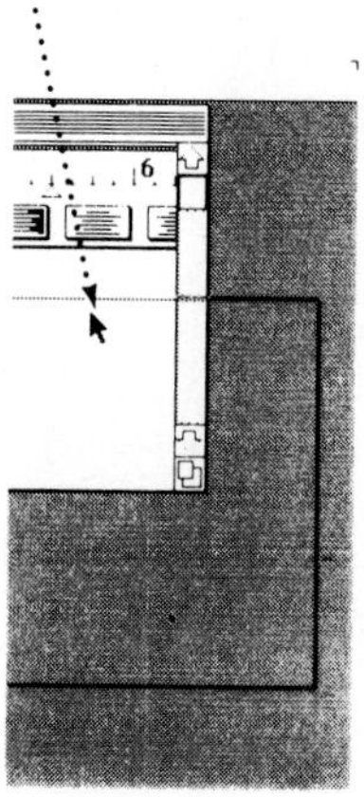

To Change the Size of a Window

■ Activate the window by clicking inside it.

■ If necessary, move the window so that the size box in the lower-right corner is visible.

■ Drag the size box until the window's outline is the size you want.

Dragging horizontally changes the width, dragging vertically changes the height, and dragging diagonally changes both.

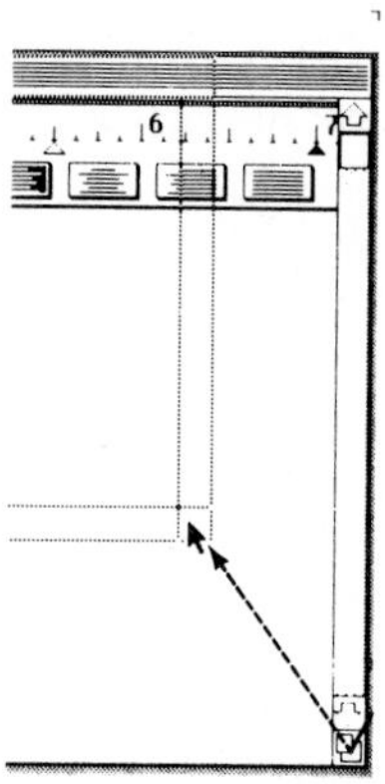

To Close a Window

■ Activate the window.

■ Choose Close from the File menu or click the close box, if any.

You can also close the Header and Footer windows with the Hide Header and Hide Footer commands in the Format menu. You can close the Clipboard with the Hide Clipboard command in the Edit menu. See "Creating Top and Bottom Margins" in this chapter.

50   CHAPTER 2 USING MACWRITE

FIGURE 1.4
Page from MacWrite Manual
Source: MacWrite Manual. Used by permission of Apple Computer Corp.

experiments or projects almost always include tables or graphs. Manuals and sets of instructions rely heavily on drawings and photographs. Feasibility reports might even include maps of sites.

In addition to the "picture" visual aid, the format of each page helps to convey meaning. Technical writers use *heads* — words or phrases that indicate the contents of each section. They also use numbered vertical lists, various marginal indentations, and white space between lines to emphasize important points or clarify difficult ones. Figure 1.4, a page from the MacWrite software program manual [(Stanton-Wyman and Espinosa 50)], uses visual aids and page design effectively to convey its message. The main heading, "Working With Windows," is larger than the three subheadings. A set-up list of key terms ("title bar," "close box," and so on) occurs in paragraph 2 of the introduction. The type of the four headings and the four columns is aligned exactly at the top. Vertical lines divide the text into units. In the square boxes, the dashed diagonal lines and the dotted parallel lines clearly indicate the actions in the text. The three pictures exactly represent the computer screen. All these elements work together to help a reader easily grasp a complex topic.

■ SUMMARY

In the workplace, people write often, and the quality of their writing affects their career advancement. Technical writing is the practical writing that people do as part of their jobs. The most common types of technical writing are memos, letters, and short reports. The most necessary skills are organizing and writing clearly. Technical writing

informs or persuades
engages a specific audience
uses plain, objective language
stresses clear organization
uses visual aids

■ EXERCISES

1. Interview a professional in your field of interest. Choose an instructor whom you know or a person who does not work on campus. Ask questions about the importance of writing to that person's job. Questions you might ask include

   How often do you write each day or week?
   How important is what you write to the successful performance of your job?

Is writing important to your promotion?
What would be a major fault in a piece of writing in your profession?
What are the features of writing (clarity, organization, spelling, etc.) that you look for in someone else's writing and strive for in your own writing?

Write a one-page memo in which you present your findings. Your instructor may ask you to read your memo to your classmates.

2. Photocopy a one- or two-page selection that you consider good technical writing. Use a textbook, an operator's manual, a technical report you may have seen, an article from a journal, a sales brochure, or some other document. Write one to three paragraphs to your instructor explaining why the selection is good technical writing. Refer to the two purposes and four characteristics of technical writing described in this chapter.

3. Photocopy two selections that treat the same subject — one technically, the other emotionally. You can find good contrasts by using poetry and textbooks — for instance, a poetic description of a bird and a field guide description of the same bird. Write several paragraphs to your instructor comparing the selections and pointing out the features that make the technical writing objective and the poetic description emotional.

◼ WORKS CITED

Barnum, Carol, and Robert Fischer. "Engineering Technologists as Writers: Results of a Survey." *Technical Communication* 31.2 (1984): 9–11.

Cronin, Frank C. "Writing Requirements in a Large Construction Firm." *The Technical Writing Teacher* 10.2 (1983): 81–85.

"Dr. Heart Plan." St. Croix Falls, WI: St. Croix Valley Memorial Hospital, 1985.

Hillerman, Tony. *A Thief of Time.* New York: Harper, 1988.

Minors Committee, English Dept., U of Wisconsin-Stout. "Final Report on Survey of Potential Employers of Graduates with Technical Writing Minors." Unpublished, 1983.

Stanton-Wyman, Pamela, and Christopher Espinosa. *MacWrite* (M1509). Cupertino, CA: Apple Computer, 1984.

Storms, C. Gilbert. "What Business School Graduates Say about the Writing They Do at Work: Implications for the Business Communication Course." *ABCA Bulletin* 46.4 (1983): 13–18.

2

The Technical Writing Process

AN OVERVIEW OF THE PROCESS

THE PREWRITING STAGE: PLANNING

THE WRITING STAGE: DRAFTING AND REVISING

THE POSTWRITING STAGE: FINISHING

GROUP WRITING

Documents don't write themselves. A writer must use a process to transform a blank sheet of paper (or a blank computer monitor screen) into a final document to send to a reader. To generate a document you need to write it, of course, but you must also perform other activities, such as researching facts, selecting visual aids, and preparing the final version. Like all processes, the writing process has stages, certain activities performed in a sequence. If you understand what to do at each stage, you can produce better documents more effectively. This chapter explains each of these stages and also introduces you to writing as a member of a group, a phenomenon common in industry and business.

AN OVERVIEW OF THE PROCESS

The goal of the writing process is to generate a clear, effective document for an audience. Experienced writers achieve this goal by performing three types of activities:

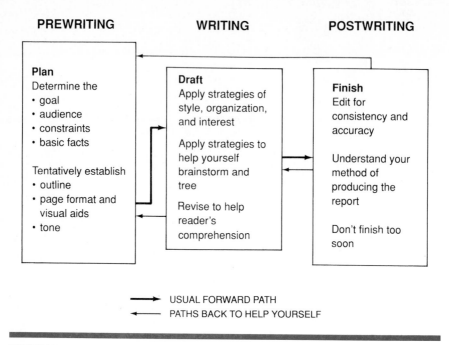

| PREWRITING | WRITING | POSTWRITING |

Plan
Determine the
• goal
• audience
• constraints
• basic facts

Tentatively establish
• outline
• page format and
  visual aids
• tone

Draft
Apply strategies of
style, organization,
and interest

Apply strategies to
help yourself
brainstorm and
tree

Revise to help
reader's
comprehension

Finish
Edit for
consistency and
accuracy

Understand your
method of
producing the
report

Don't finish too
soon

→ USUAL FORWARD PATH
← PATHS BACK TO HELP YOURSELF

FIGURE 2.1
The Technical Writing Process

Prewriting: Planning
Writing: Drafting
Postwriting: Finishing

Planning is discovering and collecting all the information you will need to start drafting. Drafting is selecting and arranging all the elements in the document. Finishing is editing the document into final form. Figure 2.1 is a flow chart of this process (adapted from Goswami 38). The main sequence is emphasized by the heavy black arrows. The light arrows, however, indicate a basic truth about this process — you should and must be ready to return to earlier stages to help yourself generate a clear document.

THE PREWRITING STAGE: PLANNING

In the prewriting stage, writers discover the dimensions of their topic. In this stage writers use a number of techniques to discover everything they need to know to write clearly. They treat this stage carefully. Experienced writers ask and clearly answer eight important questions.

1. Who is my audience?
2. What is my goal in this writing situation?
3. What constraints affect this situation?
4. What are the basic facts?
5. What is the expected final form of the document?
6. What is an effective outline?
7. What format and visual aids should I use?
8. What tone should I use?

As a writer you must ask these questions. The more you clarify the answers, the more easily you will generate the document. The following sections look at each question in more detail.

Who Is My Audience?

The *audience* is the person or people who will read your document. The more you clarify who they are, the better you can write to them. You should ask these questions about your audience:

- Who will read this document?
- How much do they know about the topic?
- Why do they need the document?
- What will they do with it or because of it?

In every writing situation, you must clarify the answers to these important questions. The answers will vary in different situations.

Suppose you are the only person in your office who has a thorough grasp of Pagemaker, a desktop publishing program. And suppose that you lose a lot of time from your own work because you have to answer questions from people who don't know how to use the program. To stop these interruptions, you decide to write a brief manual.

To begin your manual, you ask and answer the planning questions. Who will read the manual? The readers will be the people who work with the computer program in my office. How much do they know? This question has two subquestions: How much do they know about the type of computer we use? And how much do they know about Pagemaker? They all know how to operate Macintosh computers, but most don't know how to use the Pagemaker program. Why do they need the manual? They need it to give them basic information about the software program. What will

they do with or because of the manual? They will use it to guide themselves through the basic operations and because of it they will become
more independent.

What Is My Goal in This Situation?

You actually have two goals: to communicate a specific message and to
achieve a specific purpose. In other words, you ask and answer two
questions:

- What is my basic message?
- What is my purpose?

In general, your message is your content; your purpose is how you want
to affect your audience. As you continue your planning of the Pagemaker
manual, you would answer the two questions this way:

> My basic content will be specific beginner-level procedures in Pagemaker — such
> as opening a document, designing a page, or placing text.
>
> My purpose is to inform readers so that they can perform basic tasks independently.

What Constraints Affect This Situation?

Constraints are physical and psychological factors that affect your ability to
write the document and your readers' ability to read it. By thinking about
constraints, such as time and money, you achieve a clear picture of how
you can produce the document. Experienced writers think through constraints carefully in order to eliminate frustration. Basic constraints are
listed in the left-hand column below. Each constraint has a different implication for the writer and for the reader.

Constraint	*Writer's Viewpoint*	*Reader's Viewpoint*
time	time available to complete writing	length of time to read before starting to act
length	how many pages to write?	how many pages to read?
money	cost to produce the document	cost to purchase the document
physical location	place in which to write the document	place in which to read or use the document
production method	easy to use?	easy to read?

Let's consider the constraints on you as a writer of the Pagemaker manual. How much time can you devote to this project? To write a manual requires hours of actual writing, field testing to check if you described the functions correctly, rewriting, and producing the final document. If all these tasks demand, say, thirty hours, but in the next month you have only ten hours free, you have a problem. You need to drop the project, shorten the scope of the manual, find someone else to do it, or rearrange your schedule to find more time.

How long should it be? While the traditional answer is "long enough to cover the topic sufficiently," you know that people will ignore a long, complicated document, so it has to be relatively short. Will the project cost money? Probably not, if you do it at the office or lab and use all the available facilities and paper. But suppose you want to put it in a nice hard cover to protect it; you must purchase the cover. How will you produce the document? You will use a computer program and printer that you have easy access to.

Let's consider the same constraints from your readers' viewpoint. How much time will they need to spend reading before they can use the program? Since most people will claim that they are "too busy" or that they want to "get right at it," you need to make the information concise and easy to find. How long can it be? Most people tend to ignore thick manuals, so yours can't be too long. Is cost important? No, because the reader will not be buying the manual. Are there physical limits? Yes, readers must be able to handle the manual while they are sitting at the computer. So it has to lie flat and not take up too much space. Then too it must be stored near the computer. Will it be easy to read? Yes, because you will use a laser printer.

What Are the Basic Facts?

Determining the basic facts for your document is a key planning activity. You must spend time collecting these facts by reading, interviewing, or observing. For the Pagemaker manual, you must decide exactly which procedures you want to include. You also need to know the basic facts about each procedure. You would begin to list the procedures, determining the items in the list by considering your audience's needs: You might recall the operations that you have been asked to demonstrate repeatedly. You might ask some of the users what operations they would like explained, or watch to see what mistakes they make. In order to be sure you explain the steps correctly, you might have to work through all the operations, taking notes for yourself.

What Is the Expected Final Form?

Many technical writing documents require a particular final form. If you know what form is expected, you have a place to start. For instance, a report on an experiment always has a section on materials and methods, so the writer can plan that section, then go on to another. In many companies, certain reports must always use the same standard form. All trip reports, progress reports, or position papers must have certain information in the introduction followed by certain information in other specified sections. Knowing what is expected makes documents easier to write because you know which information to include and where to place it.

For the Pagemaker manual, you know that a brief manual usually has two sections (see Chapter 16). One section explains the function of each main part, and the other section gives instructions for performing basic procedures. You want to set up your manual in a similar fashion, explaining what certain items mean when they appear on the screen and then explaining the procedures.

What Is an Effective Outline?

As you begin to think about drafting, you should first construct a preliminary outline. The indented outline is very common, an informal list of major and minor points you want to make. You arrange your material into an order that will guide you as you write. Without such an order, you can easily go off on tangents or needlessly repeat material. The standard form of the document will often provide you with a broad outline, and the results of your investigations during the planning stage will provide you with the fine points. Outlines are treated in more detail in Chapter 6.

What Format and Visual Aids Shall I Use?

You need to decide how your page will look. It is important to select a format and choose visual aids that will help and not hinder your message. The two basic format elements are *margins* (the distance type is set from the left edge of the page) and *heads* (the phrases that indicate the contents of the section following). You must decide on the size of your margins and the look and placement of your heads. If you have an advanced word-processing program, you must also decide which font to use. These decisions constitute your *style sheet*, a description of the specific margins, type, and placement of heads. Most sophisticated word processing programs, Microsoft Word, for instance, allow you to determine these before you type. Then you can automatically control them as you write.

For your Pagemaker manual, you must decide how wide to set the margins, how far to indent paragraphs and lists, how to indicate levels of heads, and how to indicate major sections.

You must also choose visual aids — pictures, tables, graphs, and drawings that clarify the topic. Since visual aids will often help convey meaning more clearly than words, experienced technical writers rely heavily on them. Suppose, for instance, you need to tell people how to interact with a certain Pagemaker screen — say, the page-setup screen, on which the user arranges margins and other technical details. The easiest way to explain the process to your audience is to make a visual aid of the screen. The more visuals you can select *before* you start to write, the easier your writing will be.

What Tone Should I Use?

As you begin to draft, you must consider the tone of your document. Unfortunately, *tone* isn't a very objective term. It means what your writing sounds like. Should it sound funny or serious? Should you give silly examples or "in" jokes from work? Technical writers usually try to sound serious, but not so serious that they sound like robots. For the Pagemaker manual, you probably would choose a straightforward tone — people want to absorb the content, not be entertained.

Conclusion

The time you take to answer the planning questions will be amply repaid. Below is a checklist to help you with the planning stage.

■ CHECKLIST FOR PLANNING

Ask yourself the following questions.

☐ *Who is my audience?*
  a. Who will read this document?
  b. How much do they know about the topic?
  c. Why do they need it the document?
  d. What will they do with it or because of it?
  e. Will other people read this document? What do they need to know?

☐ *What is my goal in this writing situation?*
  a. Basic message: What do I want to tell my audience?
  b. Purpose: Why do I need to tell them that message?

☐ *What constraints affect this situation?*
  a. Time?
  b. Length?
  c. Money?
  d. Physical location?
  e. Production method?

☐ *What are the basic facts?*
  a. Where can I read about them?
  b. Who can I talk to about them?
  c. Where can I observe them?

☐ *What is the expected final form?*
  a. Does it have specified sections?
  b. What is the usual sequence of sections?
  c. What is the usual content of the sections?

☐ *What is the basic outline of topics, if none is suggested by the usual form?*

☐ *What design and visual aids should I use?*
  a. Select heads, margins, fonts.
  b. Write a style sheet.
  c. Choose visual aids that clarify essential information.

☐ *What is my tone?*

THE WRITING STAGE: DRAFTING AND REVISING

In the writing stage, you produce drafts. You already have done careful planning and produced an outline, and now you start the actual writing. You try to put on paper the words that explain the ideas in your outline. Theoretically, if you have planned thoroughly, all you need to do is flesh out the outline and describe the visual aids. But most writers don't accomplish their goals that easily. Writing the document is a difficult, often messy chore, full of false starts. You must select, reject, and reformulate your words and sentences, and find the organization that best conveys your meaning. Drafting requires concentration and energy, even if you have done extensive planning.

A Preliminary Word about Drafting

The object of drafting is to produce and to improve writing until it effectively conveys your message to the reader. Drafting has two primary aspects: clarifying and discovering. As you draft, you clarify your ideas for your reader. Like most people, you probably produce writing in bursts, often in prose that makes sense to you as you write, but makes less sense to others as they read. During drafting, you clarify this material, changing your initial, often hard-to-understand bursts into prose that is clear to readers.

But writing is also discovery. Often as you write you will suddenly conceive of new ideas or new ways to present your examples. You must evaluate these new ideas, deciding whether to use or reject them. As you write your manual, you might suddenly see that you should delete one visual aid and replace it with a better one; the replacement will lead you to write completely different but clearer instructions for the operation in question. In fact, you may discover an entirely new way to organize and approach the whole topic, one that causes you to discard much of your tentative planning. Good writers give themselves enough time to incorporate the insights that they discover in the drafting stage.

Style, Organization, and Reader Interest

Your planning will give you a good sense of your audience and your content. As a result your words will flow better. But in addition, you must have a sense of what to do as you manipulate words. Three areas which you should consider when drafting are

- style
- organization
- reader interest

Style — What Is a Good Sentence?   You should try to write shorter sentences (under twenty-five words), to use the active voice, to use parallelism, and to use words the reader understands. You will not always achieve these in the first draft, but you will be amazed how much you can achieve even in a first draft if you develop an awareness of them. These elements of style are explained more fully in Chapter 4.

Organization — What Is Clear Organization?   The strategies that make ideas easy to grasp are the "obvious organization" strategies referred to in Chapter 1 and explained at length in Chapter 4. You need to remember to make lists, repeat key terms, use heads, use definitions, and use terms the reader can understand.

Reader Interest — What Makes Writing Interesting? You create interest by using devices that help a reader "picture" the topic, thus imaginatively involving him or her in it. The picturing devices apply to both your writing and your visuals. When writing, include helpful comparisons, common examples, brief scenarios, and narratives. Visual devices include any graphic item that helps the reader visualize the topic — from pictures to drawings to tables and graphs (Duin, Slater).

Activities That Will Help When You Get Stuck

Invariably you will get stuck as you write. You need to see these stumbling blocks as challenges. When you overcome them, you enhance your confidence in your ability to do it again. Two very helpful strategies are brainstorming and treeing.

Brainstorming *Brainstorming* means listing every single item you can think of about your topic. Suppose that to write the Pagemaker manual you need to describe how to print documents with a laser printer. To write the appropriate section, you have to decide how many steps are involved in actual printing. Because you know how to print, you realize that there are enough steps to confuse an inexperienced user. You might not be sure how to start describing these steps. Brainstorming will help you. Just start a list. Write down everything, whether or not it seems relevant. For example:

computer on	file menu
print command	display box
substitute fonts	page range
use 96% for MacPaint	chooser
full size	start printer
printer power switch	Apple menu
control cursor	double click

In no time, you will generate enough material to expand into a good section. Not everything you list will be something that you can use, but much of it will be. As you continue to list, you will "warm up," and more items will come to mind. Brainstorming can help you generate ideas at any stage of the writing process, but it is especially effective when you are stuck.

Treeing When *treeing,* the writer indicates relationships by drawing lines between words arranged in descending rows. Each word represents a concept that can be broken down into subconcepts in the next row and the lines indicate the relationship. For instance, the concept *direction* could be broken down into four subconcepts: east, north, west, and south. A tree showing this relationship would look like this:

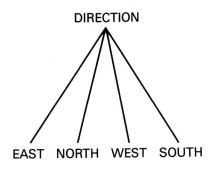

If you were to tree the ideas and steps you generated in the laser printer example, you might get something like this:

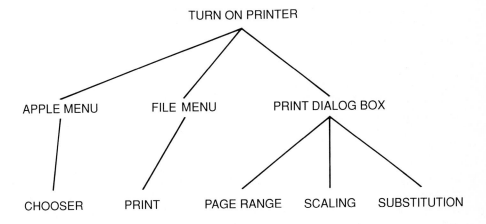

The tree shows each concept and subconcept. You can easily convert this into numbered steps.

Revising

Revising, a basic activity in this stage, is reworking the document, changing sentence structure, paragraph organization, and overall organization, and evaluating the need for effective examples. As you perform this activity, you evaluate your writing against a standard: the reader must clearly comprehend the subject matter. When you revise, ask yourself: "Will my reader understand this?" You can see this process in action in the following revisions of sentences and of paragraphs.

Changing Sentences   You must revise sentences so that a reader can easily grasp their content. First-draft sentences often reflect an author's thought process. They are a record of the ideas just as they occur. Often these sentences are too long, contain too many passive verbs, or contain strings of ideas — short bursts of content — whose relationships are not clear. Consider these sentences from the drafts of the Pagemaker manual:

> Turn on the printer and be sure you turn on the Laserwriter. You can have a real problem — you'll lose everything if you try to print before the test page prints, so wait until it prints.

This sentence contains all the basic ideas but is too long and rambling. The following more concise sentence expresses the idea clearly for a reader. Notice that the dash and the "if . . . so" construction are eliminated.

> Turn on the Laserwriter; do not begin the next step until it has printed its test page.

Revising Paragraphs   Here are two paragraphs — the original and a revision — from the Pagemaker manual. The first, organized solely in terms of the writer's thought process, presents data in terms clear only to the author. The second presents the data that the reader needs in an organization that the reader will understand.

VERSION 1

> Select Print. Click OK. Use the usual box in Imagewriter and change nothing for Laser. Ask if you don't understand.

The writing is ineffective because the details that could be convincing to the reader are locked away in the writer's head. What is the "usual box"? What is meant by "change nothing for Laser"? The reader does not have access to these details. The revised paragraph explains the process more clearly.

VERSION 2

> Select Print. A Print Dialog box will appear. After checking the appropriate boxes, select OK. For the Imagewriter, you will usually use the settings that are already selected: Faster, All, and Automatic. The Laser printer has a more complicated box to fill in. Usually you will use the settings that are already selected. If you are not sure what to do, ask a question.

This revised version names the usual boxes for the Imagewriter: faster, all, automatic. Instead of saying "do nothing," this version clearly tells the reader to accept the selection that appears on the screen.

Below is the complete "How to Print" section of the Pagemaker manual, including a visual aid.

HOW TO PRINT

Title announces goal

1. Turn on the printer. If you turn on the Laser printer, do not begin the next step until it has printed its test page.

2. Pull down the **Apple** menu.

3. Select **Choose Printer** or **Chooser**.

4. Click on the icon of the appropriate printer, Imagewriter (either top or middle icon) or Laserwriter (bottom icon). Be sure the printer icon (not the telephone icon) is selected in the port box.

5. Close the Dialog box by clicking in the box in the upper left-hand corner.

6. Pull down the **File** menu.

7. Select **Print**.

8. A Printer Dialog box will appear. After checking the appropriate boxes, select **OK**. For the Imagewriter, you will usually use the settings that are already selected: Faster, All, and Automatic. The Laser printer has a more complicated box to fill in. Usually you will use the settings that are already selected. If you are not sure what to do, ask a question.

Laser printer note:
• For documents which contain graphics, select 96% reduction.
• Always select **Substitute Fonts**.

Background for audience with limited knowledge

Boldfaced words are a format decision

Visual aid (Figure 1) illustrates difficult reality to describe

Overall tone is straightforward

Overall form reflects usual form of instructions

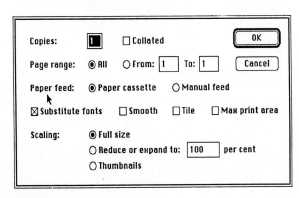

FIGURE 1
Laser Printer Dialog Box

Conclusion   Drafting is messy work, requiring a lot of energy. You must engage the reader with the writing. To do so, you must decide, "Is this best for the reader?" If the answer is no, you must revise the writing. Since your first wording often reflects only your initial thoughts, you may reject it. Developing the confidence to change is a major goal in a writing course. Of course, if you change, you must select from a range of possible options. Throughout this book, we will explain options in sentences, paragraphs, introductions, organization, and design. Many revision options are explained in Chapter 4. Below is a checklist you can use to help you with drafting.

■ CHECKLIST FOR DRAFTING

Consider the following while drafting.

☐ *Follow your initial plan and outline.*

☐ *Be aware that writing occurs in bursts.*

☐ *Write a first draft in which you express your main ideas.*

☐ *Be prepared to change your outline if you discover a new way to present the material.*

☐ *Develop a sense of*
Style strategies — try to use the active voice and parallelism.
Organization strategies — use a structure that sets up each section.
Reader-interest strategies — add comparisons, examples, or brief
     narratives, as needed.
Note: if worrying about these issues inhibits your flow of ideas,
then ignore these issues in the first draft and work at them in
a later draft.

☐ *If you are stuck, brainstorm or construct trees.*

☐ *Write second and further drafts, revising the first.*
   a. Revise sentences so they are clear to readers.
   b. Revise paragraphs so they are clear.
   c. Reorganize sentences if you discover you need to.
   d. Rework for a deductive order — definitions and overviews
      first, details second.

THE POSTWRITING STAGE: FINISHING

In postwriting, the last stage in the process, you craft the document into a product that effectively guides your reader through the topic. This stage consists of two types of actvities: editing and producing the document.

Editing

Editing means to develop a consistent, accurate text. In this step you change the document until it is right. You check spelling, punctuation, basic grammar, format of the page, and accuracy of facts. You make the text agree with various rules of presentation. When you edit, ask yourself: "Is this correct? Is this consistent?"

To help you edit, construct checklists in which you list all the possible problems that you will check for. Then read your document for all the instances of one problem. For instance, first read for apostrophes, then for heading consistency, then for spelling errors, then for consistency in the format, and so forth.

The following paragraphs demonstrate the types of decisions that you make when you edit.

VERSION 1

TECHNICAL REPROTS

The technical reprots to upper managementn will be verry detailed and will be submitted at the end of the project. It must explain; the purpos of the machine, it's operation, and the operation of it's sub systems. Assembley methods and procedures will also be presented. All design calculations for loads, stresses, velocities, and accelerations will be included. Justification for the Choice of matrials of sub systems may occurr. An example mught be; the rational for using plastic rather then steel, or using a mechanical linkage as compared to a hydraulic circuit. The material and purchased parts costs will be laid-out. Finally, Technical Writers will break down the report, and use it to develop service and user manuals.

Typos throughout

Misused semi-colon and apostrophes

Passive voice

Each sentence starts with a different subject; comprehension is difficult

Transition word indicates a process (but the paragraph is not about a process)

VERSION 2

TECHNICAL REPORTS

The technical report to upper management will be submitted at the end of the project. The report must explain, in detail, the purpose of the machine, its operation, and the operation

Typos and punc-tuation corrected

of its subsystems, as well as assembly methods and procedures. It will also include all design calculations for loads, stresses, velocities, and accelerations, as well as justification for the choice of materials in subsystems. Examples of this justification might be the rationale for using plastic rather than steel, or a comparison between a mechanical linkage and a hydraulic circuit. The report also details the cost of material and parts. Eventually, technical writers will break down the report, and use it to develop service and user manuals.	Two sentences merged into one Passive eliminated Key term repeated Transition word eliminates sense of process

A Note on Revising with a Computer   Most word-processing companies now offer various revising aids that can facilitate your editing process. Several common aids are dictionaries, grammar checkers, and thesauruses.

Dictionaries, also called spell checkers, and their closely related cousins, grammar checkers, can help you find possible problems in your writing. A spell checker will read your document and indicate any words that are not in its dictionary. If you have made a typo, such as typing "wtih," the checker will highlight the word and allow you to retype it. Some checkers even suggest several alternate spellings. A problem with these programs is that if your typo happens to be another word — such as "fist" for "first" — the program will not indicate the problem. While these devices find typos easily, they do not help with misuses of words. So if you type "to" instead of "too," the program will not indicate an error.

Grammar checkers seldom actually check grammar, such as problems in subject-verb agreement. They do, however, highlight features of your writing that you should check. For instance, the checker will highlight all the forms of "to be" in your paper, thus pointing out all the possible places that you might have used the passive voice. A good grammar checker points out sexist and racist language, overused phrases, and easily confused words. Although the checker will highlight every "your" and "you're" in your paper, you will have to decide if you have used the correct form.

A thesaurus is a collection of groups of words that have the same meaning. Like the book-version thesaurus, a computer thesaurus will suggest other words that mean the same as the one you select on your computer. You use a computer thesaurus when you want to find a sharper word than the one you have.

Producing the Document

Producing a document has two dimensions: the physical completion of the document and the psychological completion of it.

Physical Completion Physical completion means typing or printing (if you use a computer) the final document. This dimension takes energy and time. Failure to allow enough time for this stage and its problems will certainly cause frustrations. For instance, many people have discovered the difficulties of this stage when their disk crashed or their printer failed for some reason. Although physical completion is usually a minor factor in brief papers, it is a major factor in longer documents, often taking more time than the drafting stage.

Psychological Completion Psychological completion means to attend to your emotions as you near completion and to manage your time properly. Poor writing is often the result of "finishing too soon." If you prematurely decide that you are finished, you probably will not listen to other readers' suggestions (or heed your own perceptions that more revising is needed). Often these suggestions are very good, and if you acted on them, you could produce a much better document. If you have finished too soon, however, such suggestions only cause frustration because you don't want to be bothered with reworking a document you feel you have completed.

Many people enter into projects with a hidden time agenda. They decide at the beginning that they have so many hours or days to devote to a particular project. When that time is up, they must be finished. They do not want to hear any suggestions for change. A similar problem with time management often occurs in research projects. A researcher, fascinated with the reading, will continue to read "just one more" article or book, thus taking valuable time away from writing the report. When he or she begins to write the report, there is not enough time left to do the topic justice. The result is a bad report.

As you grow in your ability to generate good documents, you will also grow in your ability to estimate accurately the time that it will take you to finish a writing project. You will also develop a willingness to change the document as much as is necessary to get it right. Developing these two skills is one sure sign that you are maturing as a writer. Below is a checklist to help you with editing.

■ CHECKLIST FOR EDITING

☐ *Set up a specific time schedule. Work backward from the due date.*

☐ *Make a checklist of possible problems. Check your document for each problem.*

A possible list could include
  spelling
  grammar
  consistent word use
  head format
  typographical items (are all dashes formed the same way?)
  handling of beginning of paragraphs
  reference to figures
  Note: each project generates its own particular editing checklist.
  Formulate yours based on the actual needs of the document.

☐ *Honestly evaluate how you will produce the document.*
  Will you type or word process?
  Do you have enough time to handle unexpected difficulties?
  Will you or someone else do the production work?
  Which tasks must you finish first?
  How long will it take to complete each task?

GROUP WRITING

Many college graduates discover that once they are in business and industry, they must cowrite their documents. Often a committee or a project team — with three or more people — must produce a final report on their activities. Unless team members coordinate their activities, the give-and-take of group revision can cause hurt feelings, frustration, and an inferior report. The best way to generate an effective document is to follow a clear writing process. For each of the writing stages, you not only must plan for producing the document, but also must facilitate the group's activities.

Prewriting in a Group: Organizing

In the prewriting stage, in addition to planning about the topic, you must develop your group into a unit with a leader and a plan.

Select a Leader   The leader is not necessarily the best writer or the person most informed about the topic. Probably the best leader is the best "people" person, the one who can smooth over the inevitable personality clashes, or the best manager, the one who can best conceptualize the stages of the project.

Plan the Group's Activities   You must also think through the group's activities and develop an overall plan to resolve differences and to manage the group's activities.

Resolving differences is an inevitable part of group activity. Your group should develop a reasonable method of resolving differences that's clear to all. The usual methods are by voting, reaching a consensus, or accepting expert opinion. Voting is fast but potentially divisive. People who lose votes often lose interest in the project. Reaching a consensus is slow but affirmative. If you can thrash through your differences without alienating one another, you will maintain interest and energy in the project. Accepting expert opinion is usually, but not always, an easy way to resolve differences. If one member, who has studied citation methods closely, says that the group should use a certain format, that decision is easy to accept. Unfortunately another group member will sometimes disagree. In that case, your group will need to use one of the other methods to establish harmony.

To manage the group's activities, the group must make a work plan. The group must clarify each person's assignments and deadlines in the plan. Members should use a calendar to determine the final due date and discuss reasonable time frames for each stage in the process. The group should put everything in writing and should schedule regular meetings. At the meetings members will make many decisions — for instance, about the style sheet for head and citation format. Write up these decisions and distribute them to all members. Make — and insist on — progress reports. Help one another with problems. Tell other group members how and when they can find you.

Attention to forming the group and treating the group's activities as a project will definitely increase your chances of completing the project successfully — while still enjoying one another's company.

Writing in a Group

Before actual drafting starts, the group should have a prewriting meeting. At this meeting the group has two key tasks: to clarify assignments and deadlines and to select a method of drafting.

Clarify Assignments and Deadlines   To clarify assignments and deadlines, answer the following questions:

- What is the sequence of sections?
- Must any sections be completed before others can be started?
- What is each person's writing assignment?
- What is the deadline for each section?

Pay particular attention to deadlines. Work backward from the final deadline. If the report is due May 1, and if you need one week to type and two weeks to review and revise, then the deadline for drafting is April 9, three weeks prior to May 1.

Select a Method of Drafting   Groups can draft a document in two ways: each person writes a section, or one person writes the entire draft.

Generally each person writes a section if the document is long or if the sections are highly specialized. In a proposal, for instance, one person might write the technical description while someone else tackles the budget. If each person writes a section, the work is distributed equally. However, this method may not be efficient because of possible conflicts of style, format, or tone.

To make the document more consistent, one person often writes it, especially if it is short. A problem with this method is that the writer gets his or her ego involved, easily feeling "used" or "put upon," especially if another member suggests major revisions. The group must decide which method to use, considering the strengths and weaknesses of the group members.

Postwriting in a Group

Finishing involves two activities: editing and producing the final document.

Select an Editing Method   Groups can edit in several ways. They can edit as a group, or they can designate an editor. If they edit as a group, they can pass the sections around for comment, or they can meet to discuss the sections. Frankly, this method is cumbersome. Groups will often "overdiscuss" smaller editorial points (such as whether or not to use all capital letters for headings) and lose sight of larger issues. If the group designates one editor, that person can usually produce a consistent document. The editor should bring the edited document back to the group for review. The basic questions that the group must decide about editing include

- Who will suggest changes in drafts? one person? an editor? the group?
- Will members meet as a group to edit?
- Who will decide whether or not to accept changes?

In this phase the conflict-resolving mechanism is critical. Accepting suggested changes, for instance, is difficult for some people, especially if they are insecure about their writing.

Select a Final Production Method   The group must designate one member to oversee the final draft. Someone must collect the drafts, engage the typist, and read for typos. In addition, someone must write the introduction and attend to such matters as preparing the table of contents, the bibliography, and the visual aids. These tasks take time and require close attention to detail, especially if the document is long. Questions for the group to consider at this stage include

- Who will write the introduction?
- Who will put together the table of contents?
- Who will edit all the citations and bibliography?
- Who will prepare the final version of the visual aids?
- Who will oversee producing the final document?

Conclusion

The group writing process will challenge your skills as a writer and person. It can be a pleasant or an awful experience. Good planning will enhance your chances for a successful report and a pleasant experience. As you work with the group, remember that people's feelings are easily hurt when their writing is criticized. Be gentle. Or as one student said, "Get some tact." The checklist below summarizes the special concerns of group writing.

■ CHECKLIST FOR GROUP WRITING

Answering the following questions will help the group writing project go smoothly.

☐ *Planning Stage*
  a. Who will be the leader?
  b. How will the group manage its activities?
     Select a method to resolve differences.
     Clarify deadlines and assignments.
     Meet regularly.
     Keep records of meetings.
     Make progress reports.

☐ *Writing Stage*

    a. Ask and answer planning questions.
       What is the sequence of sections?
       Must any sections be completed before others can be started?
       What is each person's writing assignment?
       Does everyone understand the style sheet?
       What is the deadline for each section?
    b. Select a method of drafting.
       Will everyone write a section?
       Will one person write the entire draft?

☐ *Postwriting Stage*

    a. Select a method of editing.
       Who will suggest changes in drafts?
       Will the members meet as a group to edit?
       Who decides whether or not to accept changes?
    b. Oversee the production process.
       Who will write the introduction?
       Who will put together the table of contents?
       Who will edit the citations and bibliography?
       Who will prepare the final versions of the visual aids?
       Who will direct the production of the final document?

■ SUMMARY

This chapter explains the technical writing process and the special concerns of the group writing process. In the technical writing process, writers plan, draft, and finish their documents.

    Planning includes finding the answers to these eight questions:

- Who is my audience?
- What is my goal in this writing situation?
- What constraints affect this situation?
- What are the basic facts?
- What is the expected form?
- What is an effective outline?

- What format and visual aids should I use?
- What tone should I use?

Drafting consists of writing and rewriting a document in order to make it easier for the reader to grasp. Writers follow a preliminary outline but must employ effective strategies in selecting the words and developing sentences and paragraphs. They must also use such devices as brainstorming and treeing to help themselves if they get stuck.

Finishing is putting the document into its final form. Writers make the text accurate by checking the spelling, grammar, and overall consistency of similar elements in the document.

The group writing process requires special activities at each stage of the regular writing process. At the outset, groups must select a leader and devise a plan for completing the document. They also must select methods of resolving differences and clarify assignments and deadlines. As work progresses, members should meet regularly to report on activities, to make style sheet decisions, and to share information. Finally they must select a method for drafting, editing, and finishing the document, paying close attention to assignments, deadlines, and conflict resolution.

EXERCISES

1. Analyze a professional document in your field (an article, a business letter, a chapter in a textbook) to see how the writer answered the planning questions. Then assume that you are an editor at a company that publishes such documents. Use your analysis to write a letter to a prospective author explaining how to prepare such a document for your company.

2. Revise the following paragraph. The new paragraph should contain sections on reasons for writing and format. Revise sentences also.

Proposals are commonly used in the field of retail management. Proposals are written in a standard format which has a number of company parts that we often use. These are stating the problem, a section often used first in the proposal. Another one is providing a solution, of course, to the problem which you had. You write a proposal if you want to knock out a wall in your department. You also write a proposal if you want to suggest a solution that a new department be added to your store. Another section of the proposal is the explanation of the end results. You could also write a proposal if in the situation you wanted to implement a new merchandise layout. A major topic involved with writing proposals is any major physical change throughout the store. I can structure a proposal implementing any major physical changes to the upper management level, who would be my audience.

■ WRITING ASSIGNMENTS

1. Interview three people who write as part of their academic or professional work to discover what writing process they use.

   ■ A student in your major
   ■ A faculty member in your major department
   ■ A working professional in your field

   Prepare questions about each phase of their writing process. Show them the model of the process on p. 16 and ask whether it reflects the process they go through. Then prepare a one- to two-page memo to your classmates, summarizing the results of your interviews.

2. Write a description of a machine or form that might be used in your field or that you often use. Before you write the description, perform all the following activities:

   Answer all the planning questions about your document in a memo to your instructor.
   Write out a time schedule for each stage of your process.
   Hand in your checklist, schedule, and answers with your paper.

   Here is a list of suggestions. Feel free to use others.

microscope	all or part of a computer system
tax form	all or part of a stereo system
funds flow statement	microwave oven
cash register	sewing machine
oscilloscope	automobile engine
industrial robot	table saw, radial saw, other power tool
35 mm camera	waxing iron for skis
film projector	jogging shoes
horizontal camera	a drafting machine
hand calculator	oxyacetylene torch
condenser-enlarger	

3. Write a one- to two-page paper in which you describe the process you used to write a recent paper. Include a "process model" diagram of your process.

4. Your instructor will assign you to groups of three or four, based on your major or professional interest. As a group, perform assignment No. 1, above. Each person in the group should interview different people. Your goal as a group is to produce a two-page memo synthesizing the results of your interviews. Follow the steps presented in the group writing

section of this chapter. In addition to the memo, your teacher may also ask you to hand in a diary of your group activities.

 WORKS CITED

Duin, Ann Hill. "How People Read: Implications for Writers." *The Technical Writing Teacher* 15.3 (1988): 185–193.

Goswami, Dixie, et al. *Writing in the Professions: A Course Guide and Instructional Materials for an Advanced Composition Course.* Washington, DC: American Institute for Research, 1981.

Slater, Wayne H. "Current Theory and Research on What Constitutes Readable Expository Text." *The Technical Writing Teacher* 15.3 (Fall 1988): 195–206.

3

Defining Audiences

YOUR AUDIENCE'S KNOWLEDGE LEVEL

AUDIENCE ROLES

AUDIENCE'S ORGANIZATIONAL DISTANCE

AUDIENCE ATTITUDES

E very piece of technical writing has an intended audience: the reader or readers of the document. Your audience will affect all your decisions, from organization and visual aids to sentence structure and word choice. If the audience consists of consumers who will assemble a tricycle, they need clear information about the parts and the sequence of steps to follow, and they expect easy-to-follow numbered steps. Since your goal is to generate the clearest document for your audience, you must consider their needs and expectations as you plan, write, and revise. In other words, you must write your document for a specific audience in a specific situation. If either the audience or the situation changes, you must change the document.

This chapter will help you to make informed decisions about your audience as you write. You will learn to identify four basic characteristics of an audience: its knowledge level, its role in the situation, its organizational distance, and its attitude. This chapter explains these characteristics and shows you how writing changes to accommodate different audiences.

YOUR AUDIENCE'S KNOWLEDGE LEVEL

The term *knowledge level* means how much your intended audience knows about the subject matter of the document. The level ranges from layperson to expert. In technical writing, a lay audience is intelligent but uninformed about the topic, and an expert audience understands the topic's basic facts and concepts. You must plan your document to accommodate your audience's knowledge level of the subject. If your audience knows very little about the topic, you write the document to increase their knowledge. If readers know a lot about the topic, you write the document to build on what they already know.

If you write a document about subatomic particles for a nuclear physicist, you do not need to define basic terms such as *neutrino,* one of the subatomic particles. However, if you write a document on the same topic for a hotel manager, you will need to define *neutrino* and many other terms. But knowledge level is relative to the situation. If the topic is personnel regulations relating to hotel chains, the hotel manager becomes the expert and the physicist is the layperson.

Adapting to Your Audience's Knowledge Level

Your audience's knowledge level means their familiarity with the background, concepts, definitions, and terms of your topic. To accommodate your audience's knowledge level, you choose terms the audience already knows, define new terms, explain concepts by using examples or comparisons, and provide background, often in the form of the history of the subject.

Adapting to Different Knowledge Levels: An Example

Suppose that at your workplace you had to describe videocassette recorders (VCRs) to potential users. But these users have two different knowledge levels: those who know very little about these machines, and those who know a great deal. The following two examples illustrate the different strategies a writer uses to accommodate those knowledge levels.

TO A LESS KNOWLEDGEABLE AUDIENCE

To help a less knowledgeable audience understand VCRs, you write assuming they have little or no knowledge of how they work. You define any technical terms, such as *Beta, VHS, standard play,* or *extended play.* You explain concepts in common terms; for instance, you might say that "programming" is similar to setting an alarm clock that tells the machine when to wake up. You clearly indicate the start of each new section in the document by using transition words like "next" or "third"; you might supply one or more visual aids, such as pictures of the machine or its controls.

In order to use the company's video equipment effectively,   Set up
you need some background information.

First, we have videocassette recorders and videocassette   Transition term
players. A videocassette recorder (VCR) can record images
from a source, usually a TV. A VCR can also play those   Background
images back, usually through a TV. A videocassette player   explanation
(VCP) will only play prerecorded video tapes. A player cannot
record images on the tape from another source (that's why it's   Informal
a *player* and not a *recorder*). If you want to record a program   description
from our closed circuit system, you will need a VCR. **Do not
check out a VCP if you want to record.**

Second, both VCRs and VCPs have two sizes: Beta and   Transition
VHS. The two tapes are incompatible. You must match Beta
tape to Beta machines and VHS to VHS. You cannot play one
in the other. To tell what size tape you have, measure the
length and width of the cassette. A VHS plastic cassette is 4   Specific details
1/2" x 7 1/2"; Beta is about 3 1/2" x 6". We have two Beta
and four VHS machines.

Third, our tapes have different designations: 120 and   Transition
L750. These numbers refer to the number of minutes the tape
will record. VHS 120 tape will record 120 minutes of programs.   Example
A Beta L750 tape will record 90 minutes.

Fourth, tapes are recorded at different speeds. The VHS   Transition
speeds are standard (SP), long (LP), and extended (EP).
The equivalent Beta speeds are I, II, III. If you record a 120
tape at SP, it will record two hours (120 minutes) of images;   Example
LP allows a 120 tape to record four hours; EP allows six
hours. The equivalent Beta times are 90, 180, and 270 minutes.

Our VCRs will play a tape that is recorded at any of the
three speeds. The VCPs will only play VHS SP tapes. If the   Example
tape was recorded at LP or EP, the VCP will play it so fast
that the image and sound will be jumbled.

TO A MORE KNOWLEDGEABLE AUDIENCE

For the audience who knows a lot about VCRs, you would provide less
information about background, definitions, and concepts. For example,
you could tell a knowledgeable audience not to use Beta tapes because all
the machines are VHS. Since the audience understands that Beta tapes use
different technologies, they know that one will not operate in the other
kind of machine. You do not have to provide the basic details; just saying
Beta and VHS is enough.

Here are the basic facts about the Media Department's   Set up
video equipment. We have six VCRs and four VCPs. Four
of the VCRs are VHS and two are Beta. All VHS machines   Terms used with-
play SP, LP, and EP. All Beta have I, II, and III. The VCPs are   out explanation
all VHS. We do not have Beta VCPs. The VCPs will *only*   or definition
play at SP.

AUDIENCE ROLES

In any writing situation, your audience has a role. Like actors in a drama, they play a part, using the document as a "script." They perform actions after receiving the information in your document. Those who take the most active roles are users and decision makers. Users need a document that gives specific instructions for physically carrying out a process. Decision makers need documents that give them information they can use to come to an informed decision.

Elements in Role

An audience role consists of two elements: the audience's need and the audience's task. Need means why the reader is concerned with the content of the document. Task means what the reader will do after reading the document. The writer adapts to the reader's role by manipulating the document's approach and format. Approach is the way the writer presents the material; format is the way the writer arranges it on the page. A good writer will change a document to accommodate different audience roles. The topic, and even the subtopics, may be similar, but the documents will be quite different because of the different roles of the intended audiences.

Adapting to the Audience's Role: An Example

To help you understand the concept of audience role, let's look at an industrial example. MRM/Elgin, a small company, manufactures rotary piston fillers. These machines fill bottles with fluids such as shampoo, cleanser, or even perfume. The fluid is placed in a large vat. When the bottles pass under the vat, a piston mechanism draws just enough fluid out of the vat to fill the bottle. Large versions of this machine can fill thousands of bottles in an hour. As the sales manager, you might have to write to two groups: operators of the machine and their department managers. In each case the subject — the machine — is the same, but the two resulting documents are quite different.

WRITING TO AN OPERATOR

For the operator, whose role is to run the machine, your approach is to explain in detail the sequence of steps that make the piston filler fill the bottles. You would explain how to turn the machine on and off, how to increase or reduce fill speed, and how to adjust for different size bottles. You would also give troubleshooting instructions, for instance, what to do if a bottle jammed in the filler.

In this situation, the form of your document would be a manual, with brief introductions, lots of numbered how-to-do-it steps, photos or drawings of important parts, and a table of contents to help the machine operator find the relevant information in a hurry. Here is a manual section (MRM/Elgin 25) that tells an operator how to keep the machine from filling bottles unevenly. The role of the reader of this document is to take some fast action.

Troubleshooting Table			
You notice	*This may mean*	*Caused by*	*You should*
Uneven fills	Air being induced into the product	Product level in supply tank too low	Add product
		Worn piston seal	Replace (see fig. 6–202)
		O-rings in head assembly worn or missing	Replace (see fig. 6–206, 207, 214)
		Any component in the valve assembly not properly tightened	Tighten any component in the valve assembly that may be loose (see fig. 6)

From a manual by Jill Adkins. Used by permission of MRM/Elgin.

WRITING TO A DECISION MAKER

To the manager, you would write a completely different document, even though it might cover the same topics. Your approach is to explain the machine's capabilities and to show how they might benefit her staff and budget. You might point out that the machine has a variable output that can be changed to meet the changing flow of orders in the plant; you might point out that the personnel on the floor can easily maintain it so no outside help is needed for routine maintenance. And you might explain that problems such as jamming can be easily corrected.

For the manager, the document's format would be standard prose (complete sentences), divided into sections consisting of explanatory paragraphs rather than numbered how-to-do-it steps. Instead of photos, you might use a line graph that shows the effect of the variable rate of production or a table that illustrates budget, cost, or savings.

Here is a paragraph from a sales letter in which the writer wants to affect decision making. The topic is the same as that of the troubleshooting table — uneven fills — but the reader's role is different. In this case, the role is that of deciding whether or not to purchase the machine based on the information provided.

> Your staff can easily repair most of the problems that occur during operation. You will not need to suffer any lengthy downtimes waiting for our repair service to get to you. For instance, if the machine starts to fill bottles unevenly, the operator can easily replace the O-rings, a common repair for correcting such a problem. The job takes about 20 minutes.

Switching Roles

Individuals can change their role. If they do, they need to interact with the document designed for that role. If for some reason the decision maker wanted to operate the filler, she would change her audience role. She would have to use the document that was meant for the operator of the machine. She would have to fit her role to the audience expectations written into that document.

AUDIENCE'S ORGANIZATIONAL DISTANCE

Organizational distance refers to the relative positions of the reader and the writer in the hierarchy of the institution. In this regard, the audience is located above, at the same level, or below — and either near or far away. If the audience consists of people from several positions in the hierarchy, the audience is "multiple." The organization chart in Figure 3.1 shows relative positions.

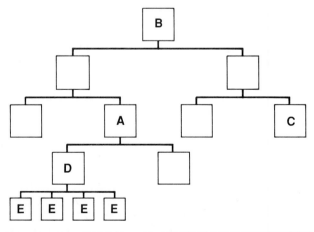

FIGURE 3.1
Organization Chart

In Figure 3.1, you are A; two levels above you is B, a vice-president. C, another superintendent, is on your level but in another section of the company. D, a supervisor, and E, floor personnel, are below you. D is near; all the others are far.

Elements of Distance

The basic elements to consider in organizational distance are formality and power. *Formality* is the degree of impersonality in the document. The more formal, the more impersonal. *Power* is the author's ability to give orders to the reader.

Generally, if you write to someone close to you, who is also the only one receiving the message, you can be relatively informal. But as you move in any direction from your place in the hierarchy, even horizontally, you probably need to become more formal. In most organizations, authority flows downward. You can issue orders to those below you but not to those above you. You can recommend or remind, however, in either direction.

Adapting to the Audience's Distance: Examples

The following examples illustrate how writers can deal with the same topic yet accommodate for the distance of the readers. While the examples have the same topic, they are different because their audiences are at different distances from them in the organization.

AUDIENCE NEAR

Suppose you want to send a note to supervisor D below you to tell him to enforce the rule that all employees must wear safety glasses. If the note is just a reminder intended only for him, then a brief handwritten message of several sentences might be the most effective document.

Feb. 14, 19XX

John,   Informal use of
    As I passed through your area last week, I noticed two   name
employees who were not wearing safety glasses. I know
we all forget or get in a hurry sometimes. The OSHA requirements   Tone is friendly
are quite explicit. All people on the floor must wear glasses.   but firm
Remember, Janeen has spare glasses for people to use if
they forget theirs. We don't want any eye accidents.

Bob

AUDIENCE FAR AND BELOW

If you write to the floor personnel at level E, some distance from you, demanding compliance with the safety glasses policy, then the document might be a formal memo with numbered paragraphs containing the pertinent statements.

To: All floor employees   February 14, 19XX   Document has a
From: Robert Hoyt   more offical
Subject: WEARING SAFETY GLASSES   structure including
     formal use of
OSHA regulations explicitly state that any employee who   names
works on a machine with moving parts must wear safety
glasses. Two incidents of noncompliance have occurred in the
last week. The following policy is now in effect:

1. All employees, including supervisors, will wear safety   Issues are stated
   glasses at all times on the floor.   precisely

2. All visitors will wear safety glasses.

3. Employees must purchase these glasses from Supply.

4. Janeen keeps spare glasses. If you forget yours, check out
   a pair from her.

5. Offenders could be subject to fines.

6. Please cooperate. We do not want anyone to damage or
   lose an eye.

AUDIENCE FAR AND ABOVE

Suppose that your audience is B, the vice-president two levels above you. If you had to write to him about eyeglasses, you would write formally both to report your action on the safety glasses and to ask his advice.

To: Joseph Pember   February 14, 19xx   Very formal
From: Robert Hoyt   document
Subject: Action on Safety Glasses

This memo details action I have taken to ensure that employees   Introduction
comply with OSHA standards and to request your aid in a   orients reader to
liability question.   actions
       After two occurrences of noncompliance, I issued a   Actions explained
written explanation of the rule, including a specific policy. A
copy is appended for your review. I hope this action is sufficient
to prevent any liability if an injury were to occur. If I need to
do more, please inform me.

Ms. Janeen Sparger, referred to in point 4 of the memo, wishes to know if she is liable if all her spare glasses are loaned out when someone needs a pair. Could you please ask the lawyers to clarify her position and ours in the problem of spare glasses?   *Request made*

Multiple Audiences

Often your audience will include readers at many levels. Suppose you have to write a document for levels B, C, and D. In this situation each person will read the document from his or her own perspective. Such documents are difficult to write because the audience includes readers with different roles and knowledge levels.

In a proposal, for instance, one group of readers might need cost figures, another technical details of the project, and a third the implications for the plant's workforce. If the group as a whole must decide whether or not to accept the proposal, an inefficient treatment of one person's area of special interest could cause the proposal to be rejected. This multiple audience is the hardest to write for, but the situation occurs frequently. For this type of audience, writers use two methods: general description and specific indication.

Choose a General Description   To describe the topic in a general manner means to write nontechnically, so that a multiple audience will understand it. To understand this concept, refer again to the VCR examples on pp. 41–42. Suppose you had to send a memo about VCRs to an entire plant. You would be sending it to all levels of distance and knowledge. To be effective, you would have to choose the longer, more formal document that assumes less knowledge. If you sent the other document, many people would not be able to understand it.

Label Sections for Specific Readers   The second method is to label sections and specifically indicate the audience for each section. In the following example, the reader has to read only the section that pertains to him or her.

RESPONSIBILITIES

I. **Receptionists**

   1. Upon receiving the alarm, the receptionist immediately pages "Dr. Heart" to *room number* and *location*, repeated 3 times, every 30 seconds x 3, then every minute until page is canceled. Cancellation will occur in ICU/CCU/ER by resetting emergency button and in the medical surgical units or nonpatient care areas by calling receptionist to cancel page.

II. **Registered Nurses**
   1. For arrests in patient care areas: all available nursing personnel respond and immediately institute CPR procedures (according to established guidelines).

III. **Nursing supervisor, ICU/CCU nurses, or other specially trained nurses (in addition to above) may:**
   1. Defibrillate as necessary.

IV. **LPNs**
   1. For arrests in patient care areas: all available nursing personnel respond and immediately institute CPR procedures (according to established guidelines).

V. **Nursing Assistants**
   1. For arrests in patient care areas: all available nursing personnel respond and immediately institute CPR procedures (according to established guidelines).

VI. **Medical Staff**
   1. All physicians in the hospital, upon hearing the page, will report to the arrest area. In cases where patient has been transported to advance life support area, check with receptionist for new location.

Secondary Audience

A *secondary audience* is someone other than the intended receiver who will also read the document. Often you must write with such a reader in mind. The secondary reader is often far from the writer, so the document must be formal. The following two examples illustrate how a writer changes the document to accommodate primary and secondary audiences.

Suppose you have to write a memo to your supervisor to request money to travel to a convention so that you can give a speech; this memo is just for your supervisor's reference since all he needs is a brief notice for his records. As an informal memo intended for a primary audience, it might read like this:

January 2, 19XX	
John, This is my formal request for $750 in travel money to give my speech about widgets to the annual Society of Manufacturing Engineers convention in San Antonio in January. Thanks for your help with this. Fred	Informal use of name No formatting to document

If this brief note is all your the supervisor needs, neither a long formal proposal with a title page and table of contents nor a formal business letter would be appropriate. The needs of the primary audience dictate the form and content of this memo.

Suppose, however, that your supervisor has to show the memo to his manager for her approval. In that case, a brief, informal memo would be inappropriate. His manager might not understand the significance of the trip or might need to know that your work activities will be covered. In this new situation, your document might look like this:

Date: January 2, 19XX To: John Jones From: Fred Johnson Subject: Travel money for speech to National Manufacturing 　　　　　Engineers convention	A more formal format used including formal use of names
As I mentioned to you in December, I will be the keynote speaker at the Annual Convention of the Society of Manufacturing Engineers in San Antonio. I would like to request $750 to defray part of my expenses for that trip.	Orients reader to background and makes request
This group, the major manufacturing engineering society in the country, has agreed to print the speech in the conference *Proceedings* so that our work in widget quality control will receive wide readership in M.E. circles. The society has agreed to pay $250 toward expenses, but the whole trip will cost about $1000.	Explains background of request
I will be gone four days, February 1–4; Warren Lang has agreed to cover my normal duties during that time. Work on the Acme Widget project is in such good shape that I can leave it for those few days. May I make an appointment to discuss this with you?	

As you can see, this document differs considerably from the first memo. It treats the relationship and the request much more formally, and presents much more data about the significance of the trip so that the manager, your secondary audience, will have all the information she needs to agree to the request.

AUDIENCE ATTITUDES

Attitude means the expectations that a reader has when he or she reads a document. Any audience will have at least two attitudes involved in the communication situation: (1) feelings about the message and the sender, and (2) expectations of form. You must accommodate your documents to meet both these expectations.

Feelings about Message and Sender

In terms of feelings, the audience's relationship to the writer and the message can be described as positively inclined, neutral, or negatively inclined. If the audience is *positively inclined*, a kind of shared community can be set up rather easily. In such a situation, many of the "small details" won't make as much difference — the form that is chosen is not as important, and the document can be brief and informal. Words that have some emotional bias can be used without causing an adverse reaction. Much the same is true of an audience that is *neutral*. A writer who has to send a neutral audience a message about a meeting or the results of a meeting might choose a variety of forms — perhaps a memo, or just a brief note with the information on it. As long as the essential facts are present, the message will be communicated.

However, if the audience is *negatively inclined*, the writer cannot assume a shared community. The small details must be attended to carefully; things like spelling, format, and word choice become even more important than usual because negatively inclined readers may react to anything that lets them vent their frustration or anger. Surprisingly, even such seemingly trivial documents as the announcement of a meeting can become a source of friction when the audience is negatively inclined.

Expectations of Form

Many audiences expect certain types of messages to have certain forms. For instance, a manager who wants a brief note to keep for handy reference may be irritated if he gets a long, detailed business letter. Or if an expert in electronics asks for information on a certain circuit, a prose discussion would be inappropriate since it is customary to give that information by supplying schematics and specifications. If an office manager has set up a form for reporting accidents, she will expect reports in that form. If she gets a different form, her attitude may change from neutral to negatively inclined. On the other hand, if she gets exactly the form that she specified, her attitude may easily turn from neutral to positively inclined. To be effective, you must provide the audience with a document in the form they expect.

■ SUMMARY

Your audience affects the document you write. You must consider your audience in terms of the following: knowledge level, role in the situation, organizational distance, and attitude. Consider carefully your audience's knowledge level. If your readers know the topic well, you may use specialized terms and refer to concepts without explaining them. If they know little, you must use terms they know, define terms they don't know, and use strategies such as comparisons to help them gain knowledge. Also consider the audiences' role, or how they will act in the situation. Will they use the document to take physical action or to make a decision? In addition consider the distance of your audience within your organization. Audiences can be above, below, or on the same level as you, as well as near or far. In most cases, you give orders to levels below you and recommendations to levels above you. Your documents will be more informal with near audiences, and more formal with far. Your audience's attitude may be positive, neutral, or negative. A positive attitude allows you to be more informal in tone and format. A negative attitude demands careful attention to tone, format, and the reader's needs.

■ WORKSHEET FOR DEFINING AUDIENCE

☐ *Who will read this document?*
Name the primary reader or readers
Name any secondary readers.

☐ *Determine the audience's knowledge level.*
What terms do they know?
What concepts do they know?
Do they need chronological background?
To find your readers' knowledge level, you must (a) ask them directly, (b) ask someone else how much they know, or (c) make an educated guess.

☐ *Determine the audience's role.*
What will they do as a result of your document?
Have you presented the document so they can easily take action?
Why do they need your document? for reference? to take to someone else for approval? to make a decision?

☐ *Determine the audience's organizational distance.*
Are they above, below, or at your level?
Are they near you? If so, you can be more informal.
Are they far from you? If so, you must be more formal.

☐ *Determine the audience's attitude.*
Are they negative or positive toward the message? toward you?

■ EXERCISES

1. In class, set up two role-playing situations. In each, let one person be the manager and two others employees in a department. In the first situation, the employees propose a change, and the manager is opposed to it. In the second, the employees propose a change, and the manager agrees but asks pointed questions because the vice-president disagrees. In each case, plan how to approach the manager, then role-play the situation. Suggestions for proposed changes include the following: switching to a four-day, ten-hour-per-day week; starting an employee recreational free time; having a drawing to determine parking spaces instead of assigning spaces closest to the building to executives.

2. Review the directions for Exercise 1. In groups of three or four, agree upon a situation like those mentioned above. As a group write a memo requesting the change to a near audience. For the next class, each person bring a memo that requests the same change but addresses a far audience. As a group select the best individual's memo and read it to the class.

3. Bring to class a piece of writing aimed at an expert or knowledgeable audience. In groups of three or four, rewrite the document for a less knowledgeable audience.

■ WRITING ASSIGNMENTS

1. Interview one or two professionals in your field whose duties include writing. Ask a series of questions to discover the kinds of audiences they write for. Write a memo summarizing your findings. Your goal is to characterize the audiences for documents in your professional area. Here are some questions that you might find helpful:

   What are two or three common types of documents that you write? (proposals? sets of instructions? informational memos? letters?)

   Do your audiences usually know a lot or a little about the topic of the document?

   Are your audiences near or far from you within the organization? above, at the same level, or below you?

   Do you ever write about the same topic to different audiences?

   Do you ever write one document aimed at a multiple audience?

   Can you give specific instances of any of the above?

2. Interview one or two professionals in your field whose duties include writing. Ask a series of questions designed to reveal the way audiences affect their writing. Prepare a memo summarizing your findings. Your goal is to describe

how professionals change their writing based on their audiences. Here are some questions you might ask:

What two or three types of documents do you commonly write?

Do you try to find out who your audience is before or as you write?

What questions do you ask about your audience before you write?

Do you change your sentence construction, sentence length, or word choice based on your audience? If so, how?

Do you ever ask someone in your intended audience to read an early draft of a document?

Can you give examples of situations in which your awareness of your audience changed your writing?

3. Write two different paragraphs about a topic that you know thoroughly in your professional field. Write the first to a person with your level of knowledge. Write the second, to a person who knows little about the topic. After you have completed these two paragraghs, make notes on the writing decisions you made to accommodate the knowledge level of each audience. Be prepared to discuss your notes with classmates on the day you hand in your paragraphs. Your topic may describe a concept, an evaluating method, a device, or a process. Here are some suggestions:

product life cycle	a direct stopwatch time study form
a balance sheet	inventory control
industrial robots	burning in
dodging	transmitting air pressure
machining processes used in wood	masers
swimming strokes	steps in fitting ski boots
bicycle gear train	accounting definitions
corporate structure	

4. Form yourselves into groups of three or four. If possible, the people in each group should have the same major or professional interest. Decide upon a short process (4 to 10 steps) that you want to describe to others so that they can carry out the process. As a group, write the process description. For the next class period, bring to class a memo that deals with the same topic but is aimed at an audience who must decide whether or not to implement the process. Agree upon the most effective memo and read it to the class.

■ WORKS CITED

"Dr. Heart Plan." St. Croix Falls, WI: St. Croix Valley Memorial Hospital, 1985.

MRM/Elgin Corp. *Rotary Piston Filler: 8 Head.* Menomonie, WI: MRM/Elgin Corp.,1985.

4

Technical Writing Style

GUIDELINES FOR WRITING CLEAR SENTENCES

GUIDELINES FOR WRITING CLEAR PARAGRAPHS

GUIDELINES FOR REVISING FOR CLARITY

GUIDELINES FOR ORGANIZING CLEARLY

To communicate well through your writing, you need to develop a clear, smooth style. While developing such a style will take time and practice, it will be well worth your while. You will especially help yourself if you understand what to do as you create drafts and what to look for as you revise drafts. In other words, you must develop standards by which to judge your own writing. This chapter explains those standards. As you internalize them, your prose will become much clearer.

The following sections present guidelines for writing clear sentences and paragraphs, for revising for clarity, and for organizing clearly.

GUIDELINES FOR WRITING CLEAR SENTENCES

As you generate your drafts, keep in mind several guidelines that will help you write clear sentences. Remember to use normal word order, write sentences of 12 to 25 words, use the active voice, use parallelism, and place the main idea first.

Use Normal Word Order

The normal word order in English is subject-verb-object. That order usually produces the clearest, most concise sentences. If you change the order, you emphasize the parts of the sentence that are out of place, but often your sentence is just wordy.

Clear	Dietitians write memos to doctors.
Wordy	Memos to doctors are kinds of writing done by dietitians.

Write Sentences of 12 to 25 Words

An easy-to-read sentence is 12 to 25 words long. Shorter or longer sentences are weaker because they become too simple or too complicated. This rule, however, is only a rule of thumb. Longer sentences, especially those containing a parallel construction, can be easy to grasp.

One sentence, 40 words long	Low-margin merchandise that is sold at discount operations like Bullseye, Quikstop, and Herms consists of items that retailers buy in large quantities so that the company can afford to sell the products to consumers at prices that are substantially reduced.
Two sentences, 17 and 12 words long	Low-margin merchandise consists of items that retailers buy in large quantities to sell at substantially reduced prices. This merchandise is sold at discount operations like Bullseye, Quikstop, or Herms.

Use the Active Voice

The active voice ("I understand it") is more direct than the passive voice ("It was understood by me"). The active voice allows you to write natural, direct, and concise sentences; it places the emphasis on the performer of the action rather than on the receiver. When the subject acts, the verb is in the active voice. When the subject is acted upon, the verb is in the passive voice. Notice that you can write the same sentence in either the active or passive voice.

Active	An insulator *surrounds* the electrode.
Passive	The electrode *is surrounded* by an insulator.

Change Passive to Active To change a verb from the passive to the active voice, follow these guidelines:

- Move the person acting out of a prepositional phrase.

  Passive The memo was sent *by the manager*.

  Active *The manager* sent the memo.

- Supply a subject (a person or an agent).

  Passive Any loose bubbles can be detected in this way.

  Active *You* can detect any loose bubbles in this way.

  Active *The worker* can detect any loose bubbles in this way.

- Substitute an active verb for a passive one.

  Passive The heated water *is sent* into the chamber.

  Active The heated water *flows* into the chamber.

Use the Passive If It Is Accurate The passive voice is sometimes more accurate; for instance, to show that a situation is typical, or usual:

Effective passive needs no agent	The robot is used in repetitive activities.
Active verb adds an unnecessary agent (*companies*)	Companies use robots in repetitive activities.

The passive voice can also be used to emphasize a certain word:

Use passive to emphasize *ink*	The ink is forced through the stencil by a squeegee.
Use active to emphasize *squeegee*	A squeegee forces the ink through the stencil.

Use Parallel Structure

Using parallel structure means using similar structure for similar elements. Careful writers use parallel structure for *coordinate elements* — elements with equal value in a sentence. Coordinate elements are connected by co-ordinating conjunctions (*and, but, or, nor, for, yet, so*) or are items in a series — words, phrases, or clauses. If coordinate elements in a sentence do not have a parallel pattern, the sentence becomes awkward and confusing.

Faulty   Management guarantees *that the old system will be replaced* and *to consider the new proposal.*

Parallel   Management guarantees *that the old system will be replaced* and *that the new proposal will be considered.*

Faulty   Use the cylinder with *a diameter of 3 ³⁄₁₆ inches* and *1½ inches high.*

Parallel   Use the cylinder with *a diameter of 3 ³⁄₁₆ inches* and *a height of 1½ inches.*
Use the cylinder that has *a diameter of 3 ³⁄₁₆ inches* and *a height of 1½ inches.*

Faulty   A successful firm is capable of *manufacturing a product*, *marketing it*, and *make a profit.*

Parallel   A successful firm is capable of *manufacturing a product*, *marketing it*, and *making a profit.*

Put the Main Idea First

Place the sentence's main idea — its subject — first. When you do so, you provide context for what follows. Notice the difference between the following two sentences. In the first, the main idea, "three pieces of equipment," comes near the end. The sentence is difficult to understand. In the second, the main idea is stated first; making the rest of the sentence easier to grasp.

A heaving line, which is a coiled line or rope, a heaving jug, which is a plastic container such as a milk or bleach bottle with a half-inch of water in it to give it buoyancy and a line or rope connected to the jug, and a ring buoy and line are three pieces of equipment used to tow victims to shore.   Main idea is last

Three pieces of equipment used to tow a victims to shore are   Main idea is first
a heaving line, which is a coiled line or rope; a heaving jug, which is a plastic container such as a milk or bleach bottle with a half-inch of water in it to give it buoyancy and with a line or rope attached; and a ring buoy and line.

GUIDELINES FOR WRITING CLEAR PARAGRAPHS

A paragraph consists of several sentences introduced by a topic sentence. The *topic sentence* expresses the paragraph's central idea, and the remaining sentences develop, explain, and support the central idea. You should generally place the topic sentence first and structure paragraphs coherently, making each sentence expand on the idea of the topic sentence.

Put the Topic Sentence First

In technical writing, almost all paragraphs begin with a topic sentence, followed by several sentences that explain its central idea. Researchers have discovered that readers comprehend a paragraph more quickly if the general or topic idea is placed first (Slater). This structure, called *deductive*, gives your paragraphs the direct, straightforward style preferred by most report readers.

To follow this guideline, state the central idea of each paragraph in a topic sentence at the beginning of a paragraph. Then follow with details that support and clarify the central idea, as in this example:

Assembly drawings are drawings that portray and explain the completely fabricated final product. Assemblies are drawn on a large sheet of paper. They generally have two or three views that relate all the information in the part to the reader. The assembly also explains different parts of the drawing through the bill of materials, a listing of all the different parts that make up the entire piece. It gives the costs, quantity, and a description of each part. The assembly drawings relay a basic overview of the entire product.	Topic sentence Supporting details

Structure Paragraphs Coherently

Coherent structure means that each sentence amplifies the point of the topic sentence. You achieve coherence by the words you choose and by the way you place sentences in paragraphs. You can indicate coherence by using words in four ways: by repeating terms, by placing key terms in the dominant position, by indicating class or membership, and by using transitions (Mulcahy 237).

Repeat Terms   Repeat terms in order to emphasize them. In the following example, "path" is repeated in the second sentence in order to provide further details.

Because fluid doesn't compress, its only path is between the gears and the housing. This path is least resistant — it allows the fluid to flow in that direction easily.

Use the Dominant Position   Placing terms in the dominant position means to repeat a key term as the subject — or main idea — of a sentence. In the following short paragraph, "contrast" is the dominant term.

> Contrast is one of the most important concepts in black-and-white printmaking. Soft contrast, just a range of grays, is used to portray a calm effect, such as a fluffy kitten. Hard contrast, on the other hand, creates sharp blacks and bright whites. It is used for dramatic effects, such as striking portraits.

"Contrast" always in dominant position

Maintain Class or Member Relationships   To indicate class or membership relationships, use words that show that the subsequent sentences are subparts of the topic sentence. In the following sentences, "store management" and "merchandising" are members of the class "career paths."

> Retailing has two career paths. Store management involves working in the store itself. Merchandising involves working in the buying office.

Provide Transitions   Using transitions means connecting sentences by using words that indicate a sequence or a pattern. Common examples are

Sequence	first . . . second . . . then . . . next . . .
Addition	and also furthermore
Contrast	but however
Cause and effect	thus so therefore

Arrange Sentences by Level   You can also develop coherence by the way you place sentences in a paragraph. In almost all technical paragraphs, each sentence has a level. The first level is the topic sentence. The second level consists of sentences that support or explain the topic sentence. The third level consists of sentences that develop one of the second-level ideas. Four sentences, for example, could have several different relationships. For instance, the last three could all expand the idea in the first:

1 First level
   2 Second level
   3 Second level
   4 Second level

Or sentences 3 and 4 could expand on sentence 2, which in turn expands on sentence 1:

1 First level
   2 Second level
      3 Third level
      4 Third level

As you write, evaluate the level of each sentence. You can do this by deciding if the idea in the sentence is level 2, a subdivision of the topic, or if it is level 3, providing details about a subdivision. Consider this example:

> (1)To develop an MRP program, our only investment will be two IBM printers. (2) One will be accessed by the line foreman for planning and controlling production orders. (3) The second printer will be used by the marketing and sales staff for customer orders. (4) The two printers will cost $1,000 each, plus a setup fee of $500, totaling $2,500.

The sentences of this paragraph have this structure:

> Level 1 To develop an MRP program, our only investment will be two IBM printers.
>   Level 2 One will be accessed by the line foreman for planning and controlling production orders.
>   Level 2 The second printer will be used by the marketing and sales staff for customer orders.
>   Level 2 The two printers will cost $1,000 each, plus a setup fee of $500, totaling $2,500.

GUIDELINES FOR REVISING FOR CLARITY

As you revise drafts, you should look for language that might cause confusion and make your writing harder to grasp. The following section will

help you develop an awareness of many constructions that cause impre-
cise, difficult-to-read sentences. Although you may produce some of
these constructions in your early drafts, you must learn to identify and
change them.

Avoid Strings of Choppy Sentences

Choppiness results from a string of short sentences. Since each idea appears as an independent sentence, the effect of such a string is to make all
the ideas seem equal to one another. As a consequence, the writing does
not clearly emphasize the important ideas. The remedy for this problem is
to combine and subordinate ideas so that only the important ones are expressed as main clauses.

Choppy	The bowl is made from stainless steel. This material makes it lightweight and durable. The weight of the bowl is four pounds. The use of stainless steel also allows for easy cleaning.
Clear	Because the bowl is made from stainless steel, it is durable, easy to clean, and lightweight — only four pounds.

Avoid Wordiness

Generally, ideas are most effective when they are expressed concisely. This
guideline does not mean that you should write sentences in telegram style
with all the short words deleted. But it does mean that you should eliminate all excess wording. Watch especially for redundancy, unnecessary
intensifiers (such as "very"), unnecessary repetition, unnecessary subordinate clauses, and unnecessary prepositional phrases.

Unnecessary subordinate clause	Two important concepts *that go along with* this field are inventory control and marketing.
Revised	Two important concepts *in* this field are inventory control and marketing.
Redundant intensifiers plus unnecessary subordinate clause	It is made of *very* thin glass *that is* milky white *in color.*
Revised	It is made of thin, milky white glass.
Unnecessary repetition	Tools can be classified into two general *categories* and be broken down further within these *categories.* The *categories* are hand *tools* and power *tools.*

Revised	Tools can be classified into two categories: hand and power tools.
Redundant	The tuning handle is a metal protrusion that can be easily grasped *hold of by the hand* to turn the gears.
Revised	The tuning handle is a metal protrusion that can be easily grasped to turn the gears.
Unnecessary repetition plus overuse of prepositions	The leadweight arms sit on each side of the specimen holder and *can be broken down into two parts. The two parts* of the arms are the abrading discs and the weights.
Revised	The lead weight arms, located on each side of the specimen holder, have two parts: the abrading discs and the weights.

Avoid Redundant Phrases

Since redundancy is such a common problem, here is a list of some common redundancies and their corrections:

Redundant phrase	*More concise word or phrase*
due to the fact that	because
employed the use of	used
basic fundamentals	fundamentals
completely eliminate	eliminate
alternative choices	alternatives
actual experience	experience
the reason is because	the reason is that
connected together	connected
final result	result
prove conclusively	prove
in as few words as possible	concisely

Avoid Abstract and General Terms

Given a choice between abstract, highly technical words and simple, concrete, specific words conveying the same meaning, always choose the concrete ones. If you must use a potentially confusing word, define it. Avoid jargon, the specialized language used and understood within a department or discipline, in reports to audiences who are distant or less knowledgeable.

Avoid or immediately clarify general terms (such as *excessive, numerous,* and *frequently*) and words of judgment (such as *effective, mediocre,* and *significant*). Specific words tell the reader far more.

Use *There Are* Sparingly

Overuse of the indefinite phrase *there are* (and its many related forms, *there is, there will be,* and so on) causes weak sentences because the effect is to bury the subject in the middle of the sentence. Most sentences are more effective if the subject is placed first.

Ineffective	*There are* three common types of splices that are used in electrical connections.
Effective	Three types of splices are commonly used in electrical connections.
Ineffective	Through research I found that *there are* six head styles that most companies carry in stock.
Effective	Through research I found that most companies stock six head styles.
Ineffective	*There are* six main parts to a small sailboat.
Effective	A small sailboat has six main parts.

Use *there are* for emphasis or to avoid the verb *exists.*

Weaker	No easy answers exist.
Stronger	There are no easy answers.

Avoid Nominalizations

Avoid using too many nominalizations — verbs turned into nouns by adding a suffix such as *-ion, -ity, -ment,* or *-ness.* Nominalizations weaken sentences by presenting the action as a static noun rather than as an active verb. Learn to find the true action in your sentences and express it with strong verbs.

Static	Research showed the *division* of waste products into biodegradable and nonbiodegradable substances.
Active	Research showed that we *can divide* waste products into biodegradable and nonbiodegradable substances.

Static	*Determination* of this percentage figure calls for an *analyzation* of various errors and failures that occurred in planned purchases and in allowances on merchandise given to employees.
Active	*Determine* this percentage figure by analyzing the errors that occurred in planned purchases and in allowances on merchandise given to employees.

Avoid Noun Clusters

Noun clusters are three or more nouns joined in a phrase. They crop up everywhere in technical writing and usually make reading difficult. Try to break them up.

Noun cluster	Allowing *individualized input variance* of data process entry will result in higher morale in the keyboarders.
Revised	We will have higher morale if we allow the keyboarders to enter data at their own rate.

Use *You* Correctly

Do not use *you* in formal reports (although writers often use *you* in informal reports). Use *you* to mean "the reader"; it should not mean "anyone" or "I."

Incorrect	I took the training course so that *you* could experience the problems firsthand.
Correct	I took the training course so that I could experience the problems firsthand.
Incorrect	The quality of performance, that is the amount of quality *you* can have and still remain competitive, is very important.
Correct	The quality of performance, that is, the amount of quality *a company* can enforce and still remain competitive, is very important.

Avoid Sexist Language

Language is considered sexist when pronouns indicate only one sex when both are intended. Careful writers rewrite sentences to avoid usages that are both insensitive and, in most cases, inaccurate. Several strategies will help you write smooth, nonsexist sentences. Avoid such clumsy phrases as *he/she* or *s/he.* An occasional *he or she* is acceptable, but a number of them in a short space destroys the sense of a passage.

Sexist	The clerk must make sure that *he* punches in.
Revise by using an infinitive	The clerk must make sure *to punch* in.
Revise by using the plural	The clerks must make sure that *they* punch in.

■ EXERCISES

Passive Sentences   Make the following passive sentences active.

1. Water from the storage drum is sent to the working drum leaving some water in the storage drum.

2. Keeping a desired overpressure is accomplished by using a pressure-drop control valve.

3. When the display has been noted and is no longer needed, press the R/S button to proceed with the program.

4. Tests are specified in the military's purchasing documents for the pouched meals; they include 27 dimensional checks on pouches.

5. Recently, it was determined that the purchase of a personal computer was needed.

Parallel Structure   Revise the following sentences, making their coordinate elements parallel.

1. It serves the purpose of pulling the sheet off the coil and to straighten or guide the sheet through the rest of the machine.

2. The two main functions of the tractor are supplying the necessary power to pull the load and to provide steering to guide the vehicle.

3. The seven steps in selling are (1) opening, (2) present merchandise, (3) handling, (4) to complete the sale, (5) suggesting, (6) record made of the sale, (7) finalize sale.

4. The student will be changing saw blades frequently for two main reasons: because of dull blades, and the need for special blades for different cutting operations.

5. The features that favor the Johnson receiver include more durability, better noise reduction and it is more efficient.

Choppiness   Eliminate the choppiness in the following sentences.

1. The impellers are jam free. They pivot on a turntable. Their name implies their shape. It is like blades of a rotor.

2. This woman is named Sue Jones. She is twenty-five years old. She is forty pounds overweight.

3. The taste test is important in food development. The test is run by lab technicians. The technicians are supervised by a sensory evaluator.

Wordiness   Revise the following sentences, removing unnecessary words.

1. The contracts also have a couple of special clauses to each one of them.

2. The soldering pencil, which is held the same way that a pencil is held, is much easier to control.

3. In caring for your blender, you should remember to care for it like you would any other electrical appliance.

4. When preparing the fabric it is wise to clean and treat the fabric, depending on the type of fabric, in the same manner as the finished garment would be cleaned, to remove this residue.

5. The plate is supported again by a thick wire that extends down into a connecting pin in the base, and the exact position of the pin is the pin next to the notch in the base.

Use of *There are*   Eliminate *there are,* or any related form, from the following sentences.

1. There are six basic requirements that blister material must fulfill.

2. Before you start to sew with the machine, there are a few steps in which one may or should follow.

3. With more shops being installed, there should be more traffic generated.

4. There are claims made by both companies that their computers are easy to learn and easy to use.

5. There are twenty-two different sizes of containers that lie in this capacity range.

Nominalization   Correct the nominalizations in the following sentences.

1. The capacity for an operator, in one day, for reconditioning plugs, is about 400.

2. For a successful business the satisfaction of the customer's needs at a profit must be done by the company.

3. The two workers accomplished the division of the material into two piles.

4. The process we use to accomplish these objectives is through the concise description of limitations, the investigation of alternatives and the establishment of communication channels.

5. Avoidance of tooling costs is accomplished if you pick a stock size.

Use of *You*   Correct the use of *you* in the following sentences.

1. The operator of the grain spout can maneuver it to deliver grain anywhere in the barge hold. You adjust the spout to compensate for changes in water level.

2. The Docu Data is also an integrated program, controlled by menus and commands. However, your flow control is poor.

3. In the microwave the aluminum heats up the oil and pops the popcorn. This process is where you get your increase in efficiency because you are using the heat to pop the popcorn.

Sexist Language   Correct the sexist language in the following sentences.

1. Each manager will direct his division's budget.

2. Every secretary will hand in her timecard on Friday.

3. If he understands the process, the machinist can improve production.

4. Each supervisor will present his/her budget to his/her vice-president at a meeting which s/he schedules.

5. We must inform each employee that she must fill out a W-4 form.

GUIDELINES FOR ORGANIZING CLEARLY

In addition to being familiar with guidelines for clear sentences and para-graphs, you should follow guidelines for clarifying your organization to your reader. These guidelines allow you to write readable documents that steer readers effortlessly through the topic. These strategies help almost all readers, especially those who are less knowledgeable or who are distant from the writer in the company structure. To clarify your organization:

Use context-setting introductions
Place important material first

Use preview lists
Use repetition and sequencing
Use structural parallelism

Use Context-Setting Introductions

Your introduction should supply an overall framework so that the reader can grasp the later details that explain and fill it. You can use an introduction to orient readers to the contents of the document in one of three ways: to define terms, to tell why you are writing, and to tell the document's purpose.

Define Terms   You can include definitions of key terms and concepts, especially if you are describing a machine or a process.

> Training is the process of preparing an individual to perform in a certain manner in a predictable situation. The training process has four steps: needs assessment, program development, delivery, and evaluation.

Tell Why You Are Writing   Although you know why you are writing, the reader often does not. To orient the reader to your topic, mention the cause of your writing. This method works well in memos and business letters. The following is a good start:

> At our recent committee meeting, you raised a number of questions about our new word processors. I couldn't answer all your questions then, but I have checked with our computer personnel, and I can now provide you with the information you requested.

State the purpose of the document   The *purpose* of the document means what the document will do for the reader. In the following one-paragraph introduction, the last two sentences state the purpose.

> Although adhesive bonding is a basic function of form/fill/seal (f/f/s) packaging, it is usually the least understood element of the package. Adhesives play a vital role in medical-device packaging today. They not only maintain package integrity but also make possible the physical combining of a wide variety of protective materials. In f/f/s packaging, adhesion must be achieved in line with a number of other packaging operations, some of which are not always conducive to good seals. This article will discuss the role played by adhesives or sealants. It will characterize their behavior in different types of packages at various stages of package formation and use. (Marotta 12)

Place Important Material First

The beginning of a section or a paragraph is an emphatic spot. Placing important material first emphasizes its importance. This strategy orients readers quickly and gives them a context so that they know what to look for as they read further. Put statements of significance, definitions, and key terms at the beginning.

The following two sentences, from the beginning of a paragraph, illustrate how a writer used a statement of significance followed by a list of key terms:

> A bill of materials (BOM) is an essential part of every MRP plan. For each product, the BOM lists each assembly, subassembly, nut, and bolt.

The next two sentences, also from the beginning of a paragraph, illustrate how a writer used a definition followed by a list of key terms:

> The assets of a business are the economic resources that it uses. These resources include cash, accounts receivable, equipment, building, land, supplies, and merchandise held for sale.

Use Preview Lists

Preview lists contain the key words to be used in the document. You can use lists in any writing situation to indicate organization. To make an effective list, you must understand the various list formats.

The basic list has three components: an introductory sentence that ends in a "control" word, a colon, and a series of items. The *control word* ("processes" in the sample below) names the items in the list and is followed by a colon. The series of items is the list itself. The items in the list are in italics in this sample:

> Wood is machined by one of two basic processes: *orthogonal cutting* and *peripheral milling.*

A more informal variation of the basic list has no colon, and the control word is the subject of the sentence. The list itself still comes at the end of the sentence. Here is an example:

> The two basic machining processes for wood are *orthogcnal cutting* and *peripheral milling.*

Writers can present lists on the page in two ways: horizontally and vertically. In a horizontal list, the items follow the introductory sentence as part of the text; the examples above are horizontal lists. In vertical lists, the items are listed in a column, thus emphasizing the items in the list. Here is an example:

> Wood is machined by one of these two basic processes:
> > orthogonal cutting
> > peripheral milling

Use Repetition and Sequencing

Repetition means restating key subject words or phrases from the preview list; *sequencing* means placing the key words in the same order in the text as in the list. The author of the following paragraph first lists the three key terms — *pneumatic, hydraulic,* and *electric.* She repeats them at the start of each sentence in the same sequence as in the list.

> The power supply required to move the segments of the manipulator can be of three types: pneumatic, hydraulic, and electric. Pneumatic power uses compressed air to supply power to certain components. It is usually used for lighter loads. Hydraulic power uses a compressed fluid for power and is usually used on heavier loads. Electric power is probably the most commonly used power source because it is easy to use and can handle most loads, light or heavy. The different kinds of power supply can be used on the same robot system.

Set up list using key words

Key words repeated at the beginning of each sentence

Use Structural Parallelism

Structural parallelism means that each section of a document has the same organization. The following two paragraphs have the same structure: first a definition, then a list of terms, then the definitions of the terms.

> **JOB DESIGN**
>
> Job design defines the specific work tasks of an individual or group. Job design focuses on four ideas: job scope, job depth, job enlargement, and job enrichment. Job scope involves the number and variety of tasks performed by the jobholder. Job depth is the amount of freedom that the jobholder has to plan and organize his work. Job enlargement gives a jobholder more tasks of a similar nature. Job enrichment increases both job scope and job depth. Adding responsibility is one of the most common ways of "enriching" a job.

Definition

List of terms

Terms defined

JOB ANALYSIS

Job analysis determines and reports pertinent information related to the nature of a specific job. Job analysis produces	Definition
both job descriptions and job specifications. A job description explains what the job is and what the duties, responsibilities,	List of terms
and general working conditions are. A job specification describes the qualifications that the jobholder must possess.	Terms defined

 SUMMARY

This chapter presents guidelines for writing effective sentences and paragraphs, and for revising sentences. In addition, the chapter presents guidelines for effectively organizing documents. These guidelines explain many skills you need to practice so that you have a repertoire of strategies to use both as you write and as you revise. Two sentence strategies of particular importance are to use the active voice and to avoid nominalizations. Following these two strategies alone will clarify much of your writing. A basic paragraph strategy is to place the topic sentence first and follow it with relevant supporting details. To make your organization clear to the reader, develop strategies for clearly indicating organization, such as context-setting introductions, preview lists, repetition, and structural parallelism. In general, remember to place important material first both in sections and paragraphs.

CHECKLIST FOR STYLE

☐ *Look for sentences with the passive voice. Change them to active.*

☐ *Look for sentences shorter than 12 or longer than 25 words. Either combine them or break them up.*

☐ *Check for parallel structure in each sentence. If coordinate elements are not parallel, make them so.*

☐ *Does the main idea come first in your sentences?*

☐ *Does each paragraph have a clear topic sentence?*

☐ *Check paragraph coherence by reviewing for*
repeated terms
key terms placed in dominant position

class or membership relationships
transitions

☐ *Evaluate the sentence levels of each paragraph. Revise sentences that do not clearly fit into a level.*

☐ *Read carefully for instances of the following potential problems.*
nominalizations
"there are" used too often
incorrect use of "you"
sexist language
choppiness
wordiness

☐ *Does your introductory material accomplish these three purposes?*
define?
tell why you are writing?
tell purpose?

☐ *Did you place the important material first?*

☐ *Did you use preview lists?*

☐ *Did you repeat key words from the preview list?*

☐ *Did you use structural parallelism?*

■ MODEL

The model that follows exemplifies many of the devices explained in this chapter.

A LAYMAN'S GUIDE TO NETWORK BAFFLEGAB (A GLOSSARY)

The world of local area networks, like any other field that is still young enough to be dominated by technologists, is full of baffling vocabulary. Not to worry. It wasn't so long ago that terms like "mainframe," "microprocessor," and "PBX" were equally mysterious. What follows will help you get up to speed in this technology and may even serve to impress your teenager (no mean trick these days).

The essential element in any network is the physical medium through which signals are transmitted. In local area networks, media are generally of three sorts: twisted pair, fiber optics, and coaxial cable. Twisted pair is the medium used for conventional telephone systems. Each path comprises two thin copper wires twisted together to form a single strand. Fiber optics is a new technology that uses pulses of light transmitted through a thin strand of glass, instead of pulses of electrical current transmitted through copper. Light waves can carry much more information than wires. Coax cable has been around for decades as the principal medium for transmitting TV signals over relatively short distances. If you have cable TV service at home, you know what this looks like.

The amount of information that can pass through a given medium in a given amount of time — its capacity — is described as its bandwidth. This is usually measured in bits per second (bps) and given in magnitudes of thousands of bps (Kbps or kilobits) or millions of bps (Mbps or megabits). Of the three media described in the preceding paragraph, twisted pair has the lowest bandwidth, optical fibers the greatest.

When coax cable is used, there are two transmission modes possible: baseband and broadband. In baseband transmission, the signals are imposed directly onto the medium, and the entire bandwidth of the cable is utilized for a single high-capacity channel. In the broadband mode, the cable's bandwidth is divided into a number of high-capacity channels enabling several transmissions to occur in parallel. Your cable TV service at home is a broadband system, which is why you can tape one program while viewing another at the same time.

Dividing bandwidth into a number of channels is accomplished through a process called multiplexing. There are two techniques: frequency division multiplexing (FDM) and time division multiplexing (TDM). In frequency division, the bandwidth is split into several parallel paths defined by bands of wave frequencies that are usually measured in millions of cycles per second (abbreviated as MHz). Look at the FM radio dial to see how this form of multiplexing is used to divide a single radio beam into several "stations." In time division, the entire bandwidth is used for every signal, but individual channels get into a queue, and each one has an allotted time slot to insert its signal into the sequence. The sequence is very short and is repeated very rapidly. It's like several lanes of traffic merging in an orderly fashion into a single high-speed tunnel.

FIGURE 4.1
Journal Article

When several communicating devices share a single channel (such as a group of terminals, each of which needs to access a computer only occasionally), there needs to be an access method which acts as the traffic cop. Two such major methods are "polling" and "random access." In local area networks these are implemented, respectively, by techniques called token-passing and CSMA/CD. Token-passing requires each device to wait its turn for a chance to use the channel. It knows when its turn comes because the network passes it a "token" in the form of a go-ahead signal. When it's through with the channel, the token is passed to the next device in a pre-determined order. CSMA/CD (Carrier Sense Multiple Access with Collision Detect) is a more random method, but one which generally yields higher utilization of the channel. It requires each device to constantly "listen" to the channel and jump on any time no other device is using it. If two devices jump on at exactly the same time (a collision), they both back off and re-try. It's really an automated party line with "politeness" programmed into it.

Most networks you've dealt with (such as AT&T's long distance telephone system) simply provide a transmission path and leave the rest to you. There are other kinds of data networks, however, that start with "raw" transmission capacity and offer other things along with it, such as speed matching, error detection, and protocol conversion. By adding these valuable functions to straightforward transmission, they have become known as value added networks (VANs). GTE's Telenet is one example of a nationwide VAN. ContelNet is an example of a local area network with VAN characteristics. (Finney 28)

FIGURE 4.1 (continued)

EXERCISES

1. In class, analyze the "Bafflegab" model on pp. 74–75 for organizational strategies. Also discuss the devices that make it interesting — comparisons, examples, brief narratives.

2. Edit the following passage, both for typos and for effective presentation:

> It has been brought to my attention by a few employeees that there is a communication gap amoung upper, middle and lower management to theri employees. I personally went to disciuss the situatio. the effects that the communication gap has on the employees is a big problem. The lack of communicationbetween the employees and managers is makeing many employees andunhappy wit their work. the workers have lowe moral, bitternes to their mangaers and the attitude of "I don't care." Anopther factor is the turnrver rate which has been increasoing in the last six months.

3. Edit the following passages for effective presentation. Put key terms and definitions first.

> a. Rolling mills are very important. They are the equipment used to reduce the heated metal into various shapes. There are 3 common types that are used to form the ingots into blooms, billets, and slabs. These are the 2-high rolling mills, reversing and nonreversing, and 3-high mills. The reversing action of the rollers allows the metal to be passed back and forth through the rollers, providing more deformation in a shorter period of time. There are the 4-high rolling mills. They have two rollers with small diameters supported by two rollers with larger diameters. These mills are used to form wide plate and sheets of metal. The added support of the larger rollers prevents variation in the thickness of the metal as it passes through the small rollers. Planetary mills allow for greater reduction of the metal in one pass. Small rollers surround a large support roller, similar to roller bearing. The metal is fed through the rollers at a much slower speed than other rolling mills. The feed rolls control the speed of the metal and the finishing rollers remove any slight scalloping defects caused by the small rollers. The planetary mill is used to reduce the thickness of plates of metal. Variations of the planetary mill are used to produce nonflat shapes. These mills have grooved rollers which control the flow of metal and produce the desired shape.
>
> b. There are two different ways pouches are cooled depending on which type of sterilization is used.
>
> When using the steam/air process a pre-cooling stage is used in which the steam is condensed, and compressed air replaces steam pressure. Cooling water is then added to condensate after which it is circulated.
>
> In using the water/air process there is no pre-cooling stage because steam pressure is not used in the process. Cooling water is circulated throughout the chamber cooling the pouches down. Once the pouches are cooled they are removed from the sterilization chamber. The pouches then go through a custom built dryer to prevent water spots.

4. In the following paragraphs, look for the strategies mentioned in this chapter. Were the strategies used well?

> A proposal is a longer report (usually several pages) that a department manager writes to her store manager or divisional manager. A department manager writes a proposal when she wants to request a major change in her department. For example, she has to write a proposal if she wants to remodel her entire department because of the major expenses it would involve. Included in the proposal are such things as sales figures previous to the change, a justification of how this proposed change will increase sales, and a pictorial plan of how the department will look after the proposed change.

> A report is a fairly brief document (usually only a few pages) that a department manager writes to the store manager after a special promotional event. A special promotional event might be a fashion show, a beauty lab, a food fest, or something similar. The sections of the report describe the special event, what happened, how smoothly it went, and how it affected sales.

5. In a professional periodical, read an article designed for experts. Analyze it in terms of the Guidelines for Organizing Clearly (pp. 68–72). Photocopy one or two pages that contain several strategies. Describe these strategies in a brief (one- to two-page) article written for the newsletter of a professional student club. Your goal is to tell fellow students how experts explain themselves to other experts.

6. From a technical encyclopedia in your field, photocopy a page that illustrates several of the audience strategies explained on pp. 69–72. In groups of three or four, read one another's photocopies to identify and discuss the strategies; list them in order of frequency. Give an oral report of your findings to the class, or as a group, prepare a one-page memo of your findings for your teacher. In either case, use specific examples.

7. Form groups of three to four. Choose a technical encyclopedia or a professional journal to investigate for writing style. Have each member bring to class a photocopy of a different page on a different topic. Analyze the sentences for frequency of the passive voice and for parallelism. Present a brief oral report of your findings to the class.

■ WRITING ASSIGNMENTS

1. You are a well-respected expert in your field. Your friend, an editor of a popular (not scholarly) magazine, has asked you to write an article describing basic terms in a new development in your field. Write a two- or three-page article, using the "Bafflegab" model (pp. 74–75) as a guide.

2. In class, write a description of a process commonly used in your field. Use three or four of the strategies named in the Guidelines for Organizing

Clearly (pp. 68–70). Name each strategy in the margin of your paper. Here are a few sample topics; however, if you are familiar with a different process, use it.

gram staining	filling out a balance sheet
analyzing nutrient intake	writing a simple computer program
figuring investment returns	testing impact resistance of corrugated board
developing film	
preparing an invitation to bid	cropping or enlarging a negative
squaring a piece of stock	setting up a tripod

 WORKS CITED

Finney, Paul R. "A Layman's Guide to Network Bafflegab (A Glossary)." *Management Technology* July 1983:28.

Marotta, Carl D. "The Vital Role That Sealants Play in Medical Device Packaging." *Packaging Technology* Dec. 1981:12–14.

Mulcahy, Patricia. "Writing Reader-Based Instructions: Strategies to Build Coherence."*The Technical Writing Teacher* 15.3 (1988): 234–243.

Slater, Wayne H. "Current Theory and Research on What Constitutes Readable Expository Text." *The Technical Writing Teacher* 15.3 (1988): 195–206.

Technical Writing Techniques

5

Researching

THE PURPOSE OF RESEARCH

QUESTIONING — THE BASIC SKILL OF RESEARCHING

METHODS OF COLLECTING INFORMATION

COLLECTING PUBLISHED INFORMATION

R esearching an issue in order to write a report is an essential part of professional life. On the job, people research everything from how high above the floor a computer screen should sit to how feasible it is to build a manufacturing plant. When you start a research project, you enter a challenging and rewarding process that culminates in a written document. Like a detective, you begin to search out facts. Your first attempts will often yield little or nothing — or even contradictory information. But as you continue to collect data, you will eventually discover the patterns in and the significance of the data. The more you conduct research, the more you will stretch your perceptions and your investigative skills. This chapter discusses the purposes of research, the essential activity of questioning, and practical methods of finding information.

THE PURPOSE OF RESEARCH

The purpose of research is to find out about a particular topic which you have perceived, and which has significance for you.

Your topic is a fact that has reached your perception. Topics can be broad general ones, such as the development of plastic oil cans, or they can be narrow specific ones, such as your office's need to purchase a new photocopier. The topic must have some significance. If your company manufactured plastic oil cans, could you make more profit? If the office obtained a new photocopier, would the office system run more smoothly?

Once you have realized its significance, you need to discover all you can about your topic. To discover new information, you research. Usually your research has a particular goal: to solve or eliminate some problem. The information — or data — turned up by your research will allow you to solve the problem. Sometimes the research attacks the problem directly: How much does a photocopier cost? What functions must it perform to help the office? Sometimes the research is more indirect: What is an effective arrangement for the room that contains the photocopier? The answer to that question might provide a basic framework to assess the most appropriate model for your office to purchase. Or you might discover information about heat or electrical needs or work flow that would cause you to purchase a different model from the one you initially had selected.

QUESTIONING — THE BASIC SKILL OF RESEARCHING

The basic skill of research is knowing how to ask questions. The answers to the questions are the facts you need. This section will explain how to discover the questions to ask and how to formulate them.

How to Discover Questions

To learn about any topic — from the role of chlorofluorocarbons in ozone-layer destruction to which photocopier to purchase for the office — you must ask questions. But what questions to ask? To formulate your questions you need to

1. Ask basic questions.
2. Ask questions about significance.
3. Consult the right sources.

Ask Basic Questions   Asking the right questions will help you learn the basic information about your topic. Basic questions include

- What are the appropriate terms and their definitions?
- What mechanisms are involved?
- What materials are involved?
- What processes are involved?

Ask Questions about Significance   Questions about significance are those which help you see the broad purposes and the context of your topic. Questions about significance include

- Who needs it and why?
- How does it relate to other items?
- How does it relate to current systems?
- What is its end goal?
- How do parts and processes contribute to the end goal?
- What controversies exist?
- What alternatives exist?
- What are the implications of that answer?

Consult the Right Sources   The right sources are the people or the printed material that have many of the facts you need. People who have the facts can answer your basic questions as well as your questions about significance. You approach such people for one of two reasons: to find out what they know or to find out what they need. An engineer, an experienced user, a salesperson — all have facts about the photocopying machines. People who already use the product have some sense of what they need. They know what they expect a photocopier to do, and they know the conditions that must be met to achieve those results. You may also consult printed sources for information. Printed sources can answer basic questions and questions of significance, often more thoroughly than people you ask.

How to Formulate Questions

There are essentially two kinds of questions: closed and open. A closed question generates a specific, often restricted answer. Technically a closed question allows only certain predetermined answers:

Closed question   How many times a week do you use the copier?
    __1-5 __ 6-10 __ 11-20 __ 21 +

An open question allows a longer, more involved answer:

Open question   Why do you use red ink?

In general, ask closed questions first in order to get basic, specific information. Then ask open questions to understand the subtleties of the topic. That sequence, however, is almost impossible to follow rigidly. Good questioners constantly switch back and forth between the two modes. If they hear a term they don't understand during the answer to an open-ended question, they ask for a definition, essentially a closed question. That definition could lead to a new open-ended question.

METHODS OF COLLECTING INFORMATION

You collect information — or find answers to your questions — in a number of ways. You can interview, survey, observe, test, and read. This section explains the first four. Collecting published information, especially in a library, is treated in a later section.

Interviewing

One convenient way to acquire information about a topic is to conduct an information interview. Your goal is to discover the appropriate facts from a person who knows them. To conduct a profitable interview, you must follow this process:

- Prepare carefully.
- Maintain a professional attitude.
- Probe.
- Record.

Prepare Carefully   To prepare carefully, inform yourself beforehand about your topic. Read background material and list questions you think will produce helpful answers.

If you are going to ask about photocopiers, for example, read about them before you interview anyone. If you do some reading, you will be in a better position to understand the significance of the answers you receive. You also need to make a list of specific questions to ask. If you don't, you may not receive all the information you want. It is easy to get off on tangents or to misappropriate time, perhaps by dwelling too long on side issues. A list will help you gather all the information you need and will allow you to pace yourself. To generate the list, brainstorm questions based on the basic questions and significance questions listed above. To interview someone about the photocopier problem, you might ask

How exactly does the photocopier malfunction?
Has that happened before? How often?
What happened just before the copier was used?

Maintain a Professional Attitude   When you interview someone, you should maintain a professional attitude. Schedule appointments for the interview beforehand. Set the context by explaining why you need to find out what they know. Be polite. Be sincere. Ask your questions as if the answers you seek are important. Most people are happy to answer questions for people who treat their answers seriously.

Be Willing to Probe   Most people know more than they say in their initial answers. You must be able to get at the material that's left unsaid. To do so, you probe by using three common probing strategies:

- Ask open-ended questions.
- Use echo technique.
- Reformulate.

The basic probe strategy is to ask an open-ended question. These questions produce answers that you can investigate further by using the other two strategies. *Echo technique* means to repeat significant words. Suppose an interviewee said, "Red really messes up a print run." In this case, an echo technique question would be "Messes up?" Almost always this technique will produce a longer, more specific answer. *Reformulating* means to repeat in your own words what the interviewee just said. The standard phrase is "I seem to hear you saying . . ." If your reformulating is accurate, your interviewee will agree, but if it is wrong, he or she will usually point out where.

Record the Answers   As you receive answers, write them down in a form you can use later. When you prepare a list of questions, put them on an 8½ x 11 sheet of paper, leaving enough room to record answers. When you record answers, write legibly and try to avoid just listing terms and abbreviations. Ask people to repeat if you didn't get the whole answer written down. After a session, go over your notes to amplify them so they will be meaningful later and to discover what you still don't understand so you can ask more questions.

Surveying

To survey is to ask people to supply written answers to your questions. You would use a survey to receive answers from many people, more than you could interview in the time you have allotted to the project. Surveys can help you determine basic facts or conditions and also the significance or importance of facts. Surveys have three elements: a context-setting introduction, closed or open questions, and a form that allows you to tabulate all the answers easily.

A context-setting introduction explains (1) why you chose this person for your survey, (2) what your goal is in collecting this information, and (3) how you will use the information. The questions may be either closed or open. The answers to closed questions are easier to tabulate. The answers to open questions can give you more insight into the situation. A general rule about survey questions is to avoid questions that require the respondent to research past records or to depend heavily on memory.

The form you use is the key to any survey. It must be well designed. Your goal is both to make it look easy to read (so people are willing to respond) and to make it easy to tabulate (so you can quickly tally the answers). For instance, if all the answers appear at the right margin of a page, you can easily transfer them to another page. Here is a sample survey using the photocopier problem.

SURVEY

In the past two weeks we have had many complaints about how difficult the new photocopier is to operate. In order to reduce frustration, we plan to develop a brief manual and to hold training sessions. To help us choose the most effective topics, would you take a moment to fill in the attached survey? Please return it to Saul Schwebs, 150 M Nutrition Building, by Friday, January 30. Thanks.

Context-setting introduction

How often do you use the copier?

Closed question

once a week	___
once every 2-3 days	___
once a day	___
several times a day	___

Do you use any of these functions:

Closed question

	Yes	No
2-sided copying	___	___
overlay copying	___	___
2-page copying	___	___
memory	___	___

Do you know how to do the following: Closed question

	Yes	No
select the proper paper key	—	—
get the copy count to zero	—	—
fill an empty paper tray	—	—
get the number of copies you wish	—	—
lay the paper on the glass with the correct orientation	—	—

Please describe your problems when you use the machine.   Open questions
Use the back of this sheet if you need more space.

Is there any topic you especially want us to cover?

Are you available at any of the times below during the week   Closed question
of 2/20–2/25 for a training session? Give first and second
preferences.

    M __   8–10__
    T __   10–12__
    W __   1–3 __
    Th__   3–5 __
    F __

How much notice do you need so that you can attend a   Closed question
training session?

    1 day __
    2 days __
    1 week__

Observing and Testing

In both observing and testing, you are carrying out a questioning strategy.
You are interacting with the machine or process yourself.

   Observing   Observing is watching intentionally in order to discover
the elements in a situation. You place yourself in the situation to observe
and record your observations. When you observe in order to collect information, you do so with the same questions in mind as when you interview:

- What are the basic facts?
- What is their significance?

To discover more about the office photocopier problems, you could learn a lot simply by watching people use the machine. You would notice where people stand, where they place their originals, how carefully they read instructions, which buttons they push, how they read the signals sent by the control panel, and so forth. If you discover that all steps move along easily except reading the control panel, you may have found a possible source of the complaints. By observation — looking in a specific way for facts and significances — you would find the data you need to solve the office problem.

Testing   To test is to change elements in a situation, to notice any differences, and to record the results. Testing, of course, is the heart of many scientific and technical disciplines. As such, it is much broader and more complex than this discussion of it. Nevertheless, simple testing often is a useful method of collecting information. Before you begin a test, you must decide what type of information you are looking for. In other words, what questions should the test answer for you? You must formulate these questions, devise your test, and then record the answers.

Suppose you had to decide which of two brands of photocopiers to buy. Before buying one, you might test both on a trial basis. You would first determine — probably through interviews or surveys — the questions you want answered. These questions should reflect the users' conce. Typical questions might be

> Which one produces 100 copies faster?
> Which one has better clarity?
> Which one generates more heat?
> Which enlarging option is easier to use?

After you determine suitable questions, you have people use both machines and then record their answers to your questions. To record the answers, you will need a recording form, much like the one used for surveys. If many people are involved in the test, you might want to survey them all to collect the best possible data.

COLLECTING PUBLISHED INFORMATION

This section discusses the basic techniques for locating and collecting published information for use in a report. As with all research projects, you must plan carefully: you must develop a search strategy, review reference material, and record your findings.

Develop a Search Strategy

With its thousands of books and periodicals, the library can be a bewildering place. The problem is to locate the relatively small number of sources that you actually need. To do so, you need to plan your activity in the library.

The trick is to develop a "search strategy" (Madland). You must find out what to look for and where to look. To develop a strategy, determine your audience, generate questions, and follow search guidelines.

Determine Your Audience   As in any writing situation, you must determine your audience and their needs. Are you writing for experts? Will they already understand the concepts in the report? Will they use your report for reference or background information, or will they act on your findings? If you are writing a seminar paper for an upper-level class, your audience — the teacher and seminar members — will expect a highly technical paper. Since none of them, perhaps not even the instructor, will know as much about the topic as you do, you will have to lead them into it gradually. You will therefore build a base of definitions, then expand into a detailed technical discussion.

Generate Questions   You will facilitate your research if you generate questions about the topic and its subtopics. It is helpful to ask these questions both before and as you continue your research. These questions are the same for library research as for interviews. You need to ask the same kinds of questions — about basics and about significance:

What is it made of?	What are its causes?
How is it made?	Who makes it?
What are its major divisions?	What are its effects?
Who uses it?	What is its future?
Where is it used?	How is it regulated?
Where is it made?	
What is its history?	
Do experts disagree about any of these questions?	

Such questions will help you focus your research, enabling you to select source materials and to categorize information as you collect it.

Follow Search Guidelines

The guidelines given below will help you find relevant material quickly.

Consider the Age of the Information   Because of the length of the publishing process, the information in books is often a year or more old — and can be out-of-date. The information in recent periodicals (magazines and newspapers) is usually much more current, only several months old. Some data bases contain even more recent information. If you need the most recent information, look in periodicals and full-text data bases (if you have access to them). If you need well-established information, use books.

Consider the Technical Level of the Information   If you need information at a high technical level, use technical journals, interviews with professionals, and even sales literature. On the job, you would also use technical reports from the company's technical information department. If you need general information, use popular magazines and newspapers. Books can provide both technical and general information.

Watch for Key Documents   As you collect articles, review their bibliographies. Some works will be cited repeatedly. These documents, whether articles or books or technical reports, are *key documents*. They contain a discussion that experts agree is basic to understanding the topic. To research efficiently, you should find and read these documents early in the process.

Find Key Terms   *Key terms* are the specific words or phrases that all writers in a particular field use to discuss a topic. You need to recognize them and master their definitions since these terms are central concepts for your topic. You can find more about these terms in specialized encyclopedias, the card catalog, periodical indexes, abstracts, and data bases. Writers will repeat them in their articles. These terms will also lead you to other useful terms, for instance, in cross references and in indexes.

Review Reference Aids

To locate ideas and material, you can use all of the following reference aids: encyclopedias, traditional card catalogs, computerized catalogs, periodical indexes, abstracts, and data bases.

Read Encyclopedias in Your Field   Encyclopedias can frequently serve as points of departure. They give background information, and many of their articles contain bibliographies giving sources of more specific and detailed data. As brief introductions, they provide basic frameworks for research topics. As you read an article in an encyclopedia, you will often find the standard terms and subdivisions of your topic. Learning those terms and subdivisions will enable you to use indexes and data bases more efficiently.

As a professional, you need to learn and use the specialized encyclopedias in your field. To list all these would take much more space than is available here. A representative sample includes:

Encyclopedia of Chemical Technology
Encyclopedia of Science and Technology
Encyclopedia of Chemistry
Encyclopedia of Chemical Process Equipment
Encyclopedia of Electronics
Encyclopedia of Management

Most fields have specialized encyclopedias. Learn the encyclopedias in your field and consult them regularly.

Use the Card Catalog   The card catalog lists every book in the library. For each book, the catalog contains (1) an author card, (2) a title card, and (3) one or more subject cards, as shown in Figure 5.1. Except for their main headings, the cards are exactly alike. If you know how to read a card, you can glean important information from it. When you start to search for information, you can use subject cards to find the books available about your topic. Think of subject words under which you might find pertinent information; then examine the cards in the subject section of the card catalog. In Figure 5.1, the subject card is titled "population." Subject cards are heavily cross-referenced so they often refer you to another subject, as here you are referred to pollution and human ecology.

While the cards contain much information, several items are especially helpful. The *call number* explains where the book can be found on the shelves. The author's name gives you a clue to other possible sources. To see if he or she has written other books that might be useful, check the author cards. The year of publication is important if rapid advances are being made in the subject. The bibliographic notation tells you if the book contains a bibliography, which could lead you to additional sources of information.

Use the Computerized Catalog   Instead of card catalogs, many libraries now have computerized catalogs, which users access through a computer terminal. These systems show the catalog information on screens, which the user accesses by keyboard, following usually simple instructions shown on the screen.

Given on pp. 93–94 are four screens from a computerized system. Figure 5.2 shows a sample circulation screen. Note the author, title, and publisher information on the top half, and the call number and shelf status on the bottom half.

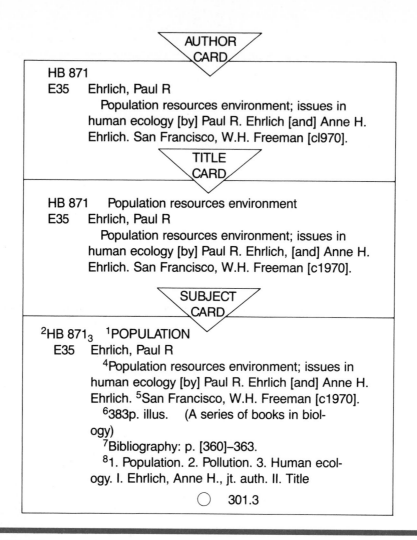

FIGURE 5.1
Author, Title, and Subject Cards

Figure 5.3 shows a bibliographic screen that describes the book in detail. This screen notes that the book has a bibliography and points out relevant subject headings. Figure 5.4, an author bibliography screen, lists other books by the same author. Figure 5.5, a subject bibliography screen, lists the three books contained in the library under the subject heading "Packaging Research." Notice, however, that the search line names only "Packaging." (The system actually listed over 50 subdivisions of Packaging, cataloging hundreds of books.)

```
                      CIRCULATION INFORMATION

        AUTHOR:  Sacharow, Stanley.
        TITLE:  Packaging regulations / Stanley Sacharow.
        EDITION:                 PUBLISHER:  AVI Pub. Co.,    YEAR:  c1979.
                                 TOTAL COPIES: 1     TOTAL HOLDS: 0

          DETAIL INFORMATION FOR UW-STOUT LIBRARY LEARNING CENTER
        1.  CALL#:  KF1665.83
        STATUS: CHECKED OUT, DUE 4/28/89 LOCATION: STACKS  MEDIA: BOOK
```

FIGURE 5.2
Circulation Screen
Source: Online Computer Library Center, Inc. Used by permission.

```
                      UW-STOUT LEARNING CENTER CATALOG
                         BIBLIOGRAPHIC INFORMATION

        AUTHOR:           Sacharow, Stanley.
        TITLE:            Packaging regulations / Stanley Sacharow
        PUBLICATION INFO:  Westport, Conn. :  AVI Publ. Co.,
                          c1979.
        PHYSICAL DESC:    x, 207 p.  : ill. ;
                          24 cm.
        NOTES:            Includes bibliographies and index.
        SUBJECT HEADINGS:  Packaging Law and legislation United States.

        REMEMBER YOU CAN ENTER H FOR HELP, Q TO START OVER, P TO PAGE BACK,
        OR YOU CAN BEGIN A NEW SEARCH FROM ANY SCREEN.
        PRESS RETURN FOR MORE INFORMATION:
```

FIGURE 5.3
Bibliography Screen
Source: Online Computer Library Center, Inc. Used by permission.

```
SEARCH:  A SACHAROW STANLEY
TOTAL MATCHES: 10

  1. AUTHOR:  Sacharow, Stanley.
     TITLE:  Handbook of Package materials /
     DATE:  c1976.
  2. AUTHOR:  Sacharow, Stanley.
     TITLE: A packaging primer /
     DATE:  c1978.
  3. AUTHOR:  Sacharow, Stanley.
     TITLE:  Packaging regulations /
     DATE:  c1979.
  4. AUTHOR:  Sacharow, Stanley.
     TITLE:  Food packaging:  a guide for the supplier, processor and distr
     DATE:  c1970.
PRESS RETURN TO CONTINUE LIST.
ENTER LINE NUMBER FOR CALL NUMBER AND CIRCULATION INFORMATION.
ENTER B FOLLOWED BY THE LINE NUMBER (EX.B4) FOR BIBLIOGRAPHIC INFORMATION
REMEMBER YOU CAN ENTER H FOR HELP, Q TO START OVER, P TO PAGE BACK,
R FOR OTHER RELATED TERMS -- OR BEGIN A NEW REQUEST FROM ANY SCREEN.
ENTER:
```

FIGURE 5.4
Author Bibliography Screen
Source: Online Computer Library Center, Inc. Used by permission.

```
SEARCH: S PACKAGING
TOTAL MATCHES: 3

  1. AUTHOR:  Container Corporation of America.
     TITLE:  An approach to packaging /
     DATE:  (197-?)
  2. AUTHOR: James, Theresa A.
     TITLE: Thermal case for Del Monte juicebars
     DATE:  c1984.
  3. AUTHOR:  Anderson, Paul W.
     TITLE:  Refrigerated distribution model
     DATE:  c1984.

PRESS RETURN TO CONTINUE LIST.
ENTER LINE NUMBER FOR CALL NUMBER AND CIRCULATION INFORMATION.
ENTER B FOLLOWED BY THE LINE NUMBER (EX.B4) FOR BIBLIOGRAPHIC INFORMATION
REMEMBER YOU CAN ENTER H FOR HELP, Q TO START OVER, P TO PAGE BACK,
R FOR OTHER RELATED TERMS -- OR BEGIN A NEW REQUEST FROM ANY SCREEN.
ENTER:
```

FIGURE 5.5
Subject Bibliography Screen
Source: Online Computer Library Center, Inc. Used by permission.

A major advantage of the computerized catalog is that it almost instantaneously gives the writer a working bibliography. For instance, if you type "S[ubject] Population," the computer will respond with a complete list of Population categories in the card catalog: Population — Demographics; Population — Major Cities; Population — Minority Statistics; Population — History of U.S. Census; or whatever. If you select one of these categories, the computer will list all the available books.

To use the system effectively, the user must use accepted subject headings. Since most of the systems are keyed to Library of Congress headings, you can use the two-volume *Library of Congress Subject Headings* to find lists of accepted subject headings. Figure 5.6 shows an entry from *Subject Headings* (2: 2323–24). The boldface word (in Figure 5.6, "Packaging") is used in the catalog. If you enter it into the computer, the system will respond with a list of books on that subject. Notice that the other symbols will help you find the correct subject heading:

sa see also; related headings that may be used in the catalog.

xx other related subjects.

— subdivision of the boldface heading; in the catalog they appear after the boldface head: **Packaging – Law and Legislation.**

See — use this heading instead of the one directly above it.

> **Packaging**
> *sa* Aluminum in packaging
>   Child-resistant packaging
>   Cigarette package labels
>   Container industry
>   *subdivision* Packaging *under subjects,*
>   e.g. Confectionery – Packaging
>
> *xx* Advertising
>   Color in advertising
>   Containers
> — Law and legislation (*Indirect*)
>   *xx* Consumer protection – Law and
>   legislation
> — Research
>   See Packaging research

FIGURE 5.6
Sample Subject Heading Entry

Use Periodical Indexes   Periodical indexes will help you find magazine articles in the same way that catalogs assist in locating books. Periodicals are published for all technical areas, and as mentioned earlier, they are especially good sources for keeping professionals informed of innovations in their fields. You should become familiar with the indexes that cover periodicals in your field.

The following are a few of the many periodical indexes for technical subjects:

Applied Science and Technology Index: subject index to articles in technology, engineering, science, and trade, including the areas of automation, construction, electronics, materials, telecommunication, and so forth

Business Periodicals Index: subject index to articles in all areas of business, including automation, labor, management, finance, marketing, public relations, communication, and so forth

Cumulated Index Medicus: subject-author index to articles that deal with biomedical topics

Engineering Index: subject-author index to publications of industrial organizations, government, research institutes, and engineering organizations; includes annotations (descriptions) of the articles

Index to U.S. Government Periodicals: subject-author index to publications of more than one hundred agencies of the United States government; includes information on hundreds of technical topics

Microcomputer Index: subject index to articles about microcomputers; includes lists of authors, companies, and products; also includes descriptive abstracts

Most indexes work the same way: they list articles beneath subject headings, as shown in Figure 5.7. Subheadings appear beneath many subject headings to identify more specific areas of information. The subheadings (not shown in Figure 5.7) beneath "Diffusion" include "Computer Simulation," "Mathematical Models," and "Tables, Calculations."

Figure 5.8 explains the first entry in Figure 5.7, an article on diffusion by M. Jansson and P. Stilbs.

A recent development in periodical indexing is a computerized index like *Infotrac*. The vendor company, Infotrac, sells the library a large compact disk that is attached to a special computer. The disk contains thousands of article titles. These titles can be accessed by key words such as *chlorofluorocarbons* or *proposals*. After you type in the key word, the computer lists all the relevant titles. These titles of articles are updated regularly so you get the most recent information available. These computerized indexes contain the same information as the regular paper copy of the in-

Design
Circulation induced by coastal diffuser discharge. E. E. Adams and J. H. Trowbridge. bibl il diags *J Waterw Port Coast Ocean Eng* 111:973-84 N '85

Diffusion
See also
Mass transfer
Semiconductors—Doping

A comparative study of organic counterion binding to micelles with the Fourier transform NMR self-diffusion technique. M. Jansson and P. Stilbs. bibl *J Phys Chem* 89:4868-73 O 24 '85

Diffusion in the stably stratified atmospheric boundary layer. J. C. R. Hunt. bibl diags *J Clim Appl Meteorol* 24:1187-95 N '85

Diffusion of benzene, toluene, napthalene, and phenanthrene in supercritical dense 2,3-dimethylbutane. C. K. J. Sun and S. H. Chen. bibl diag *AIChE J* 31:1904-10 N '85

FIGURE 5.7
Applied Science & Technology Index
Source: Applied Science & Technology Index. Copyright © 1986 by the H. W. Wilson Company. Material reproduced with permission of the publisher.

dexes. Their value is that they contain several indexes so you don't have to search each index individually.

Use Abstracts   Hundreds of abstracting services provide information about journal articles in particular fields. These abstracts are generally descriptive rather than informative; that is, they state the scope of articles but do not summarize them. While they do not serve as substitutes for the articles, they do allow you to reject inappropriate articles immediately. You should become familiar with the relevant abstracting services in your field. The following is a brief list of technical abstracts:

Aeronautical Engineering Index
Chemical Abstracts
Engineering Abstracts
Geological Abstracts
Metallurgical Abstracts
Mineralogical Abstracts

Abstracts of Instruction and Research Materials
Abstracts of Health Care Management Studies
Agricultural Index
Current Packaging Abstracts

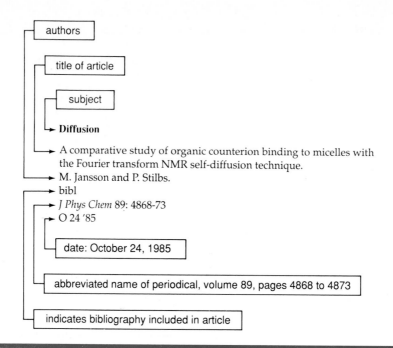

FIGURE 5.8
Explanation of Entry from Applied Science & Technology Index
Source: Applied Science & Technology Index. Copyright © 1986 by the H. W.
Wilson Company. Material reproduced with permission of the publisher.

Use Data Bases   Through computerized data bases, you can identify
many sources relevant to your topic. A *data base* is a collection of information on a specific subject. Usually the data base is a huge annotated bibliography. If you access it correctly, it will list the relevant articles on a
particular subject. You then have a customized bibliography without the
time-consuming labor of searching through volumes of abstracts and
indexes.

Data bases are particularly helpful for obtaining current information;
sometimes entries are available within a week of their appearance in print.
Since data bases are accessed by phone, using a modem, they are expensive; the costs include the phone line charge and a service access charge,
often figured in fractions of a minute. A typical charge is $100 an hour or
$1.66 a minute. At those rates you need to prepare carefully and thor-
oughly before you access the data base.

There are three kinds of data bases: full-text, bibliographic, and statistical (Miller). A *full-text data base* will provide you with an exact copy of the
article you are seeking. This type of data base, often used by financial managers, is useful for keeping abreast of important developments in a field.

Bibliographic data bases provide the researcher with publication data — and often with abstracts of articles on a certain subject. *Statistical data bases* provide economic data and financial profiles of companies.

The bibliographic data base is probably the most helpful for student researchers. To access this type of data base, you provide subject words — called *descriptors* — which are sent from your computer terminal, via a telephone line, to the data-base computer. The data-base computer selects all the articles that contain your descriptor words in the title. Depending on your commands, it will tell you how many articles it has found, display entries and abstracts on the screen, and print out the bibliography.

To search a data base effectively, you must choose your descriptors carefully. If you pick a common term, say "manufacturing" or "packaging" or "retail," the computer might tell you it has found 10,000 items. To narrow the choices, you need to combine a term like "packaging" with descriptors like "plastic" and "microwaveable." It will then search for titles that contain those three words and will present you with a much smaller list of perhaps 10 to 50 items. To make a search meaningful, you need to become adept at choosing descriptors. Although some companies publish lists or present them on a "help" screen, many do not.

Data bases provide information on almost every topic. The following partial list (Dialog) presents subject areas on the left and the data bases that contain information about them on the right.

Business/Economics	ABI/INFORM, Chemical Industry Notes, Investext, Foods AdLibra, BLS Labor Force, U.S. Exports
Chemistry	CA Search, Paper Chem, Sci Search
Directories	American Men and Women of Science, Career Placement Registry/Student
Energy and Environment	Aquaculture, DOE Energy, Pollution Abstracts, Waternet
Materials and Science	Nonferrous Metals Abstracts, Textile Technology Abstracts, World Textiles
Medicine and Biosciences	Embase, International Pharmaceutical Abstracts, Zoological Record
Science and Technology	BHRA Fluid Engineering, Geoarchive, Mathfile, Microcomputer Index, Standards and Specifications, Weldasearch

Record Your Findings

As you proceed with your search strategy, you must record your findings. To do so, you construct a bibliography, take notes, consider using visual aids, and decide whether to quote or paraphrase important information.

Make Bibliography Cards As you find potential sources of information in the card catalog and periodical indexes, list them on separate 3-by-5-inch cards. These bibliography cards should contain the name of the author, the title of the article or book, and facts about the book's publication. Record this information in the same form that you will use in your bibliography. Also record the call number and any special information about the source, for instance that you used a microfiche version. Such information will facilitate finding the source at a later date.

Take Notes After completing your bibliography cards, go directly to the sources to begin reading and taking notes. On each card, write the topic from your paper that this material supports. On every card, write the author of the work and the page number from which you are recording information. Each card should contain notes on a single subject and from a single source. This practice may seem like a waste of cards, but it greatly simplifies arranging your notes when you finally organize the report.

Make Visual Aids Visual aids will enhance your report. Remember, you are presenting complex, advanced information, and visual aids will help your readers comprehend it. You have two sources of visual aids: those that you find in your research and those that you create. If a key source has a visual aid that clarifies your topic, use it, citing it as explained in the Documentation Appendix (pp. 441–459). As you read, however, be creative and construct your own visual aids. Use flow charts to show processes, tables to give numerical data, drawings to explain machines — whatever will help first you, and later the reader, grasp the topic. The process of constructing them will help you clarify your ideas. Of course, you must refer to and discuss each visual aid in your text. Chapter 7 discusses visual aids in more detail.

Quoting and Paraphrasing

Two important skills for writing research reports are knowing how and when to quote and paraphrase. *Quoting* is using another writer's words verbatim. Use a quote when the exact words of the author clearly support an assertion you have made or when they contain a precise statement of information needed for your report. Copy the exact wording of

- definitions
- comments of significance
- important statistics

Paraphrasing means conveying the meaning of the passage in your own words. Learning to paraphrase is somewhat tricky. You cannot just change a few words and then claim that your passage is not "the exact words" of the author. To paraphrase, you must express the message in your own original language. Write paraphrases that

- outline processes or machines
- give illustrative examples
- explain causes, effects, or significance

Whether you quote or paraphrase, you must cite the source (both in the text and in the bibliography). When you write reports, you must present your quotations and paraphrases in the accepted manner. The rest of this section explains some basic rules for quoting and paraphrasing. Complete rules for documenting sources appear in the Documentation Appendix (pp. 441–459). All of the examples are based on this excerpt from *Cosmos*, by Carl Sagan.

> The Cosmos was discovered only yesterday. For a million years it was clear to everyone that there were no other places than the Earth. Then in the last tenth of a percent of the lifetime of our species, in the instant between Aristarchus and ourselves, we reluctantly noticed that we were not the center and purpose of the Universe, but rather lived on a tiny and fragile world lost in immensity and eternity, drifting in a great cosmic ocean dotted here and there with a hundred billion galaxies and a billion trillion stars. We have bravely tested the waters and have found the ocean to our liking, resonant with our nature. Something in us recognizes the Cosmos as home. We are made of stellar ash. Our origin and evolution have been tied to distant cosmic events. The exploration of the Cosmos is a voyage of self-discovery. (318)

When you quote, you place quotation marks before and after the exact words of the author. Usually you precede the quotation with a brief introductory phrase:

> According to Carl Sagan, "For a million years it was clear to everyone that there were no other places than the Earth" (318).

If you wish to delete part of a quotation from the middle of a sentence, use ellipsis dots:

> Sagan points out that "in the last tenth of a percent of the lifetime of our species . . . we reluctantly noticed that we were not the center and purpose of the Universe" (318).

If you wish to insert your own words into a quotation, use brackets:

> Sagan feels that "in the last tenth of a percent of the lifetime of our species, in the instant between Aristarchus [a Greek astronomer] and ourselves" (318), we have realized the Earth's position in relation to the rest of the Universe.

When you paraphrase, you rephrase the passage using your own words. Be sure to indicate in your text the source of your idea — the author and the page number (in parentheses) on which the idea is found in the original:

> According to Sagan, in the last several thousand years humankind has dramatically revised its opinion of its place in the Universe. While once people claimed earth was the center of the universe, now many see it as a very small part of a large, complex system. However, the very immensity of the universe, like the immensity of the ocean, seems to call people forth to discover both it and themselves (318).

Remember, when you quote or paraphrase, that you have some obligations both to the original author and to the report reader:

1. When in doubt about whether an idea is yours or an author's, give credit to the author.
2. Do not quote or paraphrase in a way that misrepresents the original author's meaning.
3. Avoid stringing one quote after another. This strategy, while often requiring great care to implement, is hard to read.

■ SUMMARY

To research effectively you must ask, then answer, questions. You must ask basic questions and questions about significance. Both types will generate information about your topic, often leading you to new questions. Five methods for finding answers to your questions are by interviewing, surveying, observing, testing, and using the library. To interview, you ask individuals for information about a topic. You must prepare questions, including probes that will cause you to ask more questions. To survey, you ask a group of people to respond to written questions. You must prepare questions and an answer form that is easy to tabulate. To observe, you watch a situation in action, looking for repeated actions and basic facts — and their significance. To test, you compare two items in terms of some criterion or set of criteria, for instance, how easy it is to use the enlarging function of two brands of photocopiers. Library research is a key method of finding information. As with other research methods, writers must first devise questions, then find answers. To find answers, you must know how to use encyclopedias, card catalogs, indexes, abstracts and data bases. To be effective, you must take notes as you read.

■ CHECKLIST FOR RESEARCH

☐ *Name the basic fact or problem that you perceive.*

☐ *Who thinks that this topic is important?*
Who would be interested in this topic? What is the nature of their interest? General curiosity? Professional involvement?

☐ *What is interesting about the topic?*
What are the implications of the topic? For instance, could a reduction in refrigeration (the topic) affect the ozone layer (implication)?

☐ *List three questions that you want to have answered about this topic.*

☐ *How can you find information about this topic?*
Who has the information that you need? Are there people in the office that you can talk to? experts who publish their information? Do you need to read? interview? survey? some combination of the three?

☐ *List the steps you will follow to find the information.*
Include a time line that estimates the number of hours and days you will need for each step.

☐ *What form will you use to record the information you discover?*
Pay special attention to this if you interview or survey. Should you construct a response form on which you can record answers?

☐ *Look for indexes and abstracting services that list articles on your topic.*

☐ *List key words that describe your topic.*

☐ *Create a working bibliography — a list of authors and their works that you think will provide helpful information.*

☐ *Read the useful publications, taking notes in the form of quotes and paraphrases.*

■ EXERCISES

1. Write a report in which you analyze and evaluate an index or abstracting service. Use one that is in your field of interest, or ask your instructor to assign one. Explain which periodicals and subjects the service lists. Discuss whether or not it is easy to use. For instance, does it have a cross-reference system? Can a reader find key terms easily? At what level of knowledge are the abstracts aimed? Beginner? Expert? The audience for your report will be other class members. Your objective is to help them with their research topics.

2. Write a memo in which you analyze and evaluate a reference book in your field of interest. Explain its arrangement, sections, and intended audience. Is it aimed at a lay or technical audience? Is it introductory or advanced? Can you use it easily? Your audience will be other class members. Your intention is to help them with their research projects.

3. Examine *Ulrich's International Periodicals Directory* or *The Standard Periodical Directory* in the library. Each lists the periodicals published for all fields. Select three periodicals in your area of interest and inspect several copies of each. Then write a memo report. Analyze and evaluate the periodicals. Discuss length of articles, intended audience, and article style. Which periodical is most helpful for a student research project? Why?

4. Divide into groups of three or four. Construct a 3- or 4-item questionnaire to give to your classmates. Write an introduction, use open and closed questions, and tabulate the answers. At a later class period, give an oral report on the results of your questionnaire. Use easy topics that inquire about the class's demographics (size of native city, year in college) or else their level of knowledge of some common area in a field chosen by the group (recycling one-liter plastic soft-drink bottles).

5. Form small groups of three or four, based on your major or your interest in a particular topic (for instance, solar heating systems, aspartame, thermoset plastics, diesel engines, industrial organization, just-in-time manufacturing). Discuss the topic briefly to formulate questions about the topic. Examples might be "What kinds of processes are used to make that?" "How will this new technology affect the workplace?" "What are the opinions of experts on this topic?" "Do the experts disagree? If so, how?" Once you begin to explore, you will generate many questions. Select two questions from the list. Use library resources to find articles that contain the answers to those questions. Then write an informal report in which you explain the answers you discovered in the articles you read.

■ WRITING ASSIGNMENT

I. Write a short research report explaining a recent innovation in your area of interest. Use at least six recent sources. Use quotations, paraphrases, and one of the citation formats explained in the Documentation Appendix (pp. 441–459). Organize your material into sections that give the reader a good sense of the dimensions of the topic. Some kinds of information you might present are

- Problems and potential solutions regarding the development of the innovation
- Issues debated in the topic area
- Effects of the innovation on your field or on industry in general
- Methods of implementing the innovation

Your instructor might require that you form groups to research and write this report. If so, he or she will give you a more detailed schedule, but you must formulate questions, research sources of information, and write the report. Use the guidelines outlined in Chapter 2.

WORKS CITED

Applied Science and Technology Index 74.2 (1986): 68.

Dialog Information Services. *Subject Guide to Dialog Data Bases.* Palo Alto, CA: Dialog, 1984.

Library of Congress. *Library of Congress Subject Headings,* 10th ed. 2 vols. Washington: Library of Congress, 1986.

Madland, Denise. Presentation. U of Wisconsin-Stout. 26 March 1986.

Miller, Tim. "Increasing Your Business I.Q. with On-line Data." *Popular Computing* Dec. 1985: 57+.

Sagan, Carl. *Cosmos.* New York: Random, 1980.

6

Summarizing and Outlining

SUMMARIZING

OUTLINING

$\mathbf{I}$n a world awash in information, the ability to construct and present concise, short versions of long documents is not only helpful but essential. The ability to *summarize* — or *abstract*, the terms are nearly synonymous — is basic to technical writers. You will often summarize your own documents, as this book explains in many chapters. But you will also need to summarize documents written by others. In addition, you must know how to outline both the material you read and the documents you intend to write. This chapter explains those skills.

SUMMARIZING

This section defines summaries and abstracts, explains the various audiences that use them, and presents the skills you need to write them.

Definitions of Summaries and Abstracts

Some confusion exists about the definitions of these two words. Both words indicate a short restatement of another document. In technical writing, a *summary* is a restatement of the major findings and conclusions of a document, placed *after* the text of the body and intended to help readers comprehend and review the text. An *abstract*, however, while also a restatement, is read *before* a reader studies the text (ANSI 7). Abstracts are designed to stand alone. Abstracts, which generally are short versions of journal articles, appear in two places: with the article in the periodical and as an independent unit provided by abstracting services (discussed in Chapter 5). There are two kinds of abstracts: indicative and informative. *Indicative* abstracts list the sections of an article; *informative* abstracts briefly explain the article's main point and the main support for that point.

In practice, the terms *abstract* and *summary* are often used interchangeably. Many report writers — as later chapters explain — put summaries first, before the body. The "executive summary," a brief overview of the report, is commonly used in longer technical reports. Furthermore, the general goals of a summary and an informative abstract are the same — to present the main point or to give the gist of the contents.

Audiences for Summaries

Readers need summaries for various reasons:

- to find out the gist of the report or article without reading the entire document
- to discover if the report or article is relevant to their needs
- to get an overview before digesting the details
- to help them keep up with current developments in the field
- to decide whether to read the article

Researchers, for instance, often find only titles of periodical articles. To decide whether or not to read an entire article, they first read the abstract to see if the article is relevant to their needs. Many people, especially managers, but often coworkers at any level, need a short version of a longer document because they don't have the time or prefer not to read the longer document.

How to Write Summaries

To write summaries, you need to understand basic summarizing strategies, two methods of organizing a summary, and details of form.

Use Basic Summarizing Strategies   To summarize effectively, you must perform two separate activities.

- Read to find the main terms and concepts.
- Decide how much detail to include.

To read to find the main idea, you must look for various elements:

What are the main divisions of the document?
What are the key statements?
    Which sentence tells the overall purpose of the document?
    Which sentences tell the main ideas of each paragraph?
What details support main ideas?
    What are the key terms? Which words are repeated or emphasized?

Consider the annotation of the article on pp. 110–111. Notice how the summarizer has indicated key sentences and terms. You might find this kind of reading and annotating difficult at first, but with practice you will become more confident of your ability to find the major divisions, main points, and main support in a document.

To decide how much detail to include, always consider your audience's needs. If they need just a description of the document, you need to name only main sections. If they need just basic facts, you might present only the report's purpose and main findings. If they need to grasp the ideas well, you must provide support details. The general rule is to be as complete as your reader requires. By the end of a useful summary, a reader should understand the report's purpose; the major findings, conclusions, or recommendations; and the major facts on which the findings are based.

Choose an Organization   The two main strategies for organizing a summary are

- proportional reduction
- main point followed by support

Proportional reduction means to devote as much space to each part in the summary as is devoted to each section in the original. Suppose the original has four sections, three of which are the same length and one of which is much longer. Your summary of this piece will have the same proportions: three shorter sections all about the same length and a fourth longer section. You can make the overall summary shorter or longer, depending on how much detail you report for each section. A short summary will give the main point of each short section in just one sentence and will use two or three sentences for the longer section. A long summary might use three or four sentences for each short section — the main point and some support detail — and then six or eight sentences to explain the longer section. Schematically, the original and summaries would look like this:

Original   *Summary*

XXXXXXXXXX   XXXXXXXXXXXXX
XXXXXXXXXXXXX
XXXXXXXXXXXXX
XXXXXXXXXX   XXXXXXXXXXXXX
XXXXXXXXXXXXX
XXXXXXXXXXXXX
XXXXXXXXXX   XXXXXXXXXXXXX
XXXXXXXXXXXXX
XXXXXXXXXXXXX
XXXXXXXXXX   XXXXXXXXXXXXX
XXXXXXXXXXXXX   XXXXXXXXXXXXX
XXXXXXXXXXXXX
XXXXXXXXXXXXX
XXXXXXXXXXXXX
XXXXXXXXXXXXX

Main point followed by support is the other method for summarizing. The main idea could be the purpose of the report, or the main findings, conclusions, or recommendations. This method is generally harder to write but is often more effective for readers because you can slant the summary to meet the reader's needs.

Use the Usual Summary Form   The usual form of summaries has the following characteristics:

- length of 250 words to 1 page (If the report is very long, the summary can be longer; a 200-page report might need 2 to 5 pages for a clear, inclusive summary.)
- verbs in the active voice and present tense
- a clear reference to the report (Generally writers put the title of the report either in the first sentence or in the summary's title. If you summarize an article, provide the entire bibliographic entry.)
- no terms, abbreviations, or symbols unfamiliar to the reader (Do not define terms in a summary unless definition was the main point of the document.)
- no evaluative comments such as: "In findings related tangentially at best to the facts she presents . . . "(Report the contents of the document without bias.)
- main points first (The first sentence usually gives the purpose of the report or the main findings, followed by support.)

SAMPLE SUMMARIES

Read the following article. Then review the abstract and two types of summaries on p. 111.

NEW GENERATION OF ANTI-STATIC FOAMS

Dramatic, improved protection from electrostatic generation and uncontrolled discharge is reported for a new generation of anti-static foam packaging developed by The Dow Chemical Co., Midland, Mich.

The technology involves an evenly dispersed, non-migrating anti-static additive system that, because of its method of performance in the foam, is not dependent on relative humidity. Unlike other former anti-static additives that perform by migrating to the foam's surface and attracting a moisture layer, this new additive is permanently fixed in the foam's matrix and does not rely on humidity for its effectiveness.

Due to the different manner in which this anti-static additive performs, the foams are expected to have extremely long shelf life, in addition to providing excellent performance in low-humidity environments. Also, there is very low corrosion potential since moisture is not attracted from the atmosphere, and very low contamination since there is no chemical migration. Overall, the anti-static properties are said to be very consistent due to greater uniformity and permanence of additive concentration throughout the foam.

PATENTS APPLIED FOR

According to Dow, U.S. patents relative to this new technology have been applied for, and developmental products upon which it is based are scheduled for commercial introduction in the near future. Initial trials have been with polyurethane foams, and Dow cited two primary applications for it. One is for packaging of sub-assemblies and similar electronic components, in which optimum static discharge protection is required because of common direct exposure of sensitive items to the packaging material. The other is cushion packaging for lightweight, electronic devices needing static-discharge protection.

"The additive used in this foam will also allow us to make anti-static polyurethane products with unique density and IFD (indent force deflection) combinations," said Jeff Lee, product marketing manager for Dow.

MOBILE TESTING LABS

In addition to the new anti-static foams, Dow also announced "mobile labs" for testing package performance. The labs are available free upon request to computer and electronics companies seeking to test performance of proposed pack designs. They are staffed by Dow Technical Service and Development specialists and travel to sites in the West and Northeast from their bases in Walnut Creek, Calif. and Boston, Mass. The two mobile labs, which are contained in 24-foot converted recreational vehicles, contain equipment to analyze all aspects of package dynamics. (*Packaging* May 1986: 65)

DESCRIPTIVE ABSTRACT

"New Generation of Anti-Static Forms" (*Packaging* May 1986: 65).	Bibliographic information
This article explains that Dow Chemical has anti-static foam packaging and mobile testing labs.	List of article's contents

SUMMARY USING REDUCTION TECHNIQUE

"New Generation of Anti-Static Forms" (*Packaging* May 1986: 65).	
Dow Chemical has developed a new generation of anti-static additive that will dramatically improve foam packaging by cutting down electrostatic generation and uncontrolled discharge.	Purpose
The additive does not migrate to the foam's surface and does not rely on humidity for effectiveness. It is permanently fixed	Part 1
in the foam's matrix. This additive should cause its packages to have longer shelf life, excellent performance in low-humidity	Part 2
situations, low corrosion potential and low contamination. Two primary applications for the additive are polyethylene foam to package electronic components and to cushion lightweight	Part 3
electronic devices. Dow also has developed mobile labs, 24-foot converted RVs, available free to companies wishing to test performance of package designs.	Part 4

SUMMARY EMPHASIZING MAJOR IDEA

"New Generation of Anti-Static Forms" (*Packaging* May 1986: 65).	
Dow Chemical has developed a new generation of anti-static adhesive that will dramatically improve foam packaging.	Purpose
Basically the additive eliminates dependence on relative humidity. Instead of migrating to the surface to form a moisture	Key concept for reader
layer, the additive is permanently fixed in the matrix. Because it has greater uniformity throughout the foam, it will perform well in low-humidity environments, protect from corrosion and contamination, and increase shelf life. Dow says that the	Effects
additive, whose patent is applied for, has been tested with polyurethane foams to protect electronic components and to cushion lightweight electronic devices. It should allow users to make unique density and indent force deflection combinations.	Applications
Dow has also developed mobile labs that will test package designs on site, for free.	Brief treatment of secondary topic

OUTLINING

An outline is a map of a document's main and support points. It is not, however, a prose piece with full sentences. It is a collection of concise phrases, organized in the same sequence as the document. This section explains the uses and types of outlines, and methods for constructing them.

Uses of Outlines

Writers use outlines in two ways — as reading aids and as prewriting devices.

As a Reading Aid   As a reading aid, the outline helps the reader grasp the sequence and relationship of the ideas in a document. So constructing an outline is often a helpful way to start the summarizing process. On a large sheet of paper, jot down main points and subpoints until the pattern in the document emerges. Once you have that, you can write the summary. Sometimes, of course, your goal is to grasp the material yourself, not to write for someone else. Then the outline is all you need. As you construct it, you achieve the understanding you need, so there's no need to write out a summary.

As a Prewriting Device   Writers use outlines as discovery and planning devices. The outline helps a writer see the relationship between, and the sequence of, ideas. Constructing a prewriting outline is a messy business. Like drafting, outlining progresses in stages. In the early stages, you must move, merge, expand, and eliminate ideas. Your goal is to discover basic topics, organization, and an approach. Later, after you have discovered your main ideas and approach, your outline can become more rigid. Theoretically, if you have thought through all the ideas well enough, you should be able to write your document from your final outline. In practice, however, writing is a process of discovery, and outlines frequently change. Many word-processing programs, like WordPerfect and Microsoft Word, now include outliners, features that allow you to construct an outline on screen and then expand it into a paper.

Types of Outlines

There are two basic types of outlines: the traditional and the nucleus.

Traditional Outlines   The traditional outline puts each phrase on a line and indicates the level of the idea by indentations and a number/letter system.

    I.
      A.
         1.
         2.
            a.
            b.
      B.
         1.
         2.

This system clearly indicates the relationship of ideas. You should note, however, that usually such a complete outline is the final product of a long process of reworking a rough outline.

    Nucleus Outlines   The nucleus outline does not express relationships or a sequence of ideas as neatly as the traditional method. Instead, the nucleus outline uses a more informal, clustering approach to group similar ideas. This type of outline, which you can make with or without circles, is helpful to use as a reading aid, allowing you to cluster on your paper ideas that are separated in the original pages. It also aids prewriting because it allows you to group related ideas in the appropriate cluster. Figure 6.1 shows a nucleus outline of the article, "New Generation of Anti-Static Foams" (p. 114).

How to Develop an Outline

To develop an outline you must draft, just as you do for the final version of the document. The basic method is to

- brainstorm
- cluster
- evaluate

Repeat the process until you have an outline.

    To brainstorm a topic is simply to list everything you know about it. The list will not have any order or logically grouped sequences, but that doesn't matter. The key is to write your ideas on paper.

    After you complete your brainstorming list, the next step is to cluster. To *cluster* means to indicate which ideas go together. You can use symbols, such as stars, for one cluster, squares for another, or you can draw joined circles around similar items. After you cluster, you make a new draft that places all the similar items together.

    Next, you evaluate your clusters. To *evaluate* is to decide if you have enough useful ideas or if you need to provide more. If you need more, you

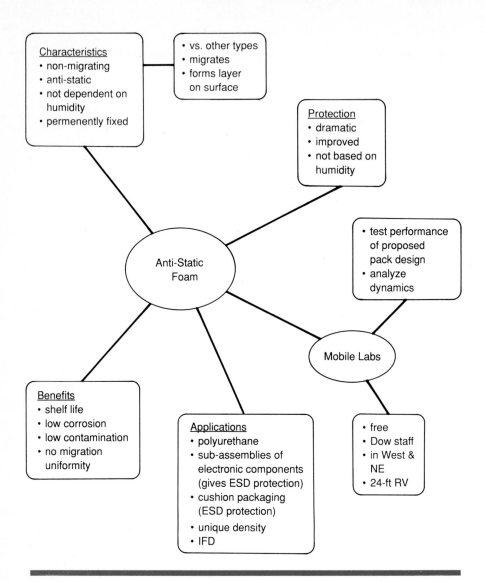

FIGURE 6.1
Nucleus Outline Based on Reading

can repeat the process of brainstorming and clustering until you are satis-
fied that you have developed a rough working outline.

◼ SUMMARY

Summaries are short (from one paragraph to one page) versions of longer
prose pieces. Used for many reasons, summaries can present the piece in
miniature, giving as much emphasis to each section as is given in the original, or they can begin with the main idea followed by pertinent supporting details. For summaries, use the active voice and present tense, and do
not use evaluative terms. Outlines are helpful devices both for reading
comprehension and for prewriting. The two common types of outlines are
the traditional and the nucleus. The traditional type, with its numbered
indented lines, indicates relationships well. The nucleus outline, with its
groups and lines, allows the easy development of subparts. Both outlines
are constructed by a drafting process that employs brainstorming and clustering to find and order ideas and concepts. Writers outline before writing
in order to find the most effective presentation for their ideas.

◼ WORKSHEET FOR SUMMARIZING

☐ *Read over the article or report you intend to summarize.*

☐ *Mark all key words and phrases.*

☐ *Arrange all the words and phrases that go together into groups or clusters.*

☐ *Write a sentence stating the main point you want to convey.*

☐ *List the order in which you will present the **groups or clusters** you have
created.*

☐ *Who is your audience for this summary?*
How much do they know about the topic? (Experts want names
of people and companies, and understand common jargon in the
field; nonexperts don't.)

☐ *How will your audience use this summary?*
Do they want an overview? Are they trying to "keep up"? Will
they use it instead of reading the entire document? Will they
make a decision based on it?

■ EXERCISES

1. Write a summary of an article you have read in a periodical related to your field. Make it a miniaturization of the original by placing your points in the same sequence as those of the original. Do not write more than one double-spaced page. Note: before you write, specify your audience's need for the summary.

2. Write a descriptive abstract of the same article.

3. Write a one-paragraph summary of the Bafflegab article in Chapter 4. Make the summary a one-to-one reduction of the original. Keep the points in the same order as in the original.

4. Read three articles about a topic in your field that interests you. Construct a nucleus outline that merges the content of the three articles. Based on the outline write a brief, one- to two-page report on the topic presented in the articles.

■ WORKS CITED

American National Standards Institute (ANSI). *American National Standard for Writing Abstracts* (Z39.14–1979). New York: ANSI, 1979.

"New Generation of Anti-Static Foams." *Packaging* May 1986: 65.

7

Formatting and Visual Aids

FORMATTING

VISUAL AIDS

TABLES AND GRAPHS

ILLUSTRATIONS

CHARTS

CHOOSING VISUAL AIDS

WRITING ABOUT VISUAL AIDS

T he personal computer has caused radical changes in the individual writer's ability to generate documents. The most noticeable change is in the way you can affect the look of a document by formatting the page and constructing and using visual aids. In the days of the typewriter, a writer had only a few limited methods for affecting the look of the page — margins, underlining, and spacing. Complicated tables, good-looking graphs, maps, photos — all were out of the question. If writers wanted them in a report, they drew them in by hand or pasted them in.

Now writers have many attractive methods available to them, methods formerly used only in expensive typeset material. Boldface, different typefaces, and different type sizes are all available in even the simplest programs. Since software allows writers to incorporate even complicated visuals into their documents with ease, the technical writer must now understand basic formatting and visual concepts. This chapter explains both.

FORMATTING

To *format* is to make choices that affect a page's appearance. The phenom-
enal growth of word processing and the ready availability of desktop publishing programs and laser printers mean that writers must understand
formatting. This section explains three basic formatting areas:

1. How to design a page
2. How to emphasize material
3. How to use heads to indicate contents

How to Design a Page

To design a page means to arrange it so that it effectively presents the
contents to the reader. Good design makes a document easier to read. Software programs such as Microsoft Word, WordPerfect, Pagemaker, and
Ready, Set, Go! now give writers the ability to manipulate the elements of
design. This section explains those elements, then gives some guidelines
for using them effectively.

The Elements of Design   The major elements of design are

rules
margins
columns
typefaces
justification
page numbers

A *rule* is a line. Its width is measured in *points,* a typographical term.
One point is 1/72 of an inch. Rules are designated by their width: "1 point,"
"2 point," and so forth.

The *margin* is the white space between the edge of the paper and the
body of text.

Columns are vertical lines of type; a normal typed page has one wide
column. Many word-processing programs allow up to twelve narrow columns; in practice, however, reports seldom need more than three columns. Some report writers are beginning to use two columns.

Every *typeface* (often called *font* in word-processing software) is named;
some frequently used typefaces are Times, Helvetica, and Palatino. Typefaces are divided into two major groups: serif and sans serif. *Serif* faces
have extenders at the ends of straight lines in letters. *Sans serif* do not have

the extenders. Serif faces impart a classical, more formal impression whereas sans serif faces appear more modern and informal.

Serif   Technical writing makes the world go around.

Sans Serif   Technical writing makes the world go around.

Justification means aligning all the first or last letters of the lines of a column. Documents in English are almost always presented "left justified" — all the first letters of each line start at the left-hand margin. *Right justified* means to align at the right-hand margin all the letters that end lines.

Page numbers appear either in the upper right corner or at the bottom middle of the page. Each piece of paper counts in the sequence of pages even if no page number appears on it.

Guidelines for Using Design Elements   Guidelines for using these design elements follow:

- Generally use 1-inch margins on the top, bottom, left, and right. Use 1½-inch margins on the left if you are going to bind the paper.
- Avoid right-justified margins. Many word-processing programs achieve right justification by inserting too many spaces between words or else by excessive hyphenation. Both methods call attention to themselves, thus distracting from the message.
- Do not use rules thicker than 2 points. Generally choose a *hairline* (or 1/2 point) rule. The thicker the rule, the more it draws the reader's eye. As a rule of thumb, thick rules appear above words, thin rules appear below.
- Generally use a single column for reports. As shown on p. 155, for design's sake you might want to use a 2- or 2½-inch left margin.
- Two columns are especially useful if you plan to use a laser printer to produce a number of graphics. You can control the visual impact of the page better.
- Use no more than two typefaces per document. In general, use the same typeface for text and heads, but you could use one typeface for text and another for heads (choose a serif face for text and a sans serif for heads).
- Use italics sparingly — italicized print is hard to read.
- Place page numbers in the same position on each page. Many word-processing programs automatically number each page. (The number "1" can appear on page one.)

To see the design elements arranged for different visual effects, turn to pages 122–123. Four models illustrate different ways of designing the same page.

Developing a Style Sheet As you generate any document, you should develop a *style sheet*, which is a list of the ways you will treat each design item.

For instance, for a three-page memo, the style sheet would be quite short:

1-inch margin on all four sides
no right justification
single-space paragraphs
double-space between paragraphs
heads flush left and underscored
triple-space above and double-space below heads
page numbers at bottom middle

For a more complicated document, you need to devise a much more complete style sheet. In addition to margins, justification, and paragraph spacing, you need to consider

a multilevel head system
page numbers
rules for page top and bottom
rules to offset visuals
captions for visuals
headers and footers — such as whether the chapter title is placed in the top (header) or bottom (footer) margins

How to Emphasize

To achieve emphasis, you highlight. *Highlighting* means to make an item look different from the items around it. You can highlight by using bold-face, all capitals, underlining, or vertical lists. The basic guidelines for highlighting are

- Highlight — but not too much.
- Use boldface as the most effective highlight (underlining is the most effective for typed material).

- Avoid long phrases written in all capital letters.
- Use vertical lists to emphasize the items in a series.

The following two examples illustrate ineffective and effective use of highlighting.

Ineffective highlighting	Caution: Do not separate the mold holders **from** the mold cavity **with** a sharp instrument. Use nothing harder than brass or wood. You could damage the mold.
Effective highlighting	**Caution**: DO NOT separate the mold holders from the mold cavity with a sharp instrument. Use nothing harder than brass or wood. You could damage the mold.

How to Use Heads to Indicate Contents

To indicate a document's contents, you develop a head system. A *head* is a word or phrase that indicates to the reader the contents of the following section or subsection. Here are four basic guidelines for developing a head system.

1. Designate levels of heads, traditionally called by letters or numbers (A, B, C, D, or level 1, level 2, level 3). Level A is a main section, level B a subsection, level C a sub-subsection, and so on.

2. Make each level look different, with the level A head more prominent than level B, and so forth. Use boldfacing or underlining. Use different print sizes. Place different levels at different positions on the page (centered or flush left). Your method of printing the document will affect your options for designating heads:

Typewriter:	Use underlining and all capital letters. Heads centered or flush left.
Laser printer:	Use boldface, all capitals, various print sizes, and different fonts.
Dot matrix:	Use either boldface or underlining (do not underline boldface). Depending on the printer and word-processing program, use various point sizes and fonts.

3. Make each level's wording parallel. Use all noun phrases or questions or *-ing* words.

4. Use content wording, not generic wording. Use phrases like "Factors in Total Cost" rather than "Cost". Use questions.

To see these guidelines in action, turn to pp. 153–156, showing a section of a report presented in several ways.

Indicating Two Levels of Head Most documents need only one or two levels of head. Most writers use a "side left" head as the level 1 head and a "paragraph" head as the level 2 head. For level 1, place the head at the left-hand margin, leave white space above and below, use no punctuation after, and either boldface or underline it. For level 2, place the head at the start of the paragraph, indented five spaces, followed by a period.

Indicating More Than Two Levels of Head The remainder of this section illustrates two different head systems, *open* and *numbered;* and both can indicate up to four or five levels of head. Examples are given for the typewriter and laser printer. Carefully note which letters are capitalized, how much space is above and below, and whether punctuation follows.

OPEN SYSTEM

---

3 Levels — Typewriter/Basic dot matrix

FIRST-LEVEL HEAD

2 spaces
XXXXXXXXXXXXXXXXXXXXXXXXXXXXXXXXXXXXXXXXXXXXX
XXXXXXXXXXXXXXXXXXXXXXXXXXXXXXXXXXXXXXXXXXXXX

3 spaces
<u>Second-Level Head</u>

2 spaces
XXXXXXXXXXXXXXXXXXXXXXXXXXXXXXXXXXXXXXXXXXXXXX
XXXXXXXXXXXXXXXXXXXXXXXXXXXXXXXXXXXXXXXXXXXXXX
     <u>Third-level Head</u>. XXXXXXXXXXXXXXXXXXXXXXXXX
XXXXXXXXXXXXXXXXXXXXXXXXXXXXXXXXXXXXXXXXXXXXX

4 Levels — Typewriter/Basic dot matrix

FIRST-LEVEL HEAD

2 spaces
XXXXXXXXXXXXXXXXXXXXXXXXXXXXXXXXXXXXXXXXXXXXXX
XXXXXXXXXXXXXXXXXXXXXXXXXXXXXXXXXXXXXXXXXXXXXX

3 spaces
<u>Second-Level Head</u>
XXXXXXXXXXXXXXXXXXXXXXXXXXXXXXXXXXXXXXXXXXXXXX
XXXXXXXXXXXXXXXXXXXXXXXXXXXXXXXXXXXXXXXXXXXXXX

3 spaces

<u>Third-Level Head</u>

2 spaces

XXXXXXXXXXXXXXXXXXXXXXXXXXXXXXXXXXXXXXXXXXXXXXXX
XXXXXXXXXXXXXXXXXXXXXXXXXXXXXXXXXXXXXXXXXXXXXXXX
     <u>Fourth-level Head</u>. XXXXXXXXXXXXXXXXXXXXXXXX

3 Levels — Laser printer/Advanced dot matrix

First-Level Head   14 pt.

  2 spaces
XXXXXXXXXXXXXXXXXXXXXXXXXXXXXXXXXXXXXXXXXXX   10 pt. text
XXXXXXXXXXXXXXXXXXXXXXXXXXXXXXXXXXXXXXXXXXX
  2 spaces
Second-Level Head   12 pt.

  2 spaces
XXXXXXXXXXXXXXXXXXXXXXXXXXXXXXXXXXXXXXXXXXX
XXXXXXXXXXXXXXXXXXXXXXXXXXXXXXXXXXXXXXXXXXX
     Third-Level Head. XXXXXXXXXXXXXXXXXXXXXX   10 pt.
XXXXXXXXXXXXXXXXXXXXXXXXXXXXXXXXXXXXXXXXXXX

4 Levels — Laser printer/Advanced dot matrix

FIRST-LEVEL HEAD   14 pt. centered

  2 spaces
XXXXXXXXXXXXXXXXXXXXXXXXXXXXXXXXXXXXXXXXXXX   10 pt. text
XXXXXXXXXXXXXXXXXXXXXXXXXXXXXXXXXXXXXXXXXXX
  2 spaces
Second-Level Head   12 pt.

  2 spaces
XXXXXXXXXXXXXXXXXXXXXXXXXXXXXXXXXXXXXXXXXXX
XXXXXXXXXXXXXXXXXXXXXXXXXXXXXXXXXXXXXXXXXXX
  2 spaces
Third-Level Head   10 pt.
XXXXXXXXXXXXXXXXXXXXXXXXXXXXXXXXXXXXXXXXXXX
XXXXXXXXXXXXXXXXXXXXXXXXXXXXXXXXXXXXXXXXXXX
     Fourth-level Head. XXXXXXXXXXXXXXXXXXXXXXXX   10 pt.
XXXXXXXXXXXXXXXXXXXXXXXXXXXXXXXXXXXXXXXXXXX

The *numbered system*, often used in more technical or more complex material, indicates level by a number before the head. Each succeeding level uses more numbers. The first level is 1.0, 2.0, etc. The second level is 1.1, 1.2; The third level 1.1.1, 1.1.2. Notice that laser-printed documents use boldface and larger type.

5 Levels — Typewriter/Basic dot matrix

1.0 FIRST-LEVEL HEAD

1.1 SECOND-LEVEL HEAD
XXXXXXXXXXXXXXXXXXXXXXXXXXXXXXXXXXXXXXXXXXXXXXXXX
XXXXXXXXXXXXXXXXXXXXXX

1.1.1 Third-Level Head
XXXXXXXXXXXXXXXXXXXXXXXXXXXXXXXXXXXXXXXXXXXXXXXXX
XXXXXXXXXXXXXXXXXXXXXX

       1.1.1.1 Fourth-level Head.   XXXXXXXXXXXXXXXXXXXXXX
XXXXXXXXXXXXXXX
       1.1.1.1.1 Fifth-level Head.   XXXXXXXXXXXXXXXXXXXXXX
XXXXXXXXXXXXXXX

5 Levels — Laser printer/Advanced dot matrix

1.0 FIRST-LEVEL HEAD

	14 pt. centered
1.1 SECOND-LEVEL HEAD	12 pt.
XXXXXXXXXXXXXXXXXXXXXXXXXXXXXXXXXXXXXXXXXXXXXX	10 pt. text
XXXXXXXXXXXXXXXXXX	
1.1.1 Third-Level Head	10 pt. head
XXXXXXXXXXXXXXXXXXXXXXXXXXXXXXXXXXXXXXXXXXXXXX	
XXXXXXXXXXXXXXXXXX	
1.1.1.1 Fourth-level Head. XXXXXXXXXXXXXXXXX	10 pt. head
XXXXXXXXXXXXXXX	
1.1.1.1.1 Fifth-level Head. XXXXXXXXXXXXXXXXXXX	10 pt. head
XXXXXXXXXXXXX	

VISUAL AIDS

Visual aids have always been an essential part of technical writing. With the advent of graphics programs for personal computers, writers have available a number of choices for using visual aids. Computers easily convert data to many kinds of visuals. The bar graphs, line graphs, and pie charts (Figures 7.1 to 7.6) shown in this chapter are all computer-generated. This chapter explains a number of different types of visual aids, shows how to discuss them in the document, and tells when to use them. Three basic types of visual aids are tables and graphs, illustrations, and charts.

Guidelines for Effective Visual Aids

These six guidelines provide a framework for incorporating visual aids into a document. Later sections of the chapter explain which types of visual aids to use and when to use them.

1. Construct high-quality visual aids, using clear lines, words, numbers, and organization. Research shows that the quality of the visual aid is the most important factor in its effectiveness.
2. Identify all visual aids as either tables or figures. Anything that is not a table is a figure — no matter what form it takes. In formal situations, place numbers and titles directly above tables and directly below figures. For informal situations, you may place the titles (and numbers if you use them) above or below, whichever seems clearer.
3. Make sure a visual conveys only one point. If you include too much data, readers cannot grasp the meaning readily.
4. Do not clutter a visual aid with too many words or lines, causing it to lose its visual quality and impact.
5. Integrate visual aids into the report at logical and convenient places. As a general rule, place illustrations in the middle of or after your discussion of them.
6. Refer to each visual aid in the text, (usually by number, e.g. "see Figure 1"), even if the illustration is right beside or below your discussion of it, and explain or interpret each one.

TABLES AND GRAPHS

Tables and graphs are the most common visual aids. Used extensively in all types of technical writing, they are easily constructed from many available software programs. This section explains how to construct them and when to use them.

Basic Definitions

The information in tables and graphs is either an independent or a dependent variable. An independent variable is the group you are discovering information about. The dependent variable is the category you use to find out about the independent variable. The dependent variable changes when the independent variable changes. In a table of weather conditions, the independent variable is the months. The dependent variables are average temperature, average precipitation, and whatever else you might wish to compare. The following sections tell you where to place these variables in tables and graphs.

Tables

Tables present information — usually numbers, but sometimes words — in columns and rows. A table shows classifications and relationships of numerical or verbal data. Here are some general principles for presenting tables.

- In general, use tables with more professional, expert audiences.
- Remember that tables are harder for less knowledgeable readers to understand.
- Refer to the table in the text.
- Point out the essential relationships you want readers to understand.
- Use a table when you must present all the numerical data to expert audiences so that they see the context of the relationships you point out.
- Use a table to compare many numbers or features (and eliminate the need for lengthy prose explanations).
- Put the items you want to compare (the independent variables) down the left side of the table in the *stub* column. Put the categories of comparison (the dependent variables) across the top in the column headings. Remember, columns are easier to compare than rows.
- Use plenty of white space between columns and above and below the table.
- Give each table a number and a clear, concise title.

The following table shows the elements and organization of a table.

TABLE 1   Number
Production Rates and Labor Costs   Title
                                                                                 Spanner

Time	No. of Harnesses Produced	Labor Costs[a]			Column heads
		Cutter[b]	Assembler[c]	Total	
1 Cycle[d]	1	$ .04	$ .21	$ .25	Data
1 Hour	33	7.00	8.50	15.50	
1 Day[e]	264	56.00	68.00	124.00	

Note: Labor contract will expire Jan. 1.   Notes
[a]Rounded to the nearest cent.
[b]Based on 1 man working at $7.00/hour.
[c]Based on 1 man working at $8.50/hour.
[d]1 Cycle = 20 seconds for cutting + 90 seconds for assembly
= 110 seconds.
[e]Based on 1 eight-hour shift.
Source: Production Department figures for first quarter.

  Guidelines for Constructing Tables   The following detailed guidelines explain how to present material in a table. The guidelines progress
through a table from top to bottom. Table 1 above illustrates all of these
guidelines.

- Number tables consecutively throughout a report with Arabic numerals
  in the order of their appearance. Put the number and title *above* the
  table. Use the "double number" method (for example, 2.3) only in
  long reports that contain chapters.
- Use the table title to identify the main point of the table. Do not
  place punctuation after the title.
- In general, use three horizontal rules and no vertical rules. If the
  report is more informal, use fewer or no rules.
- Name the independent variable in the heading of the left-hand column.
- Name the dependent variables (the categories you are comparing) in
  the column heads.
- Use a *spanner* head to name the column headings below it. Spanners
  reduce repetition in column heads.
- Use a *line heading* to identify the contents of the horizontal row of
  data to the right.

Cite the *source* of the data — unless it is obviously data collected for the paper. Either say "*Source*: Production Department figures for first quarter" or, if the table has been printed elsewhere, give a citation.

Place explanatory comments below the bottom rule. Use the word *Note* followed by the comment.

Use specific notes to clarify portions of a table. Indicate them by raised lower-case letters within the table and at the beginning of each note.

Graphs

A graph can present in simplified form the same statistical data as a table. Many software programs allow you to enter data as a table and then convert it into many different kinds of graphs. Bar graphs, pie charts, and line graphs present information more dramatically than tables, though often not as specifically. Since graphs are more dramatic but less precise, you should determine whether a table or a graph is more appropriate for your readers' needs.

Guidelines for Selecting a Graph   If you decide to present your information graphically, you must determine whether a bar, pie, or line graph is best.

- Use a *line graph* to depict trends or relationships. In a trend, the same data change over time. The population figures for one city at different points in time is an example of a trend. A relationship shows the interaction of two variables, for example, percentage of pollutant to size of filter.
- Use a *bar graph* to compare discrete items. For example, the population of three different cities at one point in time can be shown using a bar graph.
- Use a *pie chart* to represent discrete values as parts of a whole. If you want to compare profits to total income, use a pie chart.
- Research shows that bar graphs are the easiest for less knowledgeable readers to grasp.

Guidelines for Presenting Graphs   The following guidelines explain how to present the title and number of graphs clearly.

- Refer to all graphs as *figures,* and number them sequentially throughout the report, using Arabic numbers (Figure 1, Figure 2).
- Place the word *Figure* (or its common abbreviation *Fig.*) and the number at the left-hand margin.

- Treat the word *figure* consistently. Either use all capital letters (FIGURE) or capitalize just the first letter (Figure). Place a period after the number, if the title follows on the same line. If you place the title underneath the number, do not use a period.
- Place the title after the number or below it. Both methods are used. Do not underline the title; do not use a period after it; use initial capitals only, followed by lower-case letters.
- In informal reports, writers place the title either below (as in this book) or above the figure, whichever is clearer in the page makeup. (Many software programs automatically place the titles above the figure.)
- If needed, start a second line directly under the first letter of the title.
- Indicate the source of the data by placing the word *Source* under the figure, followed by the citation.
- Refer to the graph in the text.
- Point out the relationship you want readers to understand.

Bar Graphs A *bar graph* uses rectangles to indicate the relative size of several variables. Bar graphs contrast variables or show magnitude effectively. (See Figure 7.1.) A bar graph compares the items by means of the height or length of the appropriate bars. Bar graphs can be either horizontal or vertical, depending on whether the bars go up the page or across it.

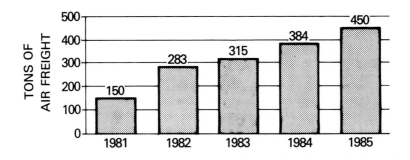

FIGURE 7.1
Bar Graph
Source: 1986 Program Strategies. Reprinted by permission of the City of Dayton, Ohio.

Vertical bar graphs (frequently called *column graphs*) are better for showing discrete values over time, such as profits at certain intervals. Either type can be used in most other instances.

The following guidelines explain how to construct bar graphs more effectively. These guidelines are for vertical bar graphs. Rearrange items accordingly for horizontal bar graphs.

- Place the names of the items you are comparing — the independent variable — under the bars (for a vertical bar graph).
- Place the units of comparison — the dependent variables (usually numbers) — at the left.
- Make the spaces between the bars one-half the width of the bars. (Many computer programs do not follow this guideline.)
- Use a *legend* — a small sample of the markings and brief phrase — to explain the meanings of the bars' markings.

Notice how the bars in Figure 7.2 are subdivided to show an additional comparison of percentages. Figure 7.3 shows a *multiple bar graph* where

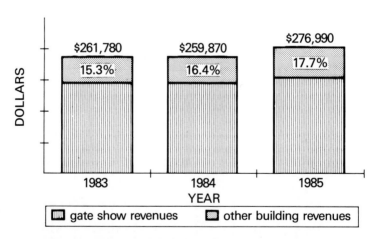

FIGURE 7.2
Divided Bar Graph

Source: 1986 Program Strategies. Reprinted by permission of the City of Dayton, Ohio.

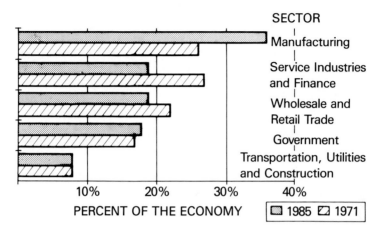

EMPLOYMENT BY SECTOR

SECTOR

Manufacturing

Service Industries
and Finance

Wholesale and
Retail Trade

Government

Transportation, Utilities
and Construction

10%   20%   30%   40%

PERCENT OF THE ECONOMY   ▨ 1985 ▧ 1971

Figures are adjusted to reflect changes in the Metropolitan Area.

FIGURE 7.3
Multiple Bar Graph
Source: 1986 Program Strategies. Reprinted by permission of the City of Dayton,
Ohio.

more than one bar is used to compare quantities; this device is useful when
the comparison must extend over several years. Notice the legends in Figures 7.2 and 7.3.

Line Graphs A *line graph* shows the relationship of two variables by
a line connecting points inside an x (horizontal) and a y (vertical) axis.
These graphs usually show trends over time, such as profits or losses from
year to year. The line connects the points, and its ups and downs illustrate
the changes — often dramatically. Choose a line graph if you want to emphasize continuity; choose a bar graph if you want to emphasize the relative size of each item.

Figure 7.4 shows the public's perception of park safety over a series of
years. The ratings of increases and decreases in park safety are easy to
grasp. Notice that percentages appear at each year to indicate precise
numbers.

The following few general guidelines show how to make line graphs
more effective.

■ Name the independent variable — the one that changes automatically,
such as years — on the horizontal axis.

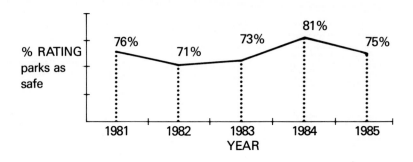

FIGURE 7.4
Line Graph
Source: 1986 Program Strategies. Reprinted by permission of the City of Dayton, Ohio.

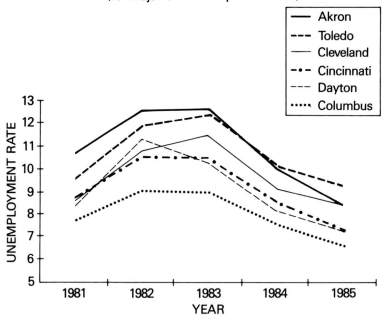

FIGURE 7.5
Multiple Line Graph
Source: 1986 Program Strategies. Reprinted by permission of the City of Dayton, Ohio.

- Name the dependent variable — the one that changes because the independent variable changes, like profits — on the vertical axis.
- If more than one line appears in the graph, use a legend to explain each.
- Don't use too many lines. If you use several lines, make them visually distinct (for example, use a dotted line and a continuous line, as in Figure 7.5).

In Figure 7.5, a *multiple line graph*, several lines compare the same information (unemployment rates in several cities) over the same time period. A legend explains what each line stands for.

Pie Charts   A *pie chart* uses segments of a circle to indicate each segment's percentage of a total. (See Figure 7.6.) The whole circle represents

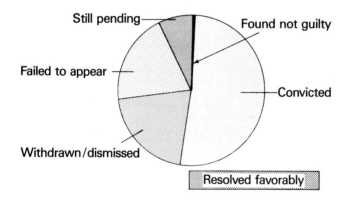

HOUSING AND ZONING CASES

TOTAL CASE LOAD

Activity	No.	% of all cases filled
Conviction	90	52%
Withdrawn/Dismissed	34	20%
Not Guilty	1	1%
Failed to Appear	34	20%
Still Pending	13	7%

FIGURE 7.6
Pie Chart
Source: 1986 Program Strategies. Reprinted by permission of the City of Dayton, Ohio.

100 percent, the segments of the circle represent each item's percentage of the total, and the legend explains the symbols used in the graph.

Pie charts show relative magnitudes best. Since pie chart segments are not very precise, most people cannot clearly distinguish between close values, such as 17 percent and 20 percent. Pie charts are effective for simultaneously comparing the components to one another and to the whole. These charts are often used to visualize financial data, for example, the relationship of operating costs to profit.

The following guidelines explain how to make pie charts more effective.

- Identify "slices" by means of a legend or callouts — phrases that name each slice. Usually cluster all the legend items in one corner of the visual, and arrange callouts around the circumference of the circle.
- In general, place percentage figures inside the segments.
- Start the "slices" at twelve o'clock and run them in sequence, clockwise, from largest to smallest. (Sometimes the data require a different arrangement: If you compare the grades in a chemistry class, the "A" slice will usually be smaller than the "C" slice. However, you would logically start the A slice at twelve o'clock.)
- For emphasis, shade the segments, or present only a few important segments rather than the whole circle.
- In general, don't divide pie charts into more than five segments. Too many segments turn into tiny slices whose size readers can't differentiate, and they require so many legends that the chart appears chaotic.

ILLUSTRATIONS

Illustrations, usually photographs or drawings, are used extensively in sets of instructions and manuals. To point out the important parts of an illustration, writers use *callouts*, letters or words connected by lines to the relevant part of the illustration. The following general guidelines explain how to make illustrations more effective.

- Use illustrations to avoid lengthy discussions. A picture of a complex part will generally be more helpful than a lengthy description.
- Use high-quality illustrations: make sure they are clear, and large enough to be effective, and set off by plenty of white space.
- Keep the illustrations as simple as possible. Show only items essential for your discussion.

Photographs are difficult to use in technical reports because, in order to be reprinted, they must be converted to *halftones*. This process changes the photograph into a series of dots, which until recently only the printing press could reproduce. Computer technology has partially solved this problem with *scanners*, devices which convert photographs into halftones electronically. These expensive devices produce acceptable but low-quality halftones. A writer inserts a photograph into a scanner, which electronically converts the photograph into dots. Using a scanner and an instant camera, many amateurs can produce acceptable halftone photographs in minutes.

Drawings are easier to reproduce. A scanner will produce good reproductions of line drawings, since they do not have to be converted to halftones. Anyone with a scanner can easily convert a line drawing on a sheet of paper into electronic dots.

Photographs

A good photograph has these advantages: it duplicates the item discussed (so audiences can be sure that they are looking at what is intended), and it shows the relationship of various parts. The disadvantages are that it reduces a three-dimensional reality to two dimensions and that it shows everything, thus emphasizing nothing.

Figure 7.7 shows a skillfully "cropped" photograph. The original contained much more visual information, but the designer blocked out a great deal of it, simplifying it so it makes only one point.

Drawings

Drawings, whether made by computer or by hand, can clearly represent an item and its relationship to other items. Since details in drawings are

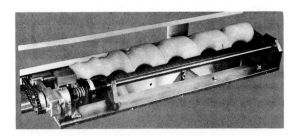

FIGURE 7.7
Cropped Photograph
Source: A manual by Jill Adkins. Reprinted by permission of MRM/Elgin.

chosen selectively by the artist, the reader can focus on just the intended object. The exploded view and the detail drawing are two commonly used types of drawings.

Exploded View   As the term implies, an *exploded view* shows the parts disconnected but arranged in the order in which they fit together, as in Figure 7.8. Exploded drawings can show the internal parts of a small and intricate object or explain how it is assembled. Manuals and sets of instructions often use exploded drawings with named or numbered parts.

Detail Drawings   Detail drawings are renditions of particular parts or assemblies. They are used in manuals and sets of instructions, usually in one of two ways. Drawings can function as an uncluttered, well-focused photograph, showing just the items that the writer wishes. They can also show cross-sections, that is, they can cut the entire assembled object in half, both exterior and interior. In technical terms, the object is cut at right angles to its axis. A cross-sectional view shows the size and relationship of all the parts. Two views of the same object, front and side views, for example, are often placed beside each other to give the reader an additional perspective of the object. (See Figure 7.9.)

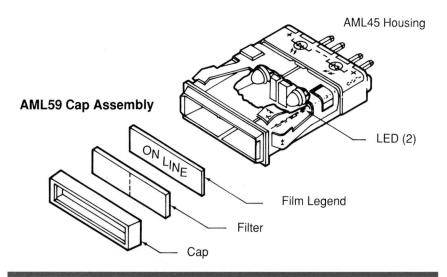

FIGURE 7.8
Exploded View
Source: Manual Controls: Independent Study Workbook by John R. Mancosky. Used by permission of Microswitch and John R. Mancosky.

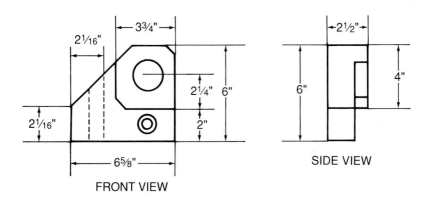

FIGURE 7.9
Detail Drawing

CHARTS

A *chart* is the catchall name for many kinds of visual aids. Charts represent the organization of something, either something dynamic like a process, or something static like a corporation. They include such varied types as troubleshooting tables, schematics of electrical systems, diagrams of the sequences of an operation, organization charts, flow charts, and decision charts. Use the same techniques to title and number these as you use for graphs. (Some software companies, like Microsoft, use *chart* to mean *graph*. You will find these terms are used inconsistently.)

Troubleshooting Tables

Troubleshooting tables in manuals identify a perceived problem and give its probable cause and cure. The problem appears at the left and the appropriate action to the right. Complicated tables, like Figure 7.10, also suggest causes. Service manuals for huge manufacturing machines and user manuals for appliances, VCRs, and automobiles use these tables. (See Figure 7.10.)

Organization Charts

Organization charts depict the flow of authority in an institution. (See Figure 7.11.) They are composed of boxes (with names and positions in them), and lines connecting the boxes. Place the most powerful position at the top and less powerful positions below.

TABLE 2
Troubleshooting Table

You Notice	This May Mean	Caused by	You Should
Containers do not center with nozzle	Infeed Starwheel is out of time	Misadjustment or loose mounting bolts	Readjust & retighten
	Machine speed is too fast	Speed not reset during change-over	Reset speed (see "Speed Adjust," p. 5)
	Wrong change parts being used		Install correct change parts (see "Change Parts Data Sheet," p. 27)
Machine vibrates	Lack of maintenance	Tight roller chains (chains that drive the conveyor) or the two chains on the end of the Spiral Screw Feed	Loosen chains
		Roller chains running dry	Lubricate chains
		Cross beam bearing dry	Lubricate (see fig. 5-148)

FIGURE 7.10
Troubleshooting Table
Source: Reprinted by permission of MRM/Elgin.

Flow charts

Flow charts symbolically depict a time sequence or a decision sequence. A flow chart has arrows that indicate the direction of the action, whatever it is, and symbols that represent steps or particular points in the action. (See Figure 7.12.) In many instances, especially in computer programming, the symbols have special shapes for certain activities. For instance a rectangle means an action to perform and an oval means the first or last action. Flow charts are especially helpful when you must to help a reader grasp a process.

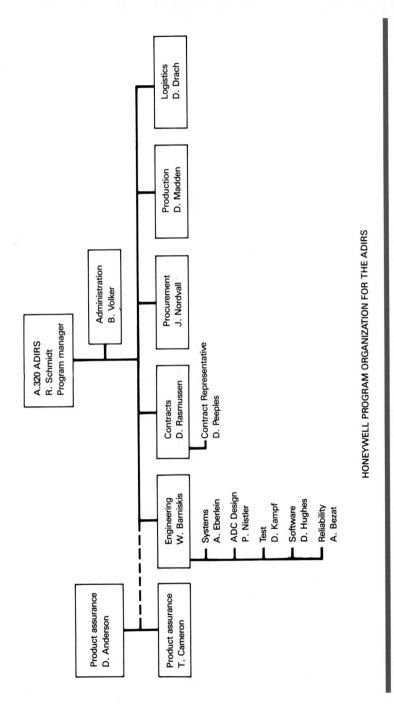

FIGURE 7.11
Organization Chart
Source: Reprinted by permission of Honeywell, Inc.

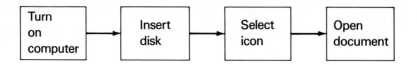

FLOW CHART OF STEPS TO OPEN A DOCUMENT

FIGURE 7.12
Flow Chart

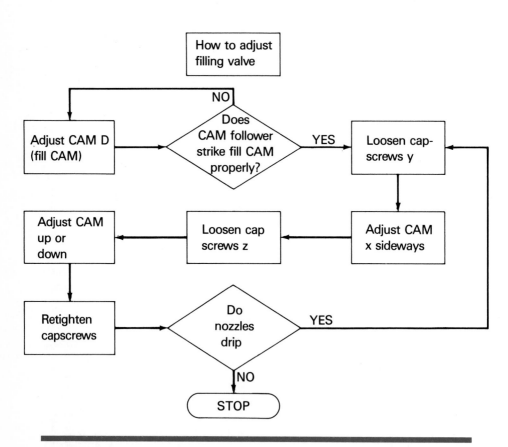

FIGURE 7.13
Decision Chart

Decision Charts

A *decision chart* (or *tree*) is a flow chart that uses graphics to explain whether or not to perform a certain action in a certain situation. At each point the reader must decide yes or no, then follow the appropriate path until the final goal is reached. (See Figure 7.13.)

CHOOSING VISUAL AIDS

Deciding which type of visual aid to use is sometimes difficult. The best way to choose is to decide what your readers need. If your readers need more scientific or "more objective" data, use tables and line graphs. Tables can classify large amounts of data so that relationships can be easily pointed out. Pie and bar graphs have more visual impact or drama. If you look back at the bar graph shown in Chapter 1 (Figure 1.1, p. 4), you will see that it presents information dramatically. The towering columns clearly reinforce the notions of "critically important" and "important" as opposed to the short column of "no importance."

    The difficulties — and challenges — of deciding what visual aid to use can be seen in the following figures. All four figures, constructed using Microsoft Works, are attempts to convey the material presented in the table in Figure 7.14. The table compares the costs of manufacturing WORM disks (a type of computer floppy disk) from 1980 to 1987 in four areas: testing, thin films, assembly, and coating. The four types of figures consist of a divided bar graph, one bar for each year (Figure 7.15); two bar graphs, one for each year (Figure 7.16); a bar graph showing bars for each component (Figure 7.17); and two pie charts, one for each year (Figure 7.18). Notice the different effects.

    The two bar graphs (Figure 7.16) emphasize changes in the relationship between items in each year. It is easy to see the relationships between

Table 1.
Components of Cost of Disks

	testing	thin films	assembly	coating
1980	$4.00	$20.00	$10.00	$4.00
1987	$5.50	$12.80	$8.00	$3.20

FIGURE 7.14
A Table Compares the Data

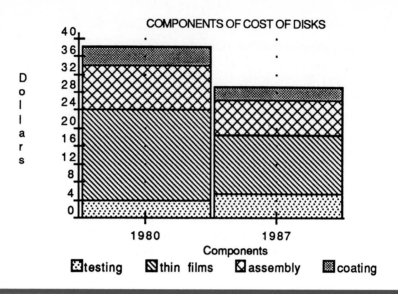

FIGURE 7.15
A Divided Bar Graph Compares the Data

testing and thin films for each year, but the two vertical axes ("Dollars") are based on different scales, so comparison is difficult.

The two pie charts (Figure 7.18) take time to figure out because they are separated from each other. Notice that they do show that the total cost has changed and that the relative percentages of thin films and testing have changed.

The divided bar graph (Figure 7.15) shows the relationships between the components and the total cost. This visual shows essentially the same information as the two pie charts. Because of the closeness of the two columns, however, the comparative relationships are easier to grasp.

The bar graph with two columns for each component (Figure 7.17) contrasts dollars, not percentages, clearly illustrating all the reasons why 1987 is cheaper overall than 1980.

If readers need to understand dollar breakdowns, Figure 7.17 would serve their needs better. If readers need percentage breakdowns, the divided bar graphs (Figure 7.15) or the pie charts (Figure 7.18) would help them more. However, if readers need precise figures and if they can work easily with relationships, the table would be a good choice (Figure 7.14).

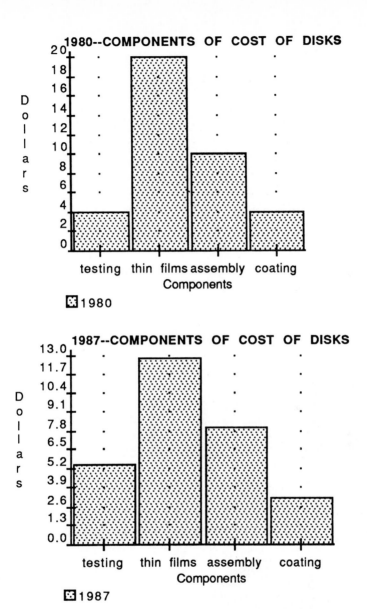

FIGURE 7.16
Two Bar Graphs Compare the Data

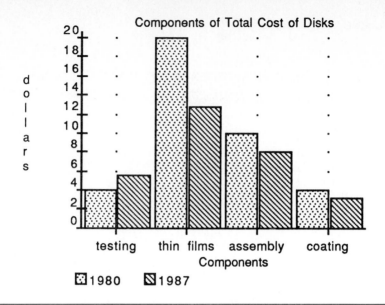

FIGURE 7.17
Bar Graph Compares the Data

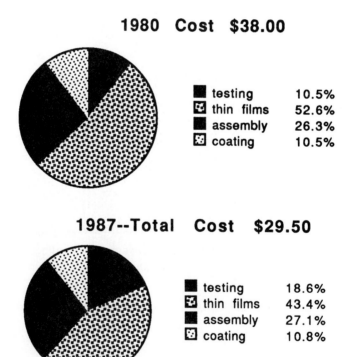

FIGURE 7.18
Two Pie Charts Compare the Data

Three Principles for Manipulating Graphs

Jan White, in his book *Using Charts and Graphs*, illustrates three ways that you can affect the perception of graphic data. Your goal is to present the graph so that it reports the data honestly.

1. Changing the width of the units on the y-axis alters the viewer's emotional perception of the data. The following graphs plot exactly the same data (Figure 7.19).

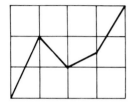 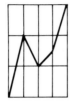 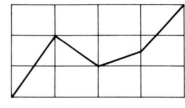

    a. normal   b. dramatic rise   c. gradual rise

FIGURE 7.19
Three Graphs Plot the Same Data
Source: Reprinted by permission of R. R. Bowker.

2. The nearer a highlighted feature appears, the more impact it has on one's consciousness. The following pies report the same data. Notice that the wedge on the left appears largest, but the wedge on the right is forced into the viewer's consciousness. The middle wedge appears unimportant because it is far away (Figure 7.20).

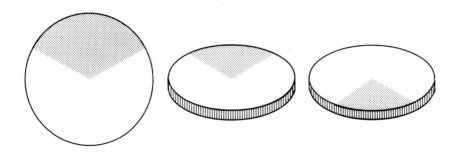

FIGURE 7.20
Three Pie Charts Plot the Same Data
Source: Reprinted by permission of R. R. Bowker.

3. A darker element seems more important to the viewer (Figure 7.21).

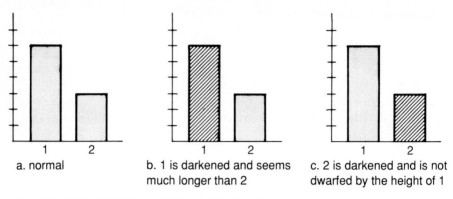

a. normal   b. 1 is darkened and seems   c. 2 is darkened and is not
                             much longer than 2   dwarfed by the height of 1

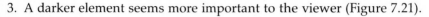

FIGURE 7.21
Three Bar Graphs Plot the Same Data
Source: Reprinted by permission of R. R. Bowker.

WRITING ABOUT VISUAL AIDS

Although visual aids can increase the impact of a report, they can also overwhelm and confuse a reader. Tables, and even graphs, may contain so much information that readers can interpret them in various ways — and perhaps not in the way you intended. As a report writer, you must learn how to guide your reader through your visual aids. This section gives the guidelines for writing about visual aids, how to refer clearly to visual aids, and how to tell readers what to notice.

Refer Clearly to the Visual Aids

Refer to the visual aid by number. If it is several pages away, include the page number in your reference. You can make the references textual or parenthetical.

Textual Reference   A *textual* reference is simply a statement in the text itself, often a subordinate clause, that calls attention to the visual aid.

As seen in Table 1 (p. 10) . . .
If you look at Figure 4 . . .
The data in Table 1 show . . .

Parenthetical Reference   The *parenthetical* reference names the visual in parentheses in the sentence in one of two ways: complete or abbreviated. The complete reference is used more in reports; the abbreviated in sets of instructions. In reports, use "see" and spell out "Table." Although "Figure" or "Fig." are both used, "Figure" is more acceptable in formal writing.

> The profits for the second quarter (see Figure 1) are . . .
> A cost analysis reveals that we must reconsider our plans for purchasing new printers (see Table 1).

In instructions, you do not need to use "see"; you can refer to figures as "Fig."

> Insert the disk into slot A (Fig. 1).
> Set the CPM readout (Fig. 2) before you go on to the next step.

Do not capitalize "see" unless the parenthetical reference stands alone as a separate sentence. In that case, also place the period inside the parentheses.

> All of this data was described above. (See Tables 1 and 2.)
> All of this data was described above (see Tables 1 and 2).

Tell the Reader What to Notice   When you discuss a visual aid, point out what the reader should look for and explain its significance. To point out an item, you simply call attention to it or name it. To explain its significance, you either give its source or discuss its implications. Examples are given in the following discussion.

Explaining a Table   Table 1 in the following example is from a complex government report on the costs of having a window constructed in a wall. But the numbers can be confusing, and the important relationships are not apparent. Should we compare the cost of a 12-foot-square window ($52.20) with the cost of 12 square feet of wall ($33.72), or to some other number or combination of numbers? Should we compare the $52.20 in the upper left-hand corner with the $216.63 in the lower right-hand corner? But why? And what would such a comparison mean?

The creators of the table (Collins et al.) were aware of this problem, so in their text they pointed out which number the readers should notice and its significance. The authors draw attention to the $18 figure in Column 1 and show its relationship first to the other figures in that column and then to the range in the "double-glazed" row. They also attempt to explain, by the statement in parentheses, an important fact that the reader must consider in interpreting their data.

Acquisition Costs

Having calculated the energy costs associated with different
window configurations, let us now turn to the costs of acquisition,
maintenance, and repair. The purchase and installation of   Background
windows in a new home are generally more expensive than
the costs of an equivalent area of nonwindowed wall. To
estimate the additional acquisition costs, window costs are
compared with wall costs in Table 1. This table shows that the   What to notice
costs of the purchase and installation of good-quality wood
windows are estimated to add between $18 and $76 to initial
building costs for single-glazed windows, and $48 to $216
for double-glazed windows, depending on their size. (Because   Important fact in
windows displace portions of the wall, they raise initial building   parentheses
costs by substantially less than their full purchase and installation
costs.)

    If management devices are used, additional acquisition
costs are incurred. Costs of venetian blinds and wooden
shutters based upon averages of currently quoted prices in
the Washington, D.C., area are given in Table 2 [not shown].

TABLE 1
Acquisition Costs of a Window in Excess of the Cost of a Nonwindowed Wall

	Dollar Costs, by Size of Area			
Component	12 ft²	18 ft²	30 ft²	60 ft²
Windows[a]				
Single Glazed	52.20	70.70	122.55	245.10
Double Glazed	81.80	109.36	192.61	385.23
Wall[b]	33.72	50.58	84.30	168.60
Window Cost Less Wall Cost[c]				
Single Glazed	18.48	20.12	38.25	76.50
Double Glazed	48.08	58.78	108.31	216.63

[a]*Purchase prices are list retail prices, reduced 10 percent to reflect a typical builder's discount, for good-quality wood double-hung windows, provided by a distributor in the Washington, D.C., area. Prices are for single and multiple units of windows of a size which comes close to providing the designated percentages of the exterior wall in glazing. The 12 ft² area is provided by a 3' × 3'11" window; the 18 ft² area, by a 3' × 6' window; the 30 ft² area by two 3' × 5' windows, and the 60 ft² area, by four 3' × 5' windows. An installation cost of $5.00 per window or pair of windows is used, based on an estimate given by a home builder in the Washington, D.C., area.*

[b]*Costs of non-windowed wall areas corresponding in size to the windowed areas are based on a price of $2.81 ft² as estimated by a home builder in the Washington, D.C., area. The wall section is assumed to be face brick veneer over 8" cinder block with building paper sheathing, 3½" of insulation, and ½" of painted interior drywall.*

[c]*The additional costs incurred for windowed areas of the building are the difference between the costs of windows and the costs of walls for the same wall area.*

Explaining a Graph These same authors also included a complex graph (Figure 1 below) in their report. Exactly which relationship their readers should focus on is not immediately clear. Should they notice just electricity or just gas points? Or north or south points? Should they compare N (elec) with S (gas)? And why? What do the comparisons show?

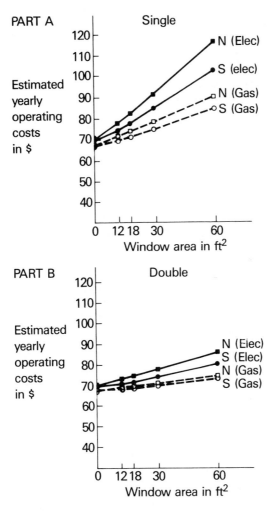

FIGURE 1
Estimated Yearly Energy Costs with North (N) or South (S) Facing Window with Gas or Electric (Elec)
Source: Belinda Collins, et al., *A New Look at Windows.* Reprinted by permission of the U.S. Department of Commerce, National Bureau of Standards, Center for Building Technology.

The authors solved the problem in their discussion. In the discussion the authors first make a general point, then give a specific example. So in the paragraph discussing Part B of the graph, the second sentence points out that "costs are lowered," and the third sentence points out an exact detail to notice — that costs are $30 lower. The $30 is the difference between N (elec) at $120 in Part A and N (elec) at $90 in Part B.

Figure 1 shows the estimated yearly energy costs for the room as a function of the window. Note that on the vertical axis, or zero window area, energy costs are given for a windowless room. To determine the operating costs attributable to just the window, the yearly operating costs for a room with a given window area must be subtracted from those for a windowless room with zero window area. Part A of Figure 1 demonstrates that, when only thermal loads are considered, estimated yearly energy costs increase for both northern and southern exposures as the size of the single-glazed window increases. A window with a northern exposure has greater energy costs, however, than one with a southern exposure, particularly when the more expensive electric heating is used. The added energy costs for the room are as much as $20 to $25 more per year for large window areas on the north wall with electric heat than for large windows on the south with gas heat.	What to notice

Significance

What to notice |
| Part B plots similar yearly energy costs for a room with a double-glazed window. When double glazing is used, energy costs are lowered for both orientations. For example, double glazing lowers energy costs by about $30 per year for the largest north-facing window in the electrically heated room. The reduction is somewhat lower for the south-facing window. | What to notice |

■ SUMMARY

As a result of the impact of computers, the visual aspect of technical writing is something that all writers must master. This chapter explains how to format pages to enhance the message and how to construct strong visual aids. Page formatting allows you to design a page that is easy to read, that calls attention to key items, and that helps the reader follow the contents. Writers must learn to construct head systems with various levels and to use these heads consistently throughout the document. The capabilities of the personal computer now allow the writer to design pages that incorporate elements — boldface type, various typefaces, type sizes, rules — than were ever used in documents formatted on the typewriter. Writers must also learn when to use and how to construct a wide array of visual aids, including tables, bar graphs, line graphs, and pie charts. Visual-aid software now enables writers to choose many attractive ways to present data.

■ WORKSHEET FOR FORMATTING

☐ *Decide on how many levels of heads you will need.*

☐ *Select a style for each level. Choose styles that reflect the descending levels of heads.*

☐ *Select outside margins.*

☐ *Select a location and format for your page numbers.*

☐ *Choose a method for distinguishing visuals from the text. Will you enclose them in a box or use a rule above and below?*

☐ *Determine the number of columns. Decide the amount of space between them.*

☐ *Do you need to place information in the header or footer area?*

■ WORKSHEET FOR VISUAL AIDS

☐ *Name the audience for this visual aid.*

☐ *What do your readers know about the topic on which this visual aid is based?*

☐ *What is your goal for them after they review your visual aid?*
Do you want them to have precise data?
Do you want them to get an overview of the topic?
Do you want them to see a picture that conveys some kind of emotional drama?

☐ *Choose a format for this visual aid:*
Determine where you will place its number and title (above? below?).

Will you have both a figure number and title, or just one? If so, which one?
Should the title be boldfaced?
If your visual is a table, how many rules should it have?
Should you include vertical lines?

☐ *What wording should you use for the title and for any legends in figures?*

☐ *How will you connect words to items in the visual — extend lines to them? use letters?*

☐ *How large will the visual aid be? Do you have room for it in your document?*

☐ *How much time do you have? Will you be able to construct a quality visual in that time?*

☐ *Have you called attention to the visual in your text. Explained it?*

■ MODELS

The following pages present the same report section for four different formats. They illustrate the variety you can use in formatting reports. Each is the result of a different style sheet.

PRINTING FEATURES

At Hobbes, we are concerned about two printing features: printing speed and automatic speed adjustment. Printing speed is the rate, measured in feet per minute, at which the system can print information clearly on a carton. This is important to Hobbes because we have an average line speed of 150 feet per minute (fpm). Therefore, a system that can print at least 150 fpm is required. Automatic speed adjustment is a feature that allows the system to automatically sense the line speed and print accordingly. This is a desirable feature because when a speed variation occurs on the line, an operator does not have to be present to make the manual adjustment. Management has given this criterion second highest priority.

Capabilities

Zorg System. As Table 2 shows, the Zorg system has a maximum printing speed of 100 fpm. This is less than the acceptable rate of speed required by the Hobbes Corporation. This system does not have an automatic speed adjustment feature (see Table 2). Thus, an operator would have to be present at all times to monitor line speeds and make necessary adjustments.

Spiff System. As Table 2 shows, the Spiff system has a maximum printing speed of 200 fpm. This exceeds Hobbes's current line speed of 150 fpm. The system has an automatic speed adjustment feature (see Table 2). This feature is economical because it saves time and reduces the frequency of illegible printing on cartons.

Conclusion. The Spiff system meets the requirements for printing speed and automatic adjustment. The Zorg system meets neither of these requirements.

TABLE 2
Printing Features of Ink Jet Printing Systems

Feature	Zorg	Spiff
Printing speed (fpm)	100	200
Automatic Speed Adjustment	No	Yes

FIGURE 7.22
Traditional Typed Page: Underlined Heads

PRINTING FEATURES

At Hobbes, we are concerned about two printing features: printing speed and automatic speed adjustment. Printing speed is the rate, measured in feet per minute, at which the system can print information clearly on a carton. This is important to Hobbes because we have an average line speed of 150 feet per minute (fpm). Therefore, a system that can print at least 150 fpm is required. Automatic speed adjustment is a feature that allows the system to automatically sense the line speed and print accordingly. This is a desirable feature because when a speed variation occurs on the line, an operator does not have to be present to make the manual adjustment. Management has given this criterion second highest priority.

CAPABILITIES

Zorg System

As Table 2 shows, the Zorg system has a maximum printing speed of 100 fpm. This is less than the acceptable rate of speed required by the Hobbes Corporation. This system does not have an automatic speed adjustment feature (see Table 2). Thus, an operator would have to be present at all times to monitor line speeds and make necessary adjustments.

Spiff System

As Table 2 shows, the Spiff system has a maximum printing speed of 200 fpm. This exceeds Hobbes's current line speed of 150 fpm. The system has an automatic speed adjustment feature (see Table 2). This feature is economical because it saves time and reduces the frequency of illegible printing on cartons.

Conclusion

The Spiff system meets the requirements for printing speed and automatic adjustment. The Zorg system meets neither of these requirements.

TABLE 2
Printing Features of Ink Jet Printing Systems

Feature	Zorg	Spiff
Printing speed (fpm)	100	200
Automatic Speed Adjustment	No	Yes

FIGURE 7.23
Page Produced with Dot Matrix or Laser Printer: Boldface Heads

Printing Features

At Hobbes, we are concerned about two printing features: printing speed and automatic speed adjustment. Printing speed is the rate, measured in feet per minute, at which the system can print information clearly on a carton. This is important to Hobbes because we have an average line speed of 150 feet per minute (fpm). Therefore, a system that can print at least 150 fpm is required. Automatic speed adjustment is a feature that allows the system to automatically sense the line speed and print accordingly. This is a desirable feature because when a speed variation occurs on the line, an operator does not have to be present to make the manual adjustment. Management has given this criterion second highest priority.

Zorg System

As Table 2 shows, the Zorg system has a maximum printing speed of 100 fpm. This is less than the acceptable rate of speed required by the Hobbes Corporation. This system does not have an automatic speed adjustment feature (see Table 2). Thus, an operator would have to be present at all times to monitor line speeds and make necessary adjustments.

Spiff System

As Table 2 shows, the Spiff system has a maximum printing speed of 200 fpm. This exceeds Hobbes's current line speed of 150 fpm. The system has an automatic speed adjustment feature (see Table 2). This feature is economical because it saves time and reduces the frequency of illegible printing on cartons.

Conclusion

The Spiff system meets the requirements for printing speed and automatic adjustment. The Zorg system meets neither of these requirements.

TABLE 2
Printing Features of Ink Jet Printing System

Feature	Zorg	Spiff
Printing speed (fpm)	100	200
Automatic Speed Adjustment	No	Yes

FIGURE 7.24
Page Produced with Laser Printer: Different Type Sizes and Indenting

does not have the ability to print a bar code. However, an option can be added to the system, at an additional cost of $700, to print readable bar codes (see Table 1).

Spiff System. As shown in Table 1, the basic Spiff system has the ability to print a readable bar code. This ability is part of the system and it requires no additional investment.

Conclusion. Both systems have the ability to print a readable bar code; however, the Zorg system requires an additional expenditure for this capability.

TABLE 1
Printing Capability of an Ink Jet Printing System

Feature	Kiwi	Diagraph
Ability to print a bar code	Yes[a]	Yes
Ability to print a readable bar code	Yes[a]	Yes

[a]Available as an option at an extra cost of $700.

Printing Features

At Hobbes, we are concerned about two printing features: printing speed and automatic speed adjustment. Printing speed is the rate, measured in feet per minute, at which the system can print information clearly on a carton. This is important to Hobbes because we have an average line speed of 150 feet per minute (fpm). Automatic speed adjustment is a feature that allows the system to automatically sense the line speed and print accordingly. Management has given this criterion second highest priority.

Zorg System. As Table 2 shows, the Zorg system has a maximum printing speed of 100 fpm. This is less than the acceptable rate of speed required by the Hobbes Corporation. This system does not have an automatic speed adjustment feature (see Table 2).

Spiff System. As Table 2 shows, the Spiff system has a maximum printing speed of 200 fpm. This exceeds Hobbes's current line speed of 150 fpm. The system has an automatic speed adjustment feature (see Table 2).

TABLE 2
Printing Features of Ink Jet Printing Systems

Feature	Zorg	Spiff
Printing speed (fpm)	100	200
Automatic Speed Adjustment	No	Yes

FIGURE 7.25
Page Showing Two-Column Format

■ EXERCISES

1. Find a line graph and convert it to a bar graph, or vice versa. Write a brief paragraph explaining each one. Emphasize the trend in the line graph; emphasize the discrete data in the bar graph.

2. Photocopy a table from the *Statistical Abstract of the United States,* available in your library's reference room. Convert the data in the table into a graph that illustrates a relationship you can see in the table. Write a paragraph explaining the relationship shown in the table and another explaining the same relationship in the graph.

3. Photocopy a table or graph, and its accompanying explanation, from a report or a journal. Bring it to class, and in small groups discuss the strengths and weaknesses of the explanations.

4. If you have access to a computer graphics program, make several different graphs of the same data. In a brief paragraph, explain the type of reader and the situation for which each graph would be appropriate.

5. Divide into groups of three or four, by major if possible. Select a process you are familiar with from your major or from your campus life. Possibilities include constructing a balance sheet, leveling a tripod, focusing a microscope, constructing an isometric projection, threading a film projector, finding a periodical in the library, or making a business plan. As a group construct a flow chart of the process. For the next class meeting each person should write a paragraph explaining the chart. Compare paragraphs within your group; then discuss the results with the class.

6. Select one of the visual aids from Figures 7.1 to 7.6. Write a brief (one to two paragraphs) news release about it for a local newspaper. Assume that you are the public relations manager in a large city government.

7. Convert the following paragraph into a table. Then rewrite the paragraph for your manager so that you give only the essential information.

  This store sells two brands of jewelry, High Fashion and Golden Moment. For each line we carry earrings, necklaces, bracelets, rings, and pins. The High Fashion jewelry lists the following prices for these pieces, respectively: $4.00, $7.00, $5.00, $35.00, and $8.00. The Golden Moment line lists the following prices, respectively: $8.00, $10.00, $7.00, $22.00 and $12.00. Except for the rings, High Fashion is cheaper by these amounts, respectively: $4.00, $3.00, $2.00, and $4.00. Golden Moment's rings are $13.00 cheaper.

8. Divide the class into three sections. Have individuals in each section convert the following numbers into visual aids. Have Section 1 make line graphs, section 2 make bar graphs, and section 3 make pie charts. Have

one person from each section put their visual on the board. Discuss them for effectiveness. Here are the figures:

Respondents to a survey were asked if they would pay more for a tamper-evident package. 8.2% said they would pay up to $.15 more; 25.8% were unwilling to pay more; 51.6% would pay $.05 more; and 14.4% would pay $.10 more.

9. Make a pie chart about some aspect of your class. An easy topic is the percentage of students from cities of various sizes — over 100,000, between 50,000 and 100,000, and so on. You will have to collect all the data and compute the percentages in class. Outside of class, make a bar graph of the same data. Write a story of several paragraphs for the school newspaper, explaining the diversity or homogeneity of your class. Refer to specific parts of your visual aid.

10. Read over the next four paragraphs from a report. Then construct a visual aid that will support the writer's conclusion.

Appleworks.   The Appleworks reference manual is very well written, with a detailed table of contents and numerous helpful examples. It has "Appleworks Tips," offering helpful hints for efficient approaches to common tasks, and "warning" messages for problem areas. For experienced users, the manual has a list of command shortcuts to speed up the processes.

Appleworks also includes a tutorial, which consists of a series of carefully designed, interactive learning experiences. The tutorial leads the user through the basic commands, keystrokes, and features, providing explanations and sample tasks. Packaged on two disks and covering key topics, it is designed to allow the user to proceed at his or her own pace. Topics can be started, stopped, repeated, or skipped as the user sees fit.

Big Worker.   The Big Worker manual serves its intended purpose fairly well, with a few exceptions. It contains no mention of the sample data diskette, nor does it mention that the examples shown in the manual are contained on the sample data diskette. It has many boldface "remember" messages that explain what will happen if the user performs certain actions. This feature is an attempt to prevent user mistakes. Overall, its reference section contains inadequate detail and no detailed system overview. No disk tutorial is available.

Conclusion.   The documentation supplied with Appleworks is far superior in all respects.

■ WORKS CITED

Collins, Belinda, et al. *A New Look at Windows.* NBSIR 77–1388. Washington, DC: National Bureau of Standards, 1978.

Honeywell, Inc. *Air Data/Inertial Reference System (ADIRS): A Proposal to British Aerospace for the BAE 146-300.* Minneapolis: Honeywell, 1984.

MRM/Elgin. *Rotary Piston Filler: Eight Head.* Menomonie, WI: MRM/Elgin, 1985.

Riordan, Timothy. *City of Dayton: 1986 Program Strategies.* Dayton, OH: City of Dayton, 1986.

White, Jan. *Using Charts and Graphs.* New York: Bowker, 1984.

8

Definition

FORMAL DEFINITIONS
INFORMAL DEFINITIONS
EXTENDED DEFINITIONS

A definition gives the precise meaning of a term. In most technical writing situations, you have to define terms. The need for definition increases when you communicate with those at a distance from you in any bureaucratic hierarchy and when you write to people with limited knowledge of your topic.

Since both situations are common, defining is an important skill. This chapter explains formal definitions, informal definitions, and extended definitions.

FORMAL DEFINITIONS

A *formal definition* is one sentence that contains three parts: the *term* that needs defining, the *class* to which the item belongs, and the *differentiation* of the item from all other members of its class. Here are some examples:

Acceleration is the rate of change of velocity with respect to time.
Resistance is any force that tends to oppose or retard motion.
A carburetor is a mixing chamber used in gasoline engines to produce an efficient explosive vapor of fuel and air.

Classify the Term

To define a term, you first place the term in a class. The class is the large group to which the term belongs. The group can be either broad or narrow. For instance, a pen can be classed as a "thing" or as a "writing instrument." A carburetor can be a "part" or a "mixing chamber." The narrower the class, the more meaning it conveys, and the less that needs to be said in the differentiation.

Differentiate the Term

To *differentiate* the term, you explain those characteristics that belong only to it and not to the other members of the class. If the differentiation applies to more than one member of the class, the definition lacks precision. For instance, if a writer says, "*evaporation* is the process of water disappearing from a certain area," the definition is too broad; water can disappear for many reasons, not just from evaporation. The differentiation must explain the characteristics of evaporation that make it unlike any other process — the change from a liquid to a vapor.

Here are five common methods for differentiating a term:

- Name its essential properties — the characteristic features possessed by all individuals of this type.
- Explain what it does.
- If the term is an object, describe what it looks like and what it is made of.
- If the term is a process, explain how to make or do it.

In the following examples, notice that the classification is often deleted. In many cases, it would be a broad statement such as ". . . is a machine."

NAME THE ESSENTIAL PROPERTIES

Engineering is the application of scientific principles to practical ends, such as the design, construction, and operation of efficient and economical structures, equipment, and systems.

Electricity is a physical phenomenon arising from the existence and interactions of charged particles.

EXPLAIN WHAT IT DOES

The microprocessor-based 7920 Multicontroller [is a machine that] automatically controls multiple-step operations in industrial production and processes (Veeder-Root 18).

DESCRIBE WHAT IT IS MADE OF

Model RS-1 Switch is a simple electronic device consisting of two cantilevered ferromagnetic blades which are hermetically sealed in a controlled atmosphere glass capsule (Humphrey 1).

A double end stepping motor [is a machine that] has an output shaft that extends from both ends of the motor (Superior 8).

DESCRIBE WHAT IT LOOKS LIKE

A well-focused photograph is a print of a negative in which all lines demarking contrast are sharp.

A sine curve is a figure that has semicircles of identical size alternating above and below a horizontal axis.

EXPLAIN HOW TO MAKE OR DO IT

Simple staining is [a process of] coloring bacteria by applying a stain, methylene blue, to a fixed smear.

Avoid Circular Definitions

Do not use circular definitions, which repeat the word being defined. You will not help a reader understand *capacitance* if you use the word *capacitor* in the differentiation. Noncrucial words, such as *writing* in the term *technical writing*, may, of course, be repeated.

INFORMAL DEFINITIONS

For specialized or technical terms that your readers will not know, you can provide an informal definition. Though they are less thorough than formal definitions, operational definitions and synonyms are quite acceptable if they furnish the necessary information.

Operational Definitions

An operational definition gives the meaning of an abstract word for one particular time and place. Scientists and managers use operational definitions to give measurable meanings to abstractions. The operational definition "creates a test for discriminating in one particular circumstance" (Fahnestock and Secor 84). For instance, to determine whether or not a marketing program is a success, managers need to define *success*. If their operational definition of success is "to increase sales by 10 percent" and if the increase occurs, the program is successful. In this sense, the operational definition is an agreed-upon criterion. If everyone agrees with it, it will facilitate discussion and evaluation of any topic.

Synonyms

A *synonym* is a word that means the same as another word. It is effective as a definition only when it is better known than the term being defined. People are more familiar with *loudspeaker* than with *electroacoustic transducer,* the technical term that has the same meaning. If your audience knows less about the topic than you do, use common words to clarify technical terms.

When using synonyms, put the common word or the technical term in parentheses, or set one of them off with dashes, as in the following examples. Notice that writers often highlight the term that they will define.

Parentheses	The maps are drawn in a projection that enlarges the areas near the edges (called limbs) of the moon, which are otherwise too foreshortened to be seen clearly.
Dash	At the edges of some of the maria we find indentations, each known as a bag — or a *sinus.*
Dash	Many of the craters have *central peaks* — round regions at their centers (Menzel and Pasachoff 323).

EXTENDED DEFINITIONS

Extended definitions are expanded explanations of the term being defined. After reading a formal definition, a less knowledgeable reader often needs more explanation to understand the term completely. Eight methods for extending definitions follow.

Explain Derivation

To explain the *derivation* of a term is to explain its origin. For example, you can show how some words were combined from other words, or can spell out acronyms. *Ammeter,* for example, is derived from the combination of *ampere* and *meter*; it is a meter that measures amperes. *COBOL* is an acronym for *common business-oriented language. Inductosyn* is a combined word, derived from *inductance* and *synchronous. Inductance,* which stems from *induce,* refers to the ability of an electric current in one circuit to cause voltage in another circuit. *Synchronous* means happening at the same time. An *inductosyn* performs actions based on simultaneous measurement of inductance in two circuits.

Explicate Terms

In this context, to *explicate* means to define difficult words contained in the formal sentence definition. Many readers would need definitions of words such as *fixed smear* and *cantilevered ferromagnetic blades* in the formal definitions on pp. 161–162. When explicating, you can often provide an informal definition rather than another formal one. Notice that *speed* defines *velocity* in the following example.

> Acceleration is the rate of change of velocity (or speed) with respect to time.

Use an Example

An extended example gives readers something concrete to help them understand a term. You should select common examples that are familiar to a wide range of readers. For example, you might amplify the definition of *weightlessness* by referring to the astronauts' experiences in space. In the following paragraph, a formal sentence definition of *order point* is amplified by an extended example.

> Order point, simply stated, is a set of procedures used to   Definition
> ensure the availability of inventory when demand is uncertain.
> Suppose the production planner must determine how many   Example
> pen barrels, ink tubes, and caps to order and when to assemble
> them to meet production requirements. Under the order point
> system, the inventory of pen barrels, ink tubes, and caps
> will be "watched." When the supply of each item falls below a   Common terms
> predetermined quantity, the order point, a "signal" will be
> sent to the planner telling him or her to place an order for an
> amount of parts, returning it to the proper inventory level. The   Definition
> time period between the order being placed and the order
> being received is called lead time. During lead time the
> production of pens must still continue. Therefore, the order
> point should occur at a time when sufficient parts remain
> in stock to meet production requirements through lead time.

Use an Analogy

An *analogy* points out a similarity between otherwise dissimilar things. If something is unknown to readers, it will help them if you call attention to its similarity to something they do know. Consider this example, which draws on the reader's knowledge of a warehouse to explain an unknown term: "Using a disk server is like leaving storage space in a warehouse, then filling it with all your belongings. Whenever you need to retrieve an item from storage, you must get it yourself" (Ushijiman 114). Here is another analogy that compares a computer network to a highway system:

> Imagine a network as an electronic freeway system where data comes from one interchange to another at 186,000 miles per second. From a vantage point in the breakdown lane, you might see parts of a document shoot from a Mac to a laser printer. A moment later, an "out of paper" message from the laser printer might travel in the opposite direction. Soon after, a road-weary document might pull into the network's rest area, a shared hard disk called a *file server.* (Heid 199)

Comparison to highway

Highway references continue

Compare and Contrast

A *comparison-contrast* definition shows both the likenesses of and differences between similar objects or processes, such as comparing water flowing through a pipe with electricity flowing through a wire. As with other methods of extending a definition, the comparison-contrast method takes advantage of something the readers know in order to explain something they do not know. Comparing and contrasting a *semiconductor* with a *conductor* of electricity works only if the reader knows what a conductor is.

Here is an extended definition that compares and contrasts a department store and a mass merchant:

> Department stores are regionally located and predominantly sell merchandise in the areas of home furnishings, household linens and dry goods, and apparel for men, women, and children. A key difference between a mass merchant and a department store is the price lines carried. A price line refers to the dollar amounts that a particular category of merchandise will retail for. For example, the mass merchant may carry blouses that cost $10, $15, and $20 whereas the department stores carry more expensive lines of blouses that cost $20, $30, and $40. The department store has a clientele with more expensive tastes and aims its sales at the middle to upper income levels. Dayton's and H. C. Prange are good examples of department stores.

Definition

Terms to contrast

Contrast

Contrast

Explain Cause and Effect

Some things are so elusive that they must be defined in terms of their causes and effects. For example, nobody really knows what electricity is, but the interactions that cause it can be explained. Magnetism, on the other hand, must be defined through the force it produces — its effect. In the following example, the causes and effects of pressure are described in order to extend the formal definition.

A valve actuator is a device that allows the hydraulic system's   Definition
pressure to regulate the flow of the system. As oil enters the   Cause/effect
inlet port of this actuator, a pressure builds up. When this
pressure exceeds the pre-set value, the rod extends. A cam,   Cause/effect
which is mounted on the rod, is the portion of the actuator
that delivers the work. It then comes in contact with the   Effect
desired valve to be opened or closed. (See Figure 1.)

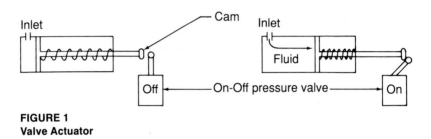

FIGURE 1
Valve Actuator

Use a Visual Aid

Drawings and diagrams can effectively reinforce definitions. For example, a small, labeled drawing of a carburetor would help explain its function, and the definition of an abstract term, such as *resistance,* would be clarified by a diagram of forces in opposition. The preceding definition of actuator is reinforced by an illustration of the actuator before and after it operates.

Analyze the Term

To *analyze* is to divide a term into its parts. Analyzing helps readers understand by allowing them to grasp the definition bit by bit. For example, *elemental times,* a term used to analyze the work of a machine operator, is easier to understand when its main parts are discussed individually:

The elemental times are the objective for taking the time   Preview
study in the first place. The elemental time slots on the time-
study form contain the following times: overall time, average
time, normal time, and standard time. *Overall time* is all times   Definition of
from start to finish, and includes total times of all elements   part 1
(each individual piece studied). *Average time* is the time   Definition of
it takes to produce or assemble each piece **into** a product.   part 2
When the overall time is divided by the number of pieces
produced, the average time per piece is obtained. *Normal*   Definition of
time is the average time per piece multiplied by a leveling   part 3
factor. This factor is just a conversion of the performance

rating found by using a time study conversion chart. This
normal time represents the time required by a qualified worker
working at the normal performance level to perform the given
task. *Standard time*, which is the time allowed to do a job, is   Definition of
obtained by adjusting the normal times according to some   part 4
allowance factor.

◼ SUMMARY

To define is to explain the meaning of a term precisely. Three ways to
explain and clarify meaning are by formal, informal, and extended definitions. Formal definitions consist of three parts: the term, the class, and the
differentiation. Informal definitions provide operational definitions or synonyms for the term. Extended definitions explain the term at some length,
giving additional information that helps readers visualize the idea. Definitions are extended by eight common methods: derivations, explications,
examples, analogies, comparison-contrast, cause and effect, visual aids,
and analyses.

◼ WORKSHEET FOR DEFINING TERMS

☐ *Name the audience for this definition.*

☐ *What do they know about the concepts on which this definition is based?*

☐ *What do they know about the topic of the document?*

☐ *What is your goal for your readers? In other words, what do you want
them to know about the concept?*

☐ *Does your audience need an overview, or do they need a technical, specific
description to enable them to make some kind of decision?*

☐ *List the visual aids that will enable a reader to grasp the topic.*

☐ *Select an approach: What devices will convey this topic?*
   Do you need a formal or informal definition?
   What method or methods of extending the definition will help
   them grasp the topic?

MODELS

The following two models begin with a formal sentence definition and use several methods to extend, or amplify, the definition. Read the models carefully and identify the methods of amplification.

RANDOM SAMPLING

The following information will orient you to basic concepts in random sampling. Random sampling is simply choosing individual items from a larger group. At our plant we use the process to determine whether or not a production run performs adequately. The example below shows how we determined that a problem existed in a run of our 12.5-ounce product. When we sample randomly, we record sample mean, sample standard deviation (Sample Std. Dev.), and cumulative data. This report explains these three concepts while providing a brief analysis of the problem.

TABLE 1
Sample Chart for a 12.5-Ounce Product

			Cumulative Data		
Sample	Sample Mean	Sample Std. Dev.	Cum. Sample	Cum. Mean	Cum. Std. Dev.
1	12.47	.015	10	12.50	.015
2	12.43	.031	20	12.49	.023
3	12.40	.055	30	12.46	.034
4	12.31	.070	40	12.42	.043
5	12.10	.080	50	12.36	.051
6	12.05	.085	60	12.31	.056
7	11.80	.104	70	12.23	.063

Sample mean

The sample mean is the average of a given sample. The second column in Table 1 reports the sample mean derived for each sample number. Each sample consists of ten different product items taken at random from the processing line. We took seven samples. The sample mean number describes the average weight (in ounces) of those ten samples. In sample 7 the average weight was considerably lower, 11.8 ounces. The data show a major problem with weight at the time when sample 7 was taken.

Sample standard deviation (Sample Std. Dev.)

The sample standard deviation column shows how much each box in the sample varied from all the others in that sample. The boxes in sample 1 did not vary a great deal from each other, only .015 ounces, but those in sample 7 varied quite a bit — .104 ounces. The Sample Std. Dev. column indicates that, during the period sample 7 was taken, the distribution of the product was most inaccurate.

FIGURE 8.1 (continues on next page)
A Definition from a Report

Cumulative data   2

Cumulative data consist of three categories: cumulative sample (Cum. Sample),
cumulative mean (Cum. Mean), and cumulative standard deviation (Cum. Std.
Dev.). The cumulative sample is a running total of all the product items used
in the sampling. For example, the number of items used in samples 1, 2, and 3
total 30; the number of items used in samples 1, 2, 3, and 4 total 40; and so on.
The cumulative mean is the mean of the cumulative sample. For example, the
mean of samples 1, 2, and 3 is 12.46 ounces. Cumulative standard deviation is
the standard deviation of the cumulative sample. The cumulative data are helpful
information when looking at the process as a whole. The last figure in each
column shows 70 pieces of data were collected for the report, the average weight
of all the data is 12.23 ounces, and the average amount each product varied
from any other was .063 ounces.

FIGURE 8.1 (continued)

Excerpted from *Better Homes and Gardens* magazine.

WHAT TYPE OF INVESTMENT

There's a slew of investments available, but stocks, bonds, and mutual funds are probably best for beginners.

When you buy **common stock**, you cast your financial fate with the company that issues the shares. If the company does well, the value of your stock stands to increase as a capital gain. Of course, the reverse is true, too: The value of your shares may plummet with the company's misfortunes or a generally bad economy. You can buy stocks in units of 100 shares; fewer, and you'll pay an extra fee for an "odd lot" trade.

Common stock issuers generally pay dividends out of overall earnings four times a year. Growth companies tend to pay lower dividends because they reinvest their earnings in the company. Established companies will distribute larger dividends, but the capital gains potential may be less.

Holders of **preferred stocks** are entitled to receive dividends — sometimes at a fixed rate — before any funds are distributed to common stockholders. Preferred-stockholders also have a claim on company assets before common-stockholders do. As a result, preferred stock may cost more, even though their values fluctuate with the market, too, and there's no guarantee that dividends will be paid.

With **bonds**, you are lending money to the issuer with the promise that you'll receive regular interest payments — specified when the bond is issued — and get your principal back when the bond matures. (Some issuers reserve the right to "call" in their bonds before their maturity date; that's a pitfall to watch for if you're counting on the long haul.) Bonds generally are issued in units of $1,000. Municipal bonds more often start at $5,000; they're especially attractive to high-income investors because the dividends are exempt from federal (and sometimes local) taxes.

You face two types of risk with bonds: (1) The *credit risk* is that the issuing company or body will default, failing to pay you interest or return your principal. You can reduce your vulnerability by purchasing highly rated bonds ("A" or above) or government-backed issues. (2) The *market risk* is that interest rates could skyrocket, cutting the bond's value if you must sell before maturity or rates drop again.

FIGURE 8.2 (continues on next page)
A Definition from a Magazine Article

If your portfolio is too small to make a diversified strategy practical, consider [2]
mutual funds — pooled-asset funds that offer instant diversification, good
liquidity, and expert management. You can invest in some for as little as $500 or
less. Mutual fund portfolios usually consist of 50 or more securities, and all the
complex buying and selling decisions are left up to the pros. You pay an annual
fee for this service — typically, 1/2 to 3/4 percent. In a "load" fund, you'll pay a
sales commission (up to 8 1/2 percent); no-loads charge none.

Mutual funds are designed to meet specific investment goals, clearly stated
in the prospectus — a document you should read carefully before investing.
Look for fees, study the track record, analyze the net return. (Laughridge 47–48)

FIGURE 8.2 (continued)

■ EXERCISES

1. Write definitions of four terms for an uninformed reader. Begin each of them with a formal sentence definition.

2. a. Write an extended 100-word definition (one paragraph) using analogy as the major method of amplification.
   b. Write an extended 100-word definition (one paragraph) using comparison-contrast as the major method.
   c. Write an extended 100-word definition (one paragraph) using cause and effect as the major method.

   To write these definitions, select a term from your major, if possible. However, you might want to use one of the following.

band saw	standard deviation
bicycle gear train	mutual funds
keyline	annuity
electron microscope	magnetron
elements of a camera lens	dichroic enlarger
word processing function	inventory
voltage	an organizational chart
inductance	halftone
electrical phase	

■ WRITING ASSIGNMENT

1. Write a 200-word definition (two or more paragraphs) using four of the eight methods of extended definition discussed in the chapter. Immediately after you have used a method, identify it in brackets.

   In writing the 200-word definition, you will find that using one method leads you to another. Consider providing synonyms and other informal definitions in each of the four methods, but do not count them as one of the four methods in your extended definition. Use format devices, especially headings and boldface or underlining, to highlight the contents.

   Try to select a term, probably from your technical area, for which you do not need to look up the definition. If you use sources, however, give credit to them. Some possible topics are the following:

hypothesis	Styrofoam	annualize
compression	robotics	depreciation
microwave oven	horsepower	CAD/CAM
fashion cycle	calibration	network
equation	feedback	layout

■ WORKS CITED

Fahnestock, Jeanne, and Marie Secor. *A Rhetoric of Argument.* New York: Random, 1982.

Heid, Jim. "Getting Started with Networking." *Macworld* 4.9 (1987): 119 + .

Humphrey Products. *Humphrey Air Cleaners* (C82A). Kalamazoo, MI: Humphrey, 1982.

Laughridge, Jamie. "Investigating Basics — Getting a Smart Start." *Better Homes and Gardens* May 1985: 47–48.

Menzel, Donald H., and Jay M. Pasachoff. *A Field Guide to the Stars and Planets.* 2nd ed. Boston: Houghton, 1983.

Superior Electric. *SloSyn DC Stepping Motors.* Bristol, CT: Superior Electric, 1979.

Ushijama, David. "Apple Share — Multifaceted Networking." *Macworld* 4.3 (1987): 108–115.

Veeder-Root Co. *Electronic Counters, Controls, Digital Instruments, Inductive Proximity Switches, Photoelectric Sensors.* 9361. Hartford, CT: Veeder-Root, n.d.

9

Description

THE COMMON ELEMENTS OF DESCRIPTION

PLANNING THE MECHANISM DESCRIPTION

WRITING THE MECHANISM DESCRIPTION

PLANNING THE PROCESS DESCRIPTION

WRITING THE PROCESS DESCRIPTION

PLANNING THE DESCRIPTION OF A PERSON IN ACTION

WRITING THE DESCRIPTION OF A PERSON IN ACTION

Description is a technique widely used in technical writing. In fact, you will probably describe something — machines, processes, or systems — in every report you write. Sometimes you will describe in intricate detail, and other times in broad outline. Regardless of the topic or the depth of detail, all descriptions share several common elements. This chapter explains those elements and shows you how to apply them when you describe a mechanism, an operation, and a process focused on a person.

THE COMMON ELEMENTS OF DESCRIPTION

Regardless of the audience or purpose, all the descriptions you write should follow a general form: start with an overview, and add necessary details (Jordan). For most descriptions, you should do the following:

- Define the mechanism or process.
- Explain its function or end goal.

- Name its subparts.
- Give relevant details:
  For mechanisms — size, shape, weight, material, method of attachment.
  For processes — quality or quantity of action, and effect of action.
- Explain the significance of the mechanism or process, if necessary.

Read over the following description of a part (of a computer printer) and the accompanying annotations to see how the author explains the part. Notice that the author starts with a definition and then adds details.

The paper advancement panel is a row of three buttons in a plastic rectangle. It allows the user to advance the paper out of the printer and to align the paper before printing. The panel is 3 inches long by 2 inches wide. It contains three ½-inch by ½-inch square buttons that control the paper advancement. These paper advancement buttons are the on-line, line feed, and form feed buttons. The on-line button controls the power to the form feed and line feed buttons. When the on-line button is on, it allows the printer to print. When the on-line button is off, the user can activate the form feed and the line feed buttons. The form feed advances the paper one full page; the line feed advances only one line at a time.	Definition Functions Size, shape, subparts Definition and function

A description of a process works the same way. In the following example, the writer starts with a definition of the end goal of the process and then explains the steps and the significance.

The quality check determines if the drawing is good enough to use. The quality control (QC) person inspects the drawing for proper standard drafting techniques and for adherence to company standards. He or she also determines whether the drawing has all the necessary information to manufacture the part. If no errors are found, the QC person signs off on the drawing, then sends it to manufacturing.	End goal Quality of action Effect of action

PLANNING THE MECHANISM DESCRIPTION

The goal of a mechanism description is to give readers all the information they need to know about the mechanism. Obviously you can't describe every part in minute detail, so you select various key parts and their functions. Most important you must plan the description before writing it.

When you plan a description of a mechanism, you must consider the audience, select an organizational principle, choose visual aids, and follow the usual form for writing descriptions.

Consider the Audience

Consider the audience's knowledge level and why they need the information. Both will determine how much detail you include. In the following mechanism description, the writer uses very specific detail to show an expert reader that the part exactly fulfills the specifications and then explains the significance of these details.

The IRU contains an Inertial Sensor Assembly (ISA), power supply, eight electronic boards, and a chassis containing a motherboard. Its form and dimensions (7.60 in. high x 12.69 in. wide x 12.76 in. long) meet the requirements of ARINC 600 10 MCU. With the implementation of second-generation electronics (14 boards reduced down to 8 boards), six empty electronic card slots are left for future growth. (Honeywell 3–10)	Overview Size Significance Significance

But in the following description, from a different document, the writer does not include any specific details, aiming instead at a reader who needs only a general understanding of the part:

The pump creates fluid flow within the system. The system has a gear-type pump made of two components: a drive gear and a driven gear, both in a closely fitted housing. The drive gear, which is powered by an electric motor, turns the driven gear in the opposite direction. As the gears turn, they mesh at a point in the housing between the inlet and the outlet ports. The fluid trapped between the teeth and the housing is pushed through the outlet port by atmospheric pressure due to the low pressure created by the rotation of the gears. This creates fluid flow.	Definition Subparts Function of part Effect of action

Select an Organizational Principle

You can choose from a number of organizational principles. For instance, you can describe an object from

- top to bottom (or bottom to top)
- outside to inside (or inside to outside)
- most important to least important (or least important to most important)

For example, if you were going to describe a secretarial chair from top to bottom, you might start with the backrest, then go to the seat, and then move down to the casters. Or you could do the reverse. If you wished to describe it from most important to least important, you might start with the seat, then describe the backrest, and then the casters.

To make your decision, consider the audience's potential use of your document. If the audience needs a general introduction, then an easy sequence, from top to bottom, is best. If your audience needs to know special details for secretaries' safety and comfort, you might start with the caster system, which prevents tipping, then go to the adjustment system, which eases back strain.

An easy way to check if your organization is working is to look for "backtracking." Your description should move steadily forward, starting with basic definitions or concepts that the audience needs to understand later statements. If your description is full of sections in which you have

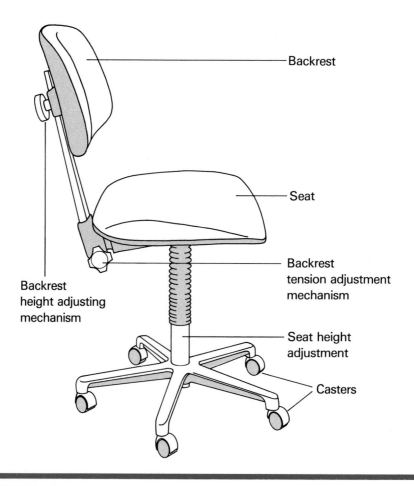

FIGURE 9.1
Secretarial Chair

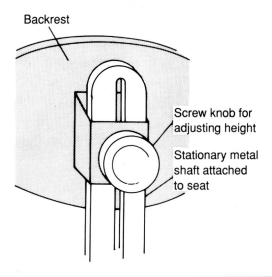

Backrest

Screw knob for adjusting height

Stationary metal shaft attached to seat

FIGURE 9.2
Detailed Drawing of Height Adjusting Mechanism

to stop and backtrack to define terms or concepts, then your sequence is probably inappropriate.

Choose Visual Aids

Use visual aids to assist your description of a complex mechanism. The type of visual you select depends on the mechanism and the reader.

If your readers need an overview of a secretarial chair so that they can see how each part is constructed, then you should use a drawing or photograph of the entire unit. (See Figure 9.1.) If, however, they need to know how the backrest can be adjusted to a comfortable height, the overall drawing is useless. Instead, provide a drawing or photograph of the adjusting assembly behind the backrest. (See Figure 9.2.)

Follow the Usual Form for Writing Descriptions

Generally, descriptions follow a similar form: an introduction or an overview followed by a body in which each part is described in turn (see pp. 180–181). Physical descriptions generally require no conclusion because a good description leaves nothing to conclude. Sometimes, however, you can place in a conclusion material that needs clarification but that does not fit elsewhere. If the operating principle of the mechanism, for instance, is complicated, you might explain it at the end of the description.

WRITING THE MECHANISM DESCRIPTION

The outline below presents the usual form for writing a mechanism description. This basic approach, with slight variations, will work in most instances:

I. Introduction
   A. Definition and purpose
   B. Overall description (size, weight, shape, material)
   C. Main parts
II. Body: Description of Mechanism
   A. Main part A (definition followed by detailed description of size, shape, material, location, and method of attachment)
   B. Main part B (definition followed by overall description, then identification of subparts)
      1. Subpart X (definition followed by detailed description of size, shape, material, location, and method of attachment)
      2. Subpart Y (same as for subpart X)
III. Optional conclusion

Introduction

The introduction gives the reader a framework for understanding the mechanism. In the introduction you should define the mechanism, tell its purpose, present an overall description, and preview the main parts. In the following introduction, the writer does all four:

To: Emily Brown   Date: July 18, 19XX
From: Steve Vande Walle
Subject: Tabletop Paper Micrometer

This memo provides the information you requested at our
July 17 meeting dealing with my department's paper micrometer.
A paper micrometer is a small measuring instrument used to   Definition and
measure the thickness of a piece of paper. The micrometer,   purpose
roughly twice as large as a regular stapler (see Figure 1), has   Overall description
four main parts: the frame, the dial, the hand lever, and the   Main parts
piston.

Body: Description of Parts

The *body* of the paper contains the detailed description. Usually you identify each main part with a heading, then describe it in a single paragraph. Each paragraph will follow the outline explained earlier in the section on The Common Elements of Description, p. 00. In the example below, notice that each section describes only one item. If necessary, you can divide a section into subsections.

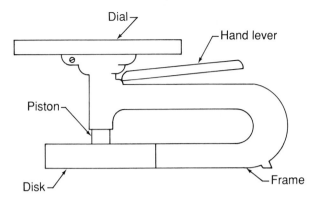

FIGURE 1
Paper Micrometer

The Frame

The frame of the paper micrometer is a cast piece of steel that provides a surface to which all of the other parts are attached. The frame, painted gray, looks like the letter *C* with a large flat disk on the bottom and a round calibrated dial on top. The disk is 4½ inches in diameter and resembles a flat hockey puck. The frame is 5⅛ inches high and 7½ inches long. Excluding the bottom disk, the frame is approximately 1¼ inches wide. The micrometer weighs 8 pounds.

Definition

Color
Analogy
Size and analogy

Weight

The Dial

The dial shows the thickness of the paper. The dial looks like a watch dial except that it has only one moving hand. The frame around the dial is made of chrome-plated metal. A piece of glass protects the face of the dial in the same way that the glass crystal on a watch protects the face and hands. The dial, 6 inches in diameter and ⅞ inches thick, is calibrated in .001 inch marks, and the face of the dial is numbered every .010 inches. The hand is made from a thin, stiff metal rod, pointed on the end.

Definition and analogy

Analogy

Size

Appearance

The Hand Lever

The hand lever, shaped like a handle on a pair of pliers, raises and lowers the piston. It is made of chrome-plated steel and attaches to the frame near the base of the dial. The hand lever is 4 inches long, ½ inch wide, and ¼ inch thick. When the hand lever is depressed, the piston moves up, and the hand on the dial rotates. When the hand lever is released and a piece of paper is positioned under the piston, the dial shows the thickness of the paper.

Analogy and definition

Relation to other parts
Effect

The Piston

The piston moves up and down when the operator depresses
and releases the hand lever. This action causes the paper's
thickness to register on the dial. The piston is ⅜ inches in
diameter, flat on the bottom, and made of metal without
a finish. The piston slides in a hole in the frame. The piston
can measure the thickness of paper up to .300 inches.

Definition
Function
Size

Relation to other
parts
Function

Other Patterns for Mechanism Descriptions

Two other patterns are useful for describing mechanisms: the function
method and the generalized method.

   The Function Method   One common way to describe a machine is to
name its main parts and then give only a brief discussion of the function
of each part. This *function method* is used extensively in manuals. The following paragraph is an example of a function paragraph.

FUNCTION BUTTONS

The four function buttons, located under the liquid crystal
display, work in conjunction with the function switches. The
four switches are hertz (Hz), decibels (dB), continuity (c), and
relative (REL). The hertz function can be selected to measure
the frequency of the input signal by pressing button 1. Press
the button again to disable. The decibel function allows you to
measure the intensity of the input signal, which is valuable
for measuring audio signals. It functions the same as the
hertz button. The continuity function allows you to turn on a
visible bar on the display, turn on an audible continuity signal,
or disable both of them. The relative function enables you to
store a value as a reference value. For example, say you
have a value of 1.00 volts stored; every signal that you measure
with this value will have 1.00 volt subtracted from it.

List subparts
Function and size
of subpart 1

Function and size
of subpart 2

Function and size
of subpart 3
Function and size
of subpart 4

   The Generalized Method   The *generalized method* does not focus on a
part-by-part description; instead the writer conveys many facts about the
machine. This method of describing is commonly found in technical journals and in technical reports. With the generalized method, writers use the
following outline (Jordan 19–22, 35):

1. General detail
2. Physical description
3. Details of function
4. Other details

General detail consists of a definition and a basic statement of the operational principle. *Physical description* explains such items as shape, size, appearance, and characteristics (weight, hardness, chemical properties, methods of assembly or construction). *Details of function* explain these features of the mechanism:

> how it works, or its operational principle
> its applications
> how well and how efficiently it works
> special constraints, such as conditions in the environment
> how it is controlled
> how long it performs before it needs service

Other details include information about

> background
> marketing
> general information, such as who makes it

The article on anti-static foam in Chapter 6 (p. 110) and the EVA hot melt model at the end of this chapter are examples of this kind of description.

PLANNING THE PROCESS DESCRIPTION

Technical writers often describe processes. Methods of testing or evaluating, methods of installing, flow of material through a plant, the schedule for implementing a proposal, and the method for calculating depreciation are all examples of processes. Manuals and reports contain many examples of process descriptions.

Processes are usually one of two types: the operations of a mechanism or system that do not involve human activity, and the operations that do involve human activity.

As with a mechanism description, the writer must consider the audience, select an organizational principle, choose visual aids, and follow the usual form for writing descriptions.

Consider the Audience

The knowledge level of audiences and their potential use of the document will vary. While most audiences for a process description have relatively little knowledge of the process, they must often make a decision based on such a description. The process description is, in effect, vital background information for the decision. For instance, a plant engineer might propose a change in material flow in a plant because a certain step is inefficient, causing a bottleneck. To get the change approved, he or she would have to describe the old and new processes to a manager, who would use that description to make a decision about whether to implement the new process or not.

Process descriptions also are used to explain in detail the implementation of a project. If a company plans to install a complicated piece of machinery in a plant, a careful schedule is written so that all affected parties understand what actions will occur at each step.

Select an Organizational Principle

The organizational principle for processes is *chronological:* the writer starts with the first action or step and continues in order until the last. Many processes have obvious sequences of steps, but others require careful examination in order to determine the most logical sequence. If you were describing the fashion cycle, you could easily determine its four parts (introduction, rise, peak, and decline). If, however, you had to describe the complex flow of material through a plant, you would want to base your sequence of steps on your audience's knowledge level and intended use of the description. You might treat "receiving" as just one step, or you might break it into several steps, like "unloading," "sampling," and "accepting." Your decision depends on how much your audience needs to know.

Choose Visual Aids

If your subject is a machine in operation, visuals of the machine in different positions will clarify the process. If you are describing a process that involves people, a flow chart can quickly clarify a sequence. For example, you might use the following flow chart to explain a hospital nutritional-assessment program.

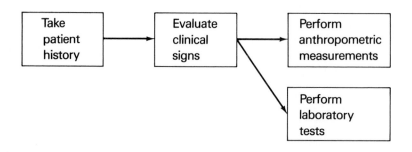

FIGURE 9.3
Mechanism Description

A *decision tree* can also explain a sequence effectively. A writer who wants to explain a grievance procedure might use this visual:

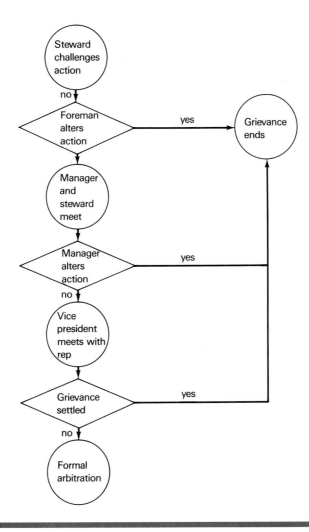

FIGURE 9.4
Process Description

Follow the Usual Form for Writing Descriptions

The process description takes the same form as the mechanism description: an introduction, which provides an overview, and the body, which treats each step in detail, usually one step to a paragraph. In each paragraph, first define the step, often in terms of its goal or end product, then describe it.

WRITING THE PROCESS DESCRIPTION

The outline below shows the usual form of a description of a process that does not involve a person. This approach will work for all such descriptions. A process description, analyzed in some detail, follows the outline.

    I. Introduction
        A. Definition of operation
        B. Principles(s) of operation
        C. Major sequences of operation
   II. Body: Description of operation
        A. Sequence A
           1. End goal of sequence
           2. Detailed description of action . . .
        B. Sequence B (same as A)
  III. Conclusion (optional)

Introduction

The introduction to a process description contains general information that prepares the reader for the specific details that will follow. In the introduction you define the process, explain its principles of operation (if necessary), and preview the major sequences. The following introduction performs all three tasks.

TO: Roger Tibbar   DATE: October 12,19XX
FROM: Brian Schmitz
SUBJECT: Conversion of AC to DC

The following is the information you requested at our last meeting. It should help give you a basic understanding of the process of converting alternating current (AC) to direct current (DC). Before explaining the conversion process, you must   Background
understand the difference between AC and DC. AC is an electrical current that continually reverses in direction at regular cycles. Its waveform is sinusoidal (see Figure 1). A typical household uses 110 volts AC. DC is an electrical current that   Common examples

flows in only one direction. A car battery produces about 12 volts DC (see Figure 1).

FIGURE 1
Comparison of AC and DC

In this memo I will describe the process of converting 110 volts AC into 12 volts DC — a very common conversion in many electronic devices today. The conversion requires three basic steps: transformation, rectification, and filtration (see Figure 2).

Preview of major sequences

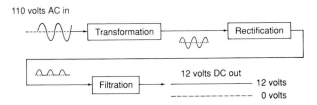

FIGURE 2
The conversion of AC to DC

Body: Description of the Operation

In the body of the paper, write one paragraph for each step of the process. Each paragraph should begin with a general statement about the end goal or main activity. Then the remainder of the paragraph explains in more detail the action necessary to achieve that goal. Notice in the following example how each paragraph starts with an overview — a statement of purpose or end goal — and how all the paragraphs are constructed in the same pattern.

Transformation

Transformation is the step in which the relatively large AC voltage (110 volts AC) is converted into a smaller AC voltage (12 volts AC). This conversion is done with a transformer, an electrical device with two parts: a primary coil and a

End goal
Tool used for activity

secondary coil. The primary coil is basically several loops of
fine wire. The secondary coil is similar to the primary coil, but   Common term
it is thicker and has fewer loops. The large AC voltage (110
AC) flows into the primary coil of the transformer. When   Action
the AC passes through this coil, a magnetic field is produced,
a phenomenon known as Faraday's law. The magnetic field
then passes through the secondary coil, producing a voltage   Action
in it (see Figure 3). The voltage produced in the secondary
coil depends on the ratio of loops of wire between the coils.   Theory of
With the correct number of loops in both the primary and   operation
secondary coils, the 110 volt AC converts into 12 volts AC.

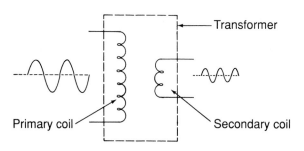

FIGURE 3
The Transformation Step

Rectification

Rectification is the step in which the small AC from the
secondary coil of the transformer is converted into DC. This   End goal
conversion is done with an electronic component known as a   Tool used for
diode (see Figure 4). A diode is like an electronic check valve;   activity
it only allows current to flow through it in one direction. When   Analogy
the AC (which by definition continually reverses in direction)
is passed through a diode, only the top half (positive half)   Common terms
of the sinusoidal waveform is allowed to pass through. The   Action
bottom half (negative half) is absorbed by the diode (see

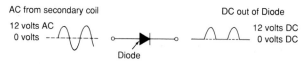

FIGURE 4
The Rectification Step

Figure 4). The current leaving the diode is traveling in only
one direction and therefore fits the definition of DC. However,
as you can see in Figure 4, the DC leaving the diode is quite
"lumpy." As a result, it is not very useful.   Common term

Filtration

Filtration smooths out the "lumpy" DC by means of a capacitor   End goal
(see Figure 5). A capacitor is a component that stores an   Tool used for
electrical current to be released later. When the DC from the   action
diode is at the top of one of the lumps, the capacitor stores   Action
some of the current. When the DC is at the bottom of the
lump, it releases it, thereby filling in the lump (see Figure 5).   Action
The DC leaving the capacitor has no lumps and is able to
perform useful work.   Common term

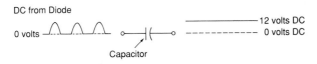

FIGURE 5
The Filtration Step

Conclusion

Conclusions to brief descriptions of operation are optional. At times writers follow the description with a discussion of the advantages and disadvantages of the process, or with a brief summary. If you have a written
relatively brief, well-constructed description, you do not need a summary.

PLANNING THE DESCRIPTION OF A PERSON IN ACTION

Describing a person going through a series of steps is a common writing
task. For instance, managers might describe the steps they take as they
review a job application or evaluate personnel. Sometimes such a description includes visual aids showing certain forms that are essential to certain
steps. A personnel manager, for example, might illustrate various forms
used in the evaluation process.

Follow the Usual Form

The usual form is similar to that of the process description shown earlier (pp. 186–189). It contains an introduction, which defines the process and its major sequences, followed by the body, which describes the process in detail.

WRITING THE DESCRIPTION OF A PERSON IN ACTION

The outline below shows the usual form for writing a description of a person in action. This approach will work for all such descriptions. A description of a person in action, analyzed in some detail, follows the outline.

    I. Introduction
       A. Definition of process
       B. Equipment needed
       C. Major sequences of process
   II. Body: Sequence of Person's Activities (same as description of operation, p. 187)

Introduction

In the introduction, writers define the process, explain the materials and the mechanisms necessary for performing the process, and list the major steps in the process. A sample introduction to the process of layout planning follows.

THE FOUR PHASES OF LAYOUT PLANNING

Layout engineers plan the spatial elements of an industrial plant in a process called *layout planning.* As shown in Figure 1, the four phases generally follow in sequence, but for the best results, the layout engineer may cause them to overlap in time.

Definition

Background

Purpose

FIGURE 1
The Four Phases of Layout Planning

The four phases are:

- Determining the location to be laid out
- Determining the overall layout
- Determining the detailed layout
- Determining the installation

Body: Sequence of a Person's Activities

In the body, the writer describes the person's actions in order, using one paragraph per step. Two notes on style seem appropriate. First, when you write a process description, do not overdo the use of the imperative (command) voice. You are trying to describe, not order, the process. So in this type of writing it's preferable to say, "We determine . . . " rather than "Determine . . ." Second, try to give the steps precise names. Notice that step 1 is named "Determining the location," a phrase that accurately describes a step. Do not give a step too concise a title, such as "The Location."

Determining The Location

Determining the location for the new layout means to select the most favorable conditions for obtaining the desired result. Determining the location does not necessarily mean to find a new site; more often, it means to decide whether to reconfigure the current location or to move to some other available space. In a typical situation, the layout engineer must decide whether to use an available storage area or a newly acquired building.

End goal

Action

Action

Determining The General Layout

Determining the overall layout means organizing the general arrangement of the area to be laid out. In this phase, the engineer works out the basic flow patterns of the product or materials in relation to the areas that have been allocated. This step allows the engineer to establish a rough idea of the general size, relationships, and configuration of each of the major areas to be laid out. Phase 2 is also called the block layout, the area allocation, or sometimes the rough layout.

End goal

Action

Significance

Other terms

Determining The Detailed Layout

Determining the detailed layout consists of locating each specific piece of machinery and equipment. In this phase, the engineer establishes the actual placement of each specific physical feature of the area to be laid out. The engineer makes detailed plans for each piece of machinery and equipment, and for aisles, storage areas, utilities, and service areas.

End goal

Action

Action

The engineer also makes a detailed plan for the overall layout
and for each of the departmental areas involved. The detailed   Background
layout is usually made of a board or sheet on which the
engineer places or draws images of the individual machines
or equipment.

Determining The Installation

Determining the installation means implementing the plans   End goal
for the new space allocation. The engineer makes final plans
for the installation, seeks the approval of the final plans by   Action
upper management, orders new equipment, and then arranges
for the necessary physical moves. The engineer does not   Action
physically move the machinery but coordinates and supervises
some other group, which does the actual installation.

Conclusion: Optional

A conclusion is optional. If you choose to include one, you might discuss
a number of topics, depending on the audience's need, including the advantages and disadvantages of the process.

◼ SUMMARY

Description is an essential technique for technical writers. All description
has common elements: the writer should define the process or mechanism,
explain its function or end goal, name its subparts, name relevant details,
and explain its significance. Three common types of description are describing mechanisms, describing operations, and describing people in action. For each type of description the writer must consider the audience,
select an organizational principle, choose visual aids, and follow the usual
form. Most descriptions begin with an introduction, which defines the
topic and previews the sections to follow. In the body of the description,
each section defines a step or part by explaining it in detail, including its
function or significance. To decide how much detail to include, writers
consider their audiences' knowledge level and how they will use the document. In most descriptions, a conclusion is optional.

◼ WORKSHEET FOR DESCRIPTION

☐ *Name the audience for this description.*

☐ *What do they know about the concepts on which this description is based?*

☐ *What do they know about the topic itself?*

☐ *What is your goal for your readers?*
What do you want them to know about the machine or process?
Parts or steps in detail or in broad outline? The function of each part?
Should they focus just on the machine or process, or grasp the broader context of the topic (such as who uses it, where and how it is regulated, who makes it, its applications and advantages)?

☐ *To write a description of mechanisms:*
Name each part.
Name each subpart.
Define each part and subpart.
List details of size, weight, method of attachment, and so forth.
Tell its function.

☐ *To write a description of processes:*
Name each step.
Name each substep.
Tell its end goal.
List details of quality and quantity of the action.
Tell significance of action.

☐ *List the visual aids that will enable a reader to grasp the topic.*

☐ *Choose a head system and your method of writing captions for each visual aid.*

☐ *Select an approach:*
What will you do first in each paragraph?
Will you use one example running through the entire paper, or will you use a different example for each paragraph?

MODELS

The models that follow describe mechanisms and processes. Based on what you know, review them carefully to discover their strengths and weaknesses. Notice that the caliper example is really an informal memo report (see Chap. 11), so the introduction contains material that differs from the usual form.

To: Nathan O'Timothy   Date: September 22, 19XX
From: Louise Esaian
Subject: Skinfold Calipers

You have expressed an interest in the skinfold caliper that our dietary department
has recently purchased. The following information should be helpful to you in
understanding the skinfold caliper and its individual parts. The skinfold caliper
(see Figure 1) is an instrument used to measure a double layer of skin and
subcutaneous fat (fat below the skin) at a specific body site. The measurement
that results is an indirect estimate of body fatness or calorie stores. The instrument
is approximately 10 inches long, made of stainless steel, and is easily held in one
hand. The skinfold caliper consists of the following parts: caliper jaws, press and
handle, and gauge.

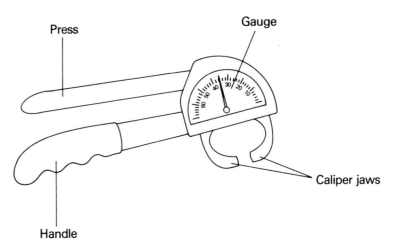

FIGURE 1
Skinfold Caliper

Caliper Jaws

The caliper jaws consist of two curved prongs. Each prong is approximately ¼
inches long. They project out from the half-moon-shaped gauge housing. The
prongs are placed over the skinfold when the measurement is taken. They clasp
the portion of the skinfold to be measured.

FIGURE 9.5
Mechanism Description

Press And Handle

The press is the lever that controls the caliper jaws. Engaging the press opens the caliper jaws so they can slip over the skinfold. Releasing the press closes the jaws on the skinfold, allowing the actual measurement. The press is 4.5 inches long and .5 inches thick. It is manipulated by the thumb while the fingers grip the caliper handle. The caliper handle is 6 inches long and .5 inches thick. The outside edge of the handle has three indentions, which make the caliper easier to grip.

Gauge

The gauge records the skinfold measurement. It is white, half-moon shaped, with 65 evenly spaced black markings and a pointer. Each marking represents 1 centimeter. The pointer projects from the middle of the straight edge of the half-moon-shaped gauge to the black markings. When the jaws tighten, the pointer swings to the marking that is the skinfold thickness.

FIGURE 9.5 (continued)

EVA HOT MELT

Hot melt is a type of adhesive whose use is rapidly growing in the field of packaging. Here at Wheeler Amalgamated, we use ethylene vinyl acetate (EVA) hot melt in a variety of packages. Since EVA is the most common hot melt adhesive used in our packaging, this report will explain it in detail. The sections that follow describe EVA, explain the role of heat, discuss EVA's three components, and review its advantages and disadvantages, and applications for use.

Description Of Eva

Ethylene vinyl acetate hot melt adhesives are polymeric thermoplastic compounds that form adhesive bonds upon cooling. Polymeric means that the material is composed of more than one compound; thermoplastic refers to a plastic that can be reheated and reformed. EVA generally comes in the form of white or brown pellets that resemble an average .22-caliber bullet in size and shape.

The Role Of Heat

The role of heat is to enhance the flow of the hot melt. In other words, the more heat that is put into the hot melt, the more fluid the hot melt becomes and the easier it is to apply to the intended surface. The heat decreases the viscosity, or thickness, of the hot melt, allowing the adhesive to be transferred to other materials more easily. In this way, EVA differs from other adhesives. Many adhesives enhance flow by dissolving or dispersing the adhesive in a volatile vehicle, such as water, which evaporates leaving a solidified adhesive to form a bond. Hot melts use no volatile solvent. Instead, they are melted prior to their application.

Eva's Three Components

Ethylene vinyl acetate hot melts belong to the class of solidifying, or wax-containing hot melts. They are generally made of three different types of materials: polymers, wax, tackifiers.

Each material has specific characteristics that help to create an effective adhesive. EVA polymers contribute to the cohesive strength, viscosity, impact resistance, and to a lesser extent, heat resistance of the final product. Heat resistance is a measure of the material's ability to hold a satisfactory bond at elevated or extreme temperatures. Wax, the most crystalline component, controls the softening point, heat resistance, open time, and the rate of set. Open and set time relate to the physical changes the adhesive goes through as it cools and forms a bond. Tackifiers promote adhesion and determine the color and odor.

FIGURE 9.6
Process Description

Different combinations of these materials produce adhesives with different characteristics. As a result, we can use EVA in many applications. For example, EVA hot melt used to coat a label would need a glossy appearance. Thus the specifications would call for a lower vinyl acetate content and a lower melt index. For such an application the hot melt would also need a paraffin and synthetic wax combined with a tackifier made from a terpene phenolic compound.

Advantages Of Eva

EVA has several advantages. It is the fastest setting of all adhesives, thus producing higher production speeds and less compression, or holding, time. Since EVA does not have a liquid solvent, it has a long storage life — six months to a year — and freeze/thaw stability. As a result, EVA can be shipped or stored for long periods of time without freezing, settling, or decomposing. EVA also offers a safer working environment, less shrinking, and better gap filling.

Disadvantages Of Eva

Perhaps the greatest shortcoming of the EVA hot melt adhesive is limited heat resistance. Thermoplastic materials lose a considerable amount of cohesiveness at even slightly elevated temperatures and will eventually remelt. They also have limited use for application on heat-sensitive materials.

Applications Of Eva

EVA has many applications. It is used to seal corrugated ("cardboard") and paperboard cases and cartons. Many products found on the shelves of stores are packaged in cartons sealed with an EVA hot melt. Examples include cereal boxes, macaroni boxes, and health-care items, such as aspirin and deodorant. EVA is also used to apply some types of labeling: the labels on beer bottles, soup cans, and detergent bottles are applied with a hot melt. The beverage industry uses hot melts for base cups (found, for instance, on plastic one-liter Coke bottles) and cap liners. Other applications include heat sealing and coating for pressure-sensitive labels and paper tape, tray forming, and bag seaming and sealing. New applications include tamper-evident and microwave packaging.

FIGURE 9.6 (continued)

■ EXERCISES

1. Find a description of a mechanism in a manual or in a journal article. Determine its intended audience by analyzing the level of terminology. Also determine from the context the use the intended audience would make of it. Write a brief (one-page) memo to your instructor explaining your findings.

2. Write two one-paragraph descriptions of a mechanism that you know well. If possible, use an item from your major. Otherwise, choose a common object like a kitchen utensil (measuring cups, whisk), a household tool (claw hammer, pliers), or an electronic appliance (tape deck, VCR). Write the first paragraph for a person who knows little about the mechanism and is curious; write the second for a knowledgeable person who is interested in the mechanism's advantages.

3. Either draw or photocopy a picture of a machine that you are familiar with (consumer manuals are a good source). Label all the parts that you think are significant. In class, exchange illustrations with a classmate. Interview your classmate to obtain details of the size, weight, color, material, as well as the use of your classmate's machine. Spend about 15 minutes writing a short (two or three paragraphs) description of your classmate's mechanism. After you have finished, critique each other's paragraphs for effectiveness.

4. In class (or in small groups, if your instructor wishes), compare a paragraph from "Layout Planning" (pp. 190–191) with a paragraph from "Skinfold Calipers" (pp. 194–195). How are the paragraphs organized? What is the function of the first sentence of each? Which paragraph seems more effective in conveying its message to the audience? Be prepared to make a brief presentation of your findings to the class.

5. Find a description of a process in a manual, a journal, or a textbook. Determine its intended audience by analyzing the level of terminology. Also determine from the context the use that the intended audience would make of the description. Write a brief memo explaining to your instructor whether or not the description would be suitable to assign to beginners.

6. Prior to class, find a brief (five to ten steps) set of instructions such as how to use a machine or system in the library. In class, rewrite the instructions into a description of a person performing the process.

7. In one paragraph, describe a person performing a very common action, such as starting a car, putting on a shirt or a blouse, purchasing a ticket for a performance, or doing an exercise routine. Start with a statement of the goal or object of the action, then describe the steps the person must take to complete the action.

8. Draw a flow chart, or a decision chart (see Chapter 7) of a process you know well. The process can be either a mechanism in operation or a person acting. Use a simple test or evaluation method in your field, or a simple procedure such as starting a computer. In class, exchange papers with another student and, for about 20 minutes, compose a draft that describes the process. Then interview each other to learn exact details and the goals of the steps. Revise the paper for the next class period.

■ WRITING ASSIGNMENTS

1. Assume that you are on the job and that your new supervisor needs detailed knowledge about a mechanism you commonly work with. The mechanism could be a mechanical object, such as a machine or a form, or it could be nonmechanical, such as a corporate structure. She will use the description as background for a series of meetings she will have with other supervisors. Write a memo of four or five paragraphs. Use the "Skinfold Calipers" memo (pp. 194–195) as a model. Use a visual aid.

2. Assume that you have a new supervisor who needs to understand a process that you use in your job. Describe the process in a memo of four or five paragraphs. Use a visual aid. Use the "Skinfold Calipers" memo (pp. 194–195) as a model for the form.

3. Write an article for a company newsletter, describing a common process on the job. Use a visual aid. Sample topics might include the route of a check through a bank, the billing procedure for accounts receivable, the company grievance procedure, the route of a job through a printing plant, or the method for laminating sheets of materials together to form a package. Your article should answer the question "Have you ever wondered how we . . .?"

4. Write several paragraphs describing a mechanism in operation; include a visual aid. The mechanism might be a machine, or it might be a system, such as the procedure for hiring new personnel. Choose a mechanism you know well, or else choose from this list:

   a sewing machine constructing a stitch

   light traveling through a camera lens to the film

   the action as a bike shifts gear

   the steps in releasing the contents of a pressurized can

   the retort process

   how an air conditioner cools air

   how a solar furnace heats a room

5. Using the "EVA Hot Melt" report (pp. 196–197) as a model, describe a mechanism in your field.

 WORKS CITED

Honeywell. *Air Data/Inertial Reference System (ADIRS).* Minneapolis: Honeywell, 1984.

Jordan, Michael P. *Fundamentals of Technical Description.* Malabar, FL: Robert E. Krieger, 1984.

10

Sets of Instructions

PLANNING THE SET OF INSTRUCTIONS

WRITING THE SET OF INSTRUCTIONS

FIELD-TESTING INSTRUCTIONS

Sets of instructions appear everywhere. Magazines and books explain how to canoe, how to prepare income taxes, how to take effective photographs; consumer manuals explain how to assemble stereo systems, how to program VCRs, how to make purchased items work. On the job you will write instructions: to explain how to perform many processes and how to run machines.

Careful planning and writing will produce clear instructions that guide readers through the task at hand. This chapter explains how to plan and write a set of useful instructions.

PLANNING THE SET OF INSTRUCTIONS

To plan your instructions, you need to determine your goal, consider your audience, identify constraints, and select an organizational principle.

Understand the Goal of Instructions

Instructions enable readers to complete a project or to learn a process. *To complete a project* means to arrive at a definite end result: the reader can complete a form or assemble a toy or make a garage door open and close on command. *To learn a process* means to master a process so that it can be performed independently of the set of instructions. The reader can paddle a canoe, log on to the computer, or adjust the camera. In effect, a set of instructions should become obsolete, either because the reader finishes the project or learns to perform the process without them.

Consider the Audience

When you analyze your audience, you estimate their knowledge level and any physical or emotional constraints they might have.

Knowledge Level   The audience will be at one of two levels:

- Absolute beginners who know nothing about the process
- Intermediates who understand the process but need a memory jog before they can function effectively

The reader's knowledge level determines how much information you need to include. Think about, for instance, telling someone to turn on a computer. If you tell beginners to "turn it on," they will not be able to do so because they will not know to look in the back — the location of the power switch on most computers. So you will also have to tell them where to find the switch. An intermediate, however, knows that the switch is at the back; all you have to say is "turn it on."

The following two examples illustrate how the audience affects the set of instructions. The first example tells a beginner how to log on to a mainframe computer; the second example tells an intermediate how to perform the same process. The first example is much longer, explaining the process in detail, with the instructions guiding the reader through the entire process. The second example does not guide at all; it simply lists the sequence of steps to jog the reader's memory.

INSTRUCTIONS FOR A BEGINNER
_____

LOGGING ON THE VAX

1. Flip on the power switch. It is on the back of the terminal to the left.   Tell beginners where to find switch

2. Press the return key until ENTER CLASS appears on the screen.   Effect of action
   Note: The computer has a 20-second time limit on the five instructions to follow, so you must move right along, or you will have to start the instructions over.   Special condition

3. Type in "3." The VAX is a class 3 option.   Explain
                                                                 significance

4. Press the return key. The computer will respond with GO.

5. Press the return key once or twice until the computer
   prints out WELCOME TO THE UW-STOUT VAX 11/80.
   The computer will then print USERNAME: on the screen.

6. Type in TS 1112220304. This is the training session **user
   name**. Ordinarily you must be enrolled in a class that uses
   the VAX to receive a **user name**; when the **user name** is   Significance
   matched with the proper password, you gain access to the
   files.

7. Press the return key. The word PASSWORD will be printed
   out on the screen.

8. Type in "ASTUDENT." This is the training session password.
   The computer will not print the password out on the screen   Unusual action
   as you type. If the password matches the user name —
   and it will if you typed it correctly — the computer will print
   WELCOME TO VAX/VMS VERSION V4.2, and a $D +   Effect
   will appear. The $ is a prompt. D + signifies that you have   Definition
   successfully entered the system.   Significance

INSTRUCTIONS FOR AN INTERMEDIATE

1. Turn the terminal on
2. Hit Return
3. Type 3, Return
4. Type Return, Return
5. Insert user name, Return
6. Password

Identify Constraints   The audience will have emotional and physical constraints in attempting to follow instructions. Many people have a good deal of anxiety about performing a task for the first time. They worry that they will make mistakes and that those mistakes will cost them their labor. What if they tighten the wrench too hard? Will the bolt snap off? What if they hit the wrong key? Will they lose the entire contents of their disk? To offset this anxiety, you should include tips about what should take place at a given step, and what to do if something else does happen. Step 8 in the first example above explains that something unusual will happen: the password will not appear on the screen. If this action is not explained, users might easily think that something has gone wrong or that they have performed the step incorrectly. The statement allays their fears.

The physical constraints are usually the materials needed to perform the process, but they might also be special environmental considerations. A Phillips screw cannot be tightened with a regular screwdriver; a three-pound hammer cannot be swung in a restricted space; in a darkroom only a red light can shine. Physical constraints also include safety concerns. If touching a certain electrical connection can cause injury, you must make that very clear.

Select an Organizational Principle

Organize the set of instructions in *chronological* order. Decide which step comes first, which second, and so on; then present them in that order. To decide where in the sequence each step belongs, you must analyze the process: you must determine the end goal, name and explain the tasks to be performed, and analyze any special conditions that the user should know. (For an example, refer to Figure 10.1 below.)

Determine the End Goal   The end goal is whatever you want the reader to achieve, the "place" the user will be at the end of the process. Suppose that your topic is to tell someone how to work a film projector. This process could end at several points. The end goal might be, "moving images will appear on the screen." Or the goal might be, "the film is re-turned to its canister and the machine left in proper condition for the next user." The end goal you choose will affect the number of steps in your document. Different end goals will require you to provide different sets of instructions, with different sections.

Analyze the Tasks   For every set of instructions you write, you must analyze the sequence of tasks, or steps, that the user takes to get to the end goal. The most effective method is to go backwards. If the end goal is

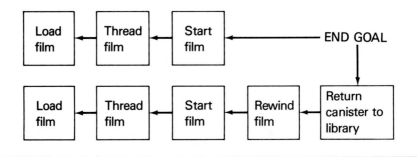

FIGURE 10.1
End Goal

to return the film to the canister, the question to ask is, "What step must the user perform immediately before putting the film into the canister?" The answer is that the film must be taken off the spool on the rewind arm. If you continue to go backwards, the next question is,"How does the film get onto the rewind spool?" As you answer that question, another will be suggested, and then another — until you are back at the beginning, taking the film out of the canister.

Name and Explain the Tasks   Once you have decided on the sequence of tasks, you should name each task and explain any subtask that accompanies it. If one of the tasks is to "thread the film through the machine," tell the user how to do that. Some projectors are easy to thread, but others require a number of substeps — levers opened, loops made a certain size. How much you say will depend on the audience's knowledge level.

Analyze Conditions   You must also analyze any special conditions that the user must know about. For instance, the projector can project only if the bulb is functioning. What if a user has to replace a burned-out bulb? You should foresee this problem and explain how to change the bulb and what size to use. Safety considerations are very important: warn the user not to touch a hot bulb and to turn off the machine before working on it.

Example of Process Analysis   The following set of instructions is the result of a careful analysis of the sequence of steps. The writer clearly names the end goal (to heat-fix the smear), clearly states the subtasks (in four steps), and accounts for special conditions (explains why not to heat too long). Notice that the following section is written for a beginner: the writer outlines the steps and substeps, and uses a visual aid. If this section were rewritten for an intermediate, the writer would only need to say, "Heat-fix the smear."

HEAT-FIXING THE SMEAR

After the smear has air-dried, heat-fix it so that the E. coli bacterial cells adhere to the glass slide.	Condition and end goal
1. Using only the right edges of the glass slide, pick up the glass slide.	Special condition placed first
2. In a rapid circular motion, pass the glass slide through the bunsen burner flame. Repeat this motion three times (see Figure 1). The bottom of the slide should become very warm but not hot. If the cells are heated too much, they will change their normal shape.	Quality of action
	Mild warning

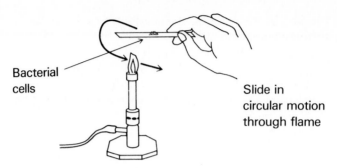

Bacterial
cells

Slide in
circular motion
through flame

FIGURE 1
Glass Slide over Flame

3. Place the heat-fixed smear on a clean paper towel.

4. Turn the bunsen burner off by pushing the handle of the
   gas terminal away from you.

Note use of *a* and
the.

Choose Visual Aids

Provide as many visual aids as you can. A visual aid can quickly clarify
and reinforce the prose explanation. Drawings and photographs are the
most effective visual aids for instructions. In most situations you can probably provide a drawing more easily than a photograph. If you have a scanner, you may want to incorporate photographs into your instructions.

   Guidelines for Choosing Visual Aids   Here are a few guidelines for
visual aids:

- The drawing or photograph should show the object or illustrate the
  step clearly.

- Place the visual aid as close as possible to the relevant discussion.

- Make each visual aid large enough. Do not skimp on size.

- Beneath each visual aid, put a label: Figure 1 (or Fig. 1) and a title.

- Refer to each visual aid clearly in the text.

- Use *callouts* — letters or words to indicate key parts. Draw a line
  or arrow from each callout to the part. Note the words "Play" and
  "Record" in the visual aid below.

Effect of a Visual Aid   Consider the difference in clarity between these two instructions.

All Words	*Words and Visual Aid*
Push the Play and Record buttons.	
The Play button is the large black button at the right end of the row of controls; it has an arrow pointing to the right.	Push the Play and Record buttons (Figure 1)
The Record button is a square orange button to the left of the Play button.	

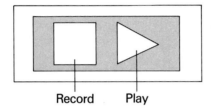

Record   Play

FIGURE 1
Control Panel

The Difference between Photographs and Drawings   A photograph shows the object realistically, often with extraneous details; a drawing can show the object more selectively.

In Figure 10.2, the photograph shows all the objects in a particular section of the tractor's motor. Notice the letter callouts.

A drawing can eliminate all the surrounding detail, allowing the reader to focus exactly on the object of the instruction. (See Figure 10.3.)

Follow the Usual Form for Instructions

The usual form for a set of instructions applies to both design features and to overall text organization. For a clear design, use white space, varied margins, underlined or boldface heads, and clear visuals. Follow these guidelines:

- Place a highlighted head (underlined or boldfaced) at the beginning of each section.

- Number each step.

FIGURE 10.2
Lawn and Garden Tractor
Source: 430 Lawn and Garden Tractor Operator's Manual. Reprinted by permission
of John Deere, Inc.

SERVICE TRACTOR SAFELY

Disconnect battery ground cable (−) before servicing if
starting engine could possibly injure operator.

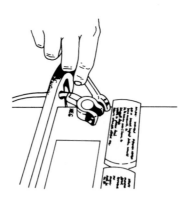

FIGURE 10.3
Lawn and Garden Tractor
Source: 430 Lawn and Garden Tractor Operator's Manual. Reprinted by permission
of John Deere, Inc.

- Start the second and following lines of each step under the first letter of the first word.
- Use margins to indicate "relative weight": show substeps by indenting to the right in outline style (see the heat-fixing example on pp. 205–206).
- Use white space above and below each step. Do not cramp the text.

A well-designed set of instructions is easy to read and thus allows the reader to concentrate on the step to be completed. The usual form for organizing the text is discussed in the next section.

WRITING THE SET OF INSTRUCTIONS

A clear set of instructions has an introduction and a body. After you have drafted them, you will be more certain that your instructions are clear if you field test them.

Write an Effective Introduction

The introduction orients the reader to the instructions. Although short introductions are the norm, you may want to include many different bits of information, depending on your analysis of the audience's knowledge level and of the demands of the process. You should always

- State the objective of the instructions for the reader.

Depending on the audience, you may also

- Define the process.
- Define important terms.
- List any necessary tools, materials, or conditions.
- Explain who needs to use the process.
- Explain where or when to perform the process.

A SAMPLE SET OF INSTRUCTIONS

In the following introduction, note that the writer states the objective ("these instructions will show you . . . "), defines terms, lists materials, explains conditions, and previews what will appear in the body.

**HOW TO PREPARE A COST OF GOODS
MANUFACTURED STATEMENT USING VISICALC
ON AN IBM PERSONAL COMPUTER**

Visicalc is a software program used on an IBM personal computer to perform various accounting and finance functions. Easy to operate, Visicalc can save its user time and paperwork.	Definition
These instructions will show you how to fill out a cost of goods manufactured statement using Visicalc.	Purpose
The IBMs are located in Room 239 of the Tech Wing along the far wall. The user must reserve time on the computer. Reservation sheets are located outside of Room 239.	Location Special conditions
These instructions tell you to push the enter key. This key is located at the right side of the keyboard with a bent-arrow symbol.	
These instructions have five parts: materials needed, loading Visicalc, entering the data, saving the data, and printing the statement.	Preview

MATERIALS NEEDED

All the materials that you need are available from the lab attendant in Room 239, Tech Wing. Just ask.	Materials

- Formatted 5¼ in. floppy disk
- DOS I.I PSHARE VISICALC Disk
- Applications for Cost and Managerial Disk

Write an Effective Body

The body consists of numbered steps arranged in chronological order. The numbered steps are the tasks that the reader must perform. To make each step clear, construct the steps carefully, place the information in the correct order, use imperative verbs, and do not omit articles (*a, an, the*) or prepositions.

Construct Steps   To make each step clear, follow these guidelines:

- Number each step.
- State only one action per number (although the effect of the action is often included in the step).
- Explain unusual effects.

- Give important rationales.
- Refer to visual aids.
- Make suggestions for avoiding or correcting mistakes.

If safety cautions are necessary, place them before the instructions. An example of how to write the body is given below.

Place Information in Useful Order   If you have to present both the instruction and an explanatory comment, you must decide in which order to place them. Generally, put the instruction first and then the explanation. If, however, the explanation is a safety warning, place it first.

1.	CAUTION: DO NOT LIGHT THE MATCH DIRECTLY OVER THE BUNSEN BURNER.	Safety warning
	Light the match and slowly bring it toward the top of the bunsen burner.	Instruction

Use Imperative Verbs   The appropriate style for instructions is the *imperative*, or order-giving, form of the verb. Give the orders clearly so that there is no mistaking what you mean. Consider the difference in these two sentences:

Precise   1. Turn the power switch to OFF.

Less precise   2. You should turn the power switch OFF.

The first is a clear imperative statement; the second is not. The difference is the word *should*, which sends an ambiguous message: you "should" do it this way, but if you're close, you'll still be all right. Use the imperative form.

Retain the Short Words   How to handle the "short" words (the articles — *the, a,* and *an,* — and prepositions, especially *of*) is an issue in writing sets of instructions. Many people think their instructions will be clearer if they use as few words as possible. So they delete the short words, making the instructions sound like a telegram. Eliminating these words often makes the sentences harder to grasp because the distinction between verbs, nouns, and adjectives is blurred.

Unclear Sentence   If paid, give patron envelope containing tickets.

SAMPLE BODY

Here is the body of the set of instructions, which follows the introduction appearing on p. 210.

Loading Visicalc   Head for step 1

1. Pick up the needed disks.
   Note: Do not touch the parts of the disk exposed in the   Special conditions
   small windows. Handle the disks only at the corners and
   edges.

2. The computer must be turned off. The power switch is
   located on the left side of the front of the desk. If it is not   How to help
   off, turn it off. If you are not sure, ask the attendant.   yourself

3. Holding the DOS I.I PSHARE VISICALC disk so that the   Special condition
   label is on the top side and towards you, insert it into drive   first; action
   A (the one on the left) and shut the door.   second

4. Turn the power on and wait for Visicalc to appear.   Action
   Note: Visicalc will appear on the screen. (See Fig. 1.)   Effect

5. When Visicalc appears on the screen, remove the DOS   Action
   disk and insert the Applications for Cost and Managerial
   Disk into drive A just as you inserted the DOS disk.

6. Type /SL COGM and then hit the enter key .

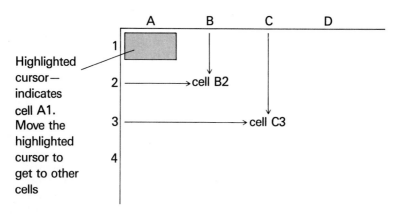

Highlighted cursor— indicates cell A1. Move the highlighted cursor to get to other cells

FIGURE 1
Visicalc Spreadsheet

Entering The Data

This step will give you practice entering data in the "cells" of   Purpose
the spreadsheet. After you have gotten the feel of it, you
can use your own spreadsheet.

Note: Use the arrow keys to move the cursor to the cell where you want to enter data (see Fig. 2).

Special condition

Caution: Do not insert commas in the numbers you are entering. Enter the numbers exactly as they are given.

Special condition

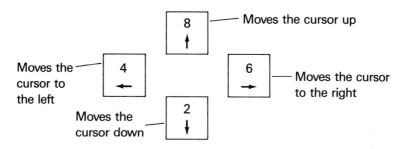

FIGURE 2
Keys to Move Cursor

1. Move to cell G5 and type: 60000. Hit the enter key.
   Note: If you make a mistake, hit the enter key and type over.
2. Move to cell F7 and type: 40000. Hit the enter key.
3. Move to cell F8 and type: 388800. Hit the enter key.
4. Move to cell Fll and type: 48000. Hit the enter key.
5. Move to cell F15 and type: 800000. Hit the enter key.
6. Move to cell F18 and type: 226000. Hit the enter key.
7. Move to cell G23 and type: 36000. Hit the enter key.

Saving The Data

1. Remove the program disk from drive A.
2. Insert the formatted personal disk into drive A in the same manner in which you inserted the first two disks.
3. Type: /SSCOGMT. The first three characters (/SS) make up the command to save the data, and the last five (COGMT) make up the name of the file. You can choose any word you wish to name the file, up to eight characters.

Significance of action

Printing The Statement

1. Make sure that the printer is on-line and ready by checking to see if both lights are lit on the printer.
2. Move the cursor to cell A1.
3. Type: /PP.

4. Move the cursor to cell G26. Hit the enter key.

5. Remove your completed statement from the printer.

Finishing

1. After you have completed all of the above steps, remove the disk from drive A and turn the power off. Return the program disks to the lab attendant.

FIELD TESTING INSTRUCTIONS

A *field test* is a method of direct observation — a method by which you can check the accuracy of your instructions. You ask someone who is unfamiliar with the process you are explaining to follow your instructions while you watch. If you have written the instructions correctly, the reader should be able to perform the entire activity without ever asking any questions. When you field test instructions, keep a record of all the places where the reader either hesitates or asks you a question.

One group of instructors recently had to write a set of instructions on how to use a computer card catalog in a library. They planned the appropriate sequence, subdivided the process into steps, and designed uncrowded pages so that users could overcome their anxiety about using the computer catalog. When they finished the instructions, they field-tested the results. The test revealed a major fault in the instructions almost immediately. The first person to use the instructions spelled an author's name wrong. The instructions gave no information on how to undo a spelling mistake. Once they were aware of it, the writers corrected the problem, but this example illustrates how easy it is to overlook a small point that can have a major impact on the effectiveness of the instructions.

■ SUMMARY

To write a useful set of instructions, you must carefully plan, write, and test the document. Planning consists of several stages. You must decide whether your audience is at a beginning or intermediate level. You must identify the constraints, emotional or physical, in these situations. You must establish a chronological sequence by determining your end goals; you should then analyze the tasks to be performed by working backwards from the end goal. You also want to tell your readers about any special conditions they should be aware of.

You need to choose visual aids (usually drawings or photographs), and you must follow clear design principles so all the steps are easy to grasp. The introduction gives the purpose of the instructions; the body contains the steps, each numbered for clarity. Effective writers field test and correct instructions before releasing them to their intended audience.

■ WORKSHEET FOR PREPARING INSTRUCTIONS

☐ *Name the audience for these instructions.*

☐ *What do your readers know about the concepts on which these instructions are based?*

☐ *What do they know about the process itself?*

☐ *What should your readers be able to do after they use the instructions and at what level of proficiency?*

☐ *List all the conditions that must be true for the end goal to occur.*
(For instance, what must be true for a document to open in a word-processing program?)

☐ *List all the words and terms that the audience might not know.*

☐ *List all the materials that a person must have in order to carry out the process.*

☐ *List the visual aids that will help a reader grasp this process.*
(For instance, use a close-up of the control panel.)

☐ *Choose your head system, margin system, method of designating individual steps, and style for writing captions for each visual aid.*

■ MODEL

The model that follows is a set of instructions written for beginners. The introduction presents essential information to orient readers; the body contains a clear sequence of steps and well-designed pages.

HOW TO MAKE STANDARD-SIZED PHOTOCOPY

The purpose of these instructions is to show you how to make a standard-sized photocopy from a standard-sized original. The instructions assume that the copier is turned on and that you have inserted an Auditron.

SINGLE-SIDED COPY

1. Place the original face down in the feed tray located on the left side of the copier. The top of the page should face left.

2. Make sure the Original Indicator Light, located on the left side of the Communications Monitor, highlights LETTER as in Fig. 1.

   a. If it doesn't highlight LETTER, lift the cover of the Special Effects Panel, located to the left of the Communications Monitor (Fig. 2).

   b. Press the Original Size Select Key one or more times until the correct direction light appears and the word LETTER shows.

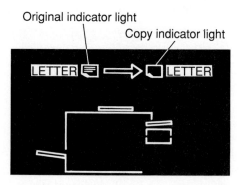

FIGURE 1
Communications Monitor

FIGURE 10.3
Set of Instructions

2

Original size

FIGURE 2
Special Effects Panel

3. Make sure the straight arrow on the Communications Monitor shows (Fig. 1). The monitor should look exactly like Figure 1.

   a. If it doesn't, press the Letter Key one time.

4. Enter the number of copies you want on the numerical key pad. If you only want one copy, skip this step.

5. Make sure the green light shows on the Print Key. If the red light shows, ask for assistance.

6. Press the Print Key. The copies will come out on the bottom left side.

FIGURE 10.3 (continued)

■ EXERCISES

1. Outside of class, write a brief (1- to 2-paragraph) process description of a person carrying out a common task, such as tying a shoe, filling out a form (include a copy of the form), or applying for a parking space. Then rewrite the process as a set of instructions on a well-designed page, using heads, numbered steps, and white space. In class, exchange the first process description for one written by a classmate. Then convert your classmate's description into a set of instructions. At the end of the class period, compare the two sets of instructions for the same process. Be prepared to comment on the effectiveness of the designs.

2. Write a set of instructions for a common activity, such as wrapping a package. Bring the necessary materials to class (it's a good idea to bring extra materials so that the reader has to choose to correct ones). Have a classmate try to perform the process by following your instructions.

3. Make a flow chart or a decision chart of a process. Choose an easy topic, such as a hobby or a campus activity or some everyday task. In class, write the instructions that a person would need to perform the process. Depending on your instructor's wishes, you may either use your own chart or exchange charts with another student and write instructions for that student's chart.

4. Divide your class into groups of three or four. Have the entire class pick a topic that everyone knows, such as checking books out of the library, applying for financial aid, reserving a meeting room in the student union, operating a microfilm reader, replacing a lost ID or driver's license, or appealing a grade. Then each team should write a set of instructions for that process. When you are finished (probably at the next class period), decide which team's set is best in terms of design, clarity of steps, and introductions.

5. Rewrite the following steps.
   From instructions for operating a computer-controlled robot:

   2. Now that you are sitting down with the CRT screen in front of you, turn it on. The switch is located on the bottom right-hand side of the keyboard. You will have to wait for the CRT to warm up, which will take from 30 to 60 seconds. You will know the screen is warmed up when the cursor appears in the top left-hand corner of the screen.

   From instructions for ringing up a sale:

   2. Take the piece of merchandise from the customer and read what the price tag states.
      a. Always take the lowest price on the ticket.
   3. Type the price stated on the ticket into the register.
      a. If the item costs 2.00, simply type 200 and the decimal point will automatically register.

    b. If an error has been made in the price, hit the clear button and start over with step 3.

4. Look at the price ticket again and in the upper left-hand corner there will be a classification number.

    a. The 4 stands for the classification and the 52 refers to the subclassification.

5. Type the classification into the register.

■ WRITING ASSIGNMENTS

1. Write a set of instructions for a process you know well. If possible, pick a process that a beginning student in your major will have to perform; otherwise, choose something that you do at a job or as a hobby — waxing skis, developing film, ringing up a sale, taking inventory. After you have written the instructions, pair up with a classmate. Outside of class, field test each other's instructions. Note every place that your classmate hesitates or asks a question, and revise your instructions accordingly. Use visual aids, and design your pages effectively.

2. Analyze several manuals for using word-processing programs (manuals for Apple Macintosh software programs are good choices, but any manual is fine, even the instructions that accompany small appliances, such as hand-held calculators). Compare them for page design, and for methods of introducing and handling individual steps, including cautions or explanations of results. Then compare one of the manuals with a technical manual for, say, an electrical generator or some other piece of equipment.

    Write a memo to a supervisor recommending one of these manuals as a model. Give clear reasons for your choice.

■ WORK CITED

John Deere, Inc. *430 Lawn and Garden Tractor Operator's Manual*. OM-M87456 Issue H4. Moline, IL: John Deere, Inc., 1985.

Technical Writing Applications

11

Memorandums and Informal Reports

THE BASIC ELEMENTS OF THE MEMO FORMAT

THE ELEMENTS OF THE INFORMAL REPORT

USES OF INFORMAL FORMATS

The day-to-day operation of a company depends on informal reports that circulate within and among its departments. These reports present the results of investigations of problems and convey information about products, methods, and equipment. The informal format makes the contents seem less threatening and more readily accessible. The basic informal format is easy to adapt to nearly any situation — and has been adapted to many purposes throughout industry. As a result, these reports are used both for readers close to the writer, such as immediate supervisors and coworkers, and for readers who are distant in the hierarchy. In addition, an informal report or memo can be used to send important information to a large group of people across a number of levels in the institutional hierarchy.

To explain this important format, this chapter has three sections: the elements of memo format, the elements of the informal report, and the uses of informal format, including analysis reports, trip reports, lab reports, and preprinted forms.

THE BASIC ELEMENTS OF THE MEMO FORMAT

Memos are an important, and frequent, job responsibility. Because memos communicate the information necessary to keep a company running smoothly, you must write them clearly and quickly. Do not be surprised if your supervisor says, "Send me a memo on that line slow-down by this afternoon." Memos test your ability to analyze a problem quickly and to write a concise, accurate solution. Your ability to handle them tells your reader a great deal about your potential as a problem solver and decision maker.

Memo Format

Memo format is easy to construct. At the top of the page, place: To, From, Subject, and Date lines. That's all there is to it. What follows below those lines is a *memo report*. Usually such a report is brief — from one or two sentences to one or two pages. Theoretically such a report has no length limit; practically, however, such reports are seldom longer than four or five pages.

Follow these guidelines to set up a memo or memo report:

1. Place the To, From, and Subject lines at the left-hand margin.
2. Place the date either to the right, without a head, or at the top of the list with a head (Date:).
3. Follow each item with a colon and the appropriate information.
4. Name the contents or main point in the subject line.
5. Place names of people who receive copies below the name of the main recipient.
6. Sign to the right of your typed name.
7. Choose a method of capitalization and placement of colons.

MEMO FORMAT: EXAMPLE 1

	February 14, 19XX	
To: E. J. Mentzer		Date on far right
c: Jane Doe		Copy
From: Judy Davis		Signature
Subject: Remodeling of Office Complex		Memo heads — only first letter capitalized

MEMO FORMAT: EXAMPLE 2

DATE: February 14, 19XX	Date line
TO: E. J. Mentzer	Memo heads in all caps
FROM: Judy Davis	
SUBJECT: REMODELING OF OFFICE COMPLEX	Subject line capitalized for emphasis

MEMO FORMAT: EXAMPLE 3

To: E. J. Mentzer   March 29, 19xx   Memo heads
From: Judy Davis   aligned on colons
Subject: Remodeling Office Complex

Whatever memo format you choose, you need to write clear, concise text to convey your information. The following report has a short introduction, briefly tells the results of a test, and recommends a course of action — all in three paragraphs.

To: Bob Mitchell   April 1, 19XX
From: Marcia L. Cody
Subject: Shurstik vs. WA tapes

      I've completed the laboratory testing on the Shurstik and the WA tapes used for Bi-pack cans. This memo report presents the results of my tests along with my recommendation. I performed the laboratory vibration testing according to ASTM packaging regulations.

Purpose of memo

Credibility of writer

      The WA tape is far superior to the Shurstik tape in preventing can separation during shipping. The Shurstik tape showed 27 out of the 60 cans tested completely separated from each other. Another 17 out of the 60 cans joined together with the Shurstik tape showed severe tape scuffing. The WA tape tested showed no cans separated, and only 5 out of the 60 cans had some scuffing.

Basic conclusion first

Data to support conclusion

      My recommendation is to use the WA rather than the Shurstik tape to prevent can separation during shipping.

Recommendation in separate paragraph for emphasis

THE BASIC ELEMENTS OF THE INFORMAL REPORT

The informal report has five basic elements, arranged in a fairly standard form. You can adapt it to many situations — from presenting background to recommending and proposing. The form has five parts:

Introduction
Summary
Background
Conclusions and Recommendations
Discussion

Informal Introductions

The object of an introduction is to orient the reader to the contents. Depending on the situation, you can choose one of four types of introductions:

- State the objective.
- State the context.
- Alert the reader to a problem.
- Use a preprinted form if one exists.

State the Objective   The basic informal introduction is a one-sentence statement of the purpose of the report.

> Objective: To report the results of the investigation of delays at Work Station 3.

Many reports start with this method; it is very common.

State the Context   An introduction that states the *context* provides enough information to orient a reader to the rest of the report. This common type of introduction, which will fit most situations, is an excellent way to begin almost all memos, letters, and informal reports.

To use an introduction that states context, include four pieces of information: cause, credibility, purpose, and preview. Follow these guidelines:

- Tell what caused you to write. Perhaps you are reporting back on an assignment, or you have discovered something the recipient needs to know.
- Explain why you are credible in the situation. You are credible either because of your actions or your position.
- Name the report's purpose. Use one clear sentence: "This report explains why Work Station 3 is inefficient."
- Preview the contents. List the main heads that will follow.

Two sample introductions that use these guidelines follow.

INTRODUCTION: EXAMPLE 1

I am responding to your recent request that I research the types of writing tasks that will be required in the course of my data processing career. In gathering this information, I interviewed John Broderick, the Data Processing Manager for	Cause for writing Source of credibility

Stephen Thomas and Associates, an architectural firm in Lexington, Kentucky. This report explains the writing responsibilities of a Data Processing Manager. These responsibilities consist of program descriptions, documentation, instruction sets, and proposals.	Purpose Preview

INTRODUCTION: EXAMPLE 2

In response to your request that I recommend an integrated software program for our office system, I have investigated two programs, Microsoft Works and Incredible Jack. I have read several detailed reviews and have talked to several current users of each. This memo presents my recommendation and the reasons for it. The decision is based on three criteria: cost, documentation, and expandability.	Cause for writing Writer's credibility Purpose Preview

Alert the Reader to a Problem   To alert the reader to a problem, you focus attention on it. To help the reader, set up a contrast between a positive and a negative, making the problem clear. Use one of the following methods:

- Contrast a general truth (positive) with the problem (negative)
- Contrast the problem (negative) with a proposed solution (positive)

In either case, you should point out the significance of the problem or the solution. If you cast the problem as a negative, show how it violates some expected norm. If you are proposing a solution, point out its positive significance.

EXAMPLE 1

The products we sell must reach our retailers in satisfactory condition. To do so, the products must remain tightly packed in their corrugated boxes. Recent complaints from trucking firms that deliver our products indicate that the flaps on our corrugated boxes are constantly opening during normal transport. The loose flaps present a serious safety threat to our products and jeopardize relations with our retailers.	General truth Problem Significance of the problem

EXAMPLE 2

Posting financial data on our balance sheets and on the forms derived from them has always been a tedious task. It is a costly, inefficient method; it is so difficult that comparative	Problem and its significance

| figures have often not been available for pricing decisions. A new spreadsheet program based on Lotus 1-2-3 could completely eliminate the tedious recopying and allow us to have many versions of the figures available for major decisions. | Proposed solution Positive significance |

Use a Preprinted Form　Some organizations commonly use a preprinted form for informal report introductions. Figure 11.1 shows such a preprinted form entitled "Technical Report Summary." The form contains spaces for content information and blanks for information management details. The content information is labeled Current Objective and Abstract. The information management details include department number, report number, employee number, security level, special information on chemicals, and a key-word glossary. Information managers use these key words to code the report into a data base so that other researchers can find and use it as needed, thus avoiding needless duplication of effort.

Summary

The summary — also called "Abstract," or sometimes "Executive Summary" — is a one-to-one miniaturization of the discussion section. If the discussion section has three parts, the summary has three statements, each giving the major point of one of the sections. In the Galaxy Foods example on pp. 243–244, notice that both the summary and the discussion contain two sections. Each statement in the summary presents the major point of the corresponding section in the discussion.

Background

The background statement gives the reader a context by explaining the project's methodology or history. If the report has only an objective statement, this section orients the reader to the material of the report. In the Galaxy Foods example, the background section orients the reader to the topic.

Conclusions and Recommendations

Informal reports often have a section called "Conclusions and Recommendation." As shown in the Galaxy Foods example (pp. 243–244), this section usually provides information that differs from the summary. Sometimes, however, this section can replace the summary. In the example on p. 230, no summary appears because the Conclusions and Recommendation section serves the same function.

TECHNICAL REPORT SUMMARY

TO: TECHNICAL COMMUNICATIONS CENTER

(Important — If report is printed on both sides of paper, send two copies to TCC.) Report Summary must be typewritten.
Guidelines on reverse side.

Division	Dept. Number
Packaging Products Division	59

Project	Project Number
Oriented Polypropylene Films	117326

Report Title	Report Number
Summary Report--July-September, 1985	005

To	Period Covered or Date
T. E. Erickson	September 30, 1985

Author(s)	Employee Number(s)
L. B. Brown	1002

Notebook Reference	No. of Pages Including Coversheet
1112	8

SECURITY ▶ ☒ Open Report & Summary ☐ Closed Report—Open Summary (Company Confidential) (Special Authorization)   **CHEMICAL REGISTRY ▶** ☐ Check box if new chemicals are reported. Use Chemical Registry Form 6092 to report all new substances.

KEYWORDS:
Lab Code

PPD

Other Keywords

Polypropylene
Polypropylene Film
Resin
Process
Properties
Oriented
   Polypropylene

CURRENT OBJECTIVE:

The overall objective of this program is to develop a fundamental understanding of oriented polypropylene film technology and its relationship to our products. Specific goals over the past quarter included:

A. Continuing evaluation of new-technology polypropylene resin from XYZ Co. to determine the differences between this material and the resin currently supplied by XYZ.

B. Continuing evaluation of ABC Co. fluoroelastomer to determine its characteristics a a polypropylene film additive.

REPORT ABSTRACT: This abstract information is distributed by the Technical Communications Center to alert Company R&D. It is Company confidential material.

Abstract:

This report covers progress in the Oriented Polypropylene Films Program for the 3rd quarter (July-September 1985). The area of new-technology polypropylene resins is discussed in detail.

The new-technology polypropylene resin from XYZ Co. exhibits significant différences in performance characteristics in comparison to the standard XYZ product. Work continues with XYZ personnel to determine how to adjust for these differences.

Wheeler polypropylene resin continues to show promising performance characteristics.

ABC Co. fluoroelastomer has shown some promising characteristics as a polypropylene film additive.

Information Liaison
Initials
_____

CONFIDENTIAL

FIGURE 11.1
Preprinted Form for Report Introduction

INTRODUCTION

This report responds to your request for setup and run hour comparisons on Part # 5008179 to determine if a proposed near net shape T15HV material is more cost effective than the CPM10V bar material currently used.

  Cause for writing memo

  A hypothetical process plan, using T15HV near net shape material for Part # 5008179, is compared to the current process plan, which utilizes CPM10V bar material. The comparison is made on the total normal labor hours per piece for each plan. The process plan having the lowest total normal labor hours per piece will reflect which material is more cost effective.

  Purpose

  Methodology
  Basic criterion

CONCLUSIONS AND RECOMMENDATION

The following conclusions were reached during this project:

  Conclusions listed

1. There is an $8.00 per part savings in material cost if the near net shape T15HV material is used.

2. Setup and run hour comparisons reveal that using CPM10V bar material presents lower total normal labor hours per piece than T15HV in order quantities over 11 pieces.

3. The T15HV material will present difficulties in quality inspection and internal spline broaching.

It would not be feasible to use a T15HV near net shape gear for the production of Part # 5008179 since the $8.00 material savings per part will be offset in larger orders by labor costs.

  Recommendation based on all evidence presented in conclusions

Discussion

The discussion section contains the more detailed, full information of the report. As the examples below show, writers use heads to subdivide this section, employ visual aids, and sometimes provide the discussion with its own introduction and conclusion. If the author has written the main introduction and summary correctly, the discussion will contain no surprises in information, just more depth. Two format concerns of the discussion section are pagination and heads.

Pagination Paginate informal reports as you would letters or reports. Follow these guidelines:

- To paginate as a letter, place the name of the recipient, the page, and the date across the top of the page:

E. J. Mentzer   2   February 14, 19XX

- To paginate as in a report, place the page number in the upper right-hand corner or in the bottom center.

To see a method of handling pagination, see the "Galaxy Foods" example (pp. 243–244); the page number is placed at the top right of each page and the date (01/21/87) and report number (OK/16/87) are at the right.

Heads   Informal reports almost always contain heads. Usually you need only one level; the most commonly used format is the "side left," though other formats are acceptable. Follow these guidelines:

- Place heads at the left hand margin, triple-spaced above and double-spaced below; use underline or boldface.
- Capitalize only the first letter of each main word (do not cap *a, an,* or *the,* or prepositions).
- Use no punctuation after heads.
- In your heads, use a word or phrase to indicate the contents immediately following.
- At times, use a question for an effective head.

<u>Kinds of Writing I Will Encounter</u>

According to Mr. Welke, I will encounter three kinds of technical writing: reports, memos, and proposals. Of the two types of reports used in the office, the most common one gives information on inventory of parts, tools, and equipment. The second type of report gives information on personnel-related matters, such as hiring, firing, layoffs, and absenteeism. Memos are used to give (or request) information or instructions. Proposals will include layout changes and equipment appropriation requests.

Side left, under-lined, no punc-tuation following

<u>For Whom Will I Write?</u>

My writing will be directed at people who will know more, less, or equal amounts of information about my subject. I will also write for people higher or lower than I am, or at my same level within the corporation.

Triple-space

Double-space

USES OF INFORMAL FORMATS

Writers frequently use informal formats, in many variations, such as pre-printed forms and outline reports. Because the format is so versatile, you can use it for nearly any kind of content. This section introduces you to several variations.

Brief Analysis Reports

Brief analysis reports are very common in industry. They present conclusions about an endless array of problems that beset normal operating procedures. Like the example below, they often have an objective, summary, introduction (optional), conclusions, and a discussion built around a visual aid.

ACE BAKING MEMORANDUM

Date: November 5, 1990
To: Tom Patterson   Memo heads
From: Ted Umentum
Subject: REDUCTION OF FATIGUE FACTOR   Subject line
                                                                capitalized for
                                                                emphasis

OBJECTIVE

To show that a modified conveyor belt will reduce employee   One sentence
fatigue and back problems.   explains purpose
                                                                  of the report

EXECUTIVE SUMMARY

The workers in Department B are sustaining reduced output   First sentence
due to fatigue and backaches. This problem will be solved by   summarizes first
raising the height of the conveyor belt, to reduce unnecessary   part of discussion;
bending for these workers.   second sentence
                                                                  summarizes part
DISCUSSION   of discussion

Recently Ace Baking Consultants conducted a study comparing   Method used
the production output of one department of workers with the
output of another department. These consultants found that
as the workday progressed, output of Department A stayed
constant whereas Department B decreased considerably.

Employee Fatigue

The workers in Department B are sustaining reduced output   Part 1 of discus-
due to fatigue and backaches. After numerous observations, I   sion: the problem
have found that these workers are suffering from these health
problems because the conveyor belt, from which they pick up
products to be packed, is too low to the ground. This poor
position of the conveyor belt causes workers to bend over
further than necessary to retrieve products. The extra bending
is the cause of numerous back problems, which result in

medical expenses paid by the company. It is also the cause of increased fatigue, resulting in a lower production output rate per 8-hour shift (see Table 1).

TABLE 1
Mean Number of Units Produced per Hour in Departments
A and B

Dept.	Hours							
	1	2	3	4	5	6	7	8
A	200	198	197	197	*	199	198	196
B	198	192	187	185	*	196	191	185

*Designated lunch hour, no output.

Table presents data that verify a problem exists.

As you can see in Table 1, the production rate of Department B, per hour, decreases considerably towards the lunch break, whereas the production rate of Department A stays relatively consistent. Both departments then receive an hour break for lunch and return to work. Once again the production rate of Department B decreases until quitting time, whereas Department A remains consistent.

Conveyor Modification

My proposal is to implement a modified conveyor belt. The proposed conveyor belt will stand 8 to 10 inches higher than the previous conveyor. This procedure can be accomplished by soldering an 8- to 10-inch extension piece to each of the existing legs consecutively and then refastening the conveyor to the floor. This modification of the conveyor will increase productivity due to reduced stress from bending over. The modification will also reduce medical expenses paid out by the company due to past backaches.

Part 2 of discussion: the solution

Another solution considered was to implement an adjustable conveyor belt that would automatically adjust to a number of different heights. This solution was rejected because of its high cost.

Rejected alternative

Trip Reports

As part of your job you may often take trips to conferences or to other sites. Many companies require that, upon your return, you file a trip report to discuss your activities and main experiences.

To write a good trip report, you must prepare before you start your trip (Reimold). You should ask your boss and colleagues what information they would like you to discover. For example, if you are planning to attend the Widget Designers International Convention, your colleagues might want to learn more about environmental concerns in recycling widgets and new software to aid in widget design and simulation. At the conference, you would look especially for that kind of information.

A good trip report includes an introduction, information on the specific concerns of your readers, and information on topics of general interest to the company. You should provide

- as much relevant, specific information as possible
- the significance of those specifics for your company

SAMPLE TRIP REPORT

Date: April 10, 19XX
To: C. M. Kirkland
From: James Hall
Subject: 1990 Widget Designers International Convention

This report discusses the Annual Widget Designers International Convention held in Jonesboro, Tennessee on April 1. The convention featured sessions on widget recycling and computerized widget design. Also featured were discussions of widget quality control and displays of other computer developments for widget design.

Purpose

Preview

Recycling

Dr. Janet Polansky reported on widget recycling bills now under consideration. A federal bill before Congress will require all manufacturers of more than one thousand widgets per hour to contribute one cent per widget to the Superfund. This fee would cause our price to rise. State laws are still a hodgepodge, but the emerging issue appears to be biodegradability. All nonbiodegradable widgets will be outlawed on the East and West coasts by 1992. Developing a biodegradable widget would give us a clear competitive edge since we would not lose customers in that area.

Data

Significance to company

Design

A new design program — Widget! — allows designers to simulate up to five simultaneous designs in any widget application. The program costs $895. I learned the basics in a one-hour workshop. The program is remarkably easy. The five designs can be produced in two hours, down from the current three days.

Data

Significance to company

Quality Control

While quality control is not an issue here at Wheeler Amalgamated, impressive reductions on waste have been achieved with special "quality teams" at Nash Incorporated. These teams make surprise inspections and, if necessary, hold training sessions immediately. They also — and this is the key — offer positive reinforcement for achieving preset goals.

Data

Computerized Displays

The displays contained exhibits from our competitors — about the same as those detailed in last year's report. A new development, though, is the use of color computer slide shows that highlight each product. Our booth did not contain such a system and so seemed old-fashioned. I recommend that we investigate this approach for next year.

Data

Significance to company

Laboratory Reports

A laboratory report communicates information gained through laboratory tests, the most rigid of all data-gathering methods. The outline below presents the usual form for writing a lab report. This basic approach will work in most situations.

> Introduction
>   Purpose
>   Problem
> Methodology
> Results (often in tabular form)
> Discussion of Results
> Conclusions
> Recommendations

A laboratory report's descriptions of methodology affirm the accuracy of the data that is discussed. Not all laboratory reports include recommendations. The following professional report attempts to determine the difference between a new resin and standard ones. Note how the introduction gives the purpose and indicates, by its structure, the contents of the body. The section entitled Results and Discussion also contains a description of methodology. Only selections from the report are presented here.

SAMPLE LABORATORY REPORT

INTRODUCTION

Work has continued in evaluation of new-technology polypropylene resins to determine the differences between these materials and standard polypropylene resins and how these differences might be exploited to provide improved product performance. Results from this work are described in the following report.

Purpose of testing activity

Purpose of report

RESULTS AND DISCUSSION

Evaluation of XYZ Co. New-Technology Polypropylene Resin

XYZ Co. supplied us with a 1000-lb sample of new-technology polypropylene resin, Lot #486. This material was prepared at its new U.S. manufacturing facility. The resin was evaluated in our laboratory in comparison to XYZ's currently supplied material (Lot #312) and also in comparison to a second sample of new-technology material from XYZ prepared from European-made flake (Lot #687). Lot #687, described in this report, is a remake of Lot #13577, which was evaluated in the Film Division last December.

Description of material

These resins were evaluated to determine melt flow-rate (before and after processing), die swell, polydispersity, thermal characteristics, processability, physical property performance, and tape properties. Initial melt flow-rate, die swell, and polydispersity results and thermal characteristics of these materials were reported previously (L.B. Brown. Technical Report Summary #004, June 1990). Processability and performance results for these materials are discussed below; the three materials were evaluated in July.

Objectives of tests

Table 1 summarizes before- and after-processing melt flow-rate and die swell results for XYZ standard-technology and new-technology polypropylene resins. All measurements were made using the melt plastometer in the laboratory.

As seen from the data in Table 1, the new-technology, U.S.-made polypropylene resin has a slightly lower melt flow-rate and lower die swell compared to the standard polypropylene resin. The melt flow-rate for this sample is also much higher than that of the previously supplied resin sample prepared from European-made flake, while the die swell is much lower. An increase in melt flow-rate from 2 to 3 g/10 minutes had been recommended by XYZ personnel for Lot #486 in order to decrease extruder gate pressure; there had been a 30% increase in gate pressure compared to the control resin, as well as an increase in extruder current-draw when processing Lot #577 in the December evaluation. Some of the previously

Text explains significance of data in visual aid

Note key terms repeated from objectives paragraph

observed differences in gate pressure and current draw for processing the new-technology resin may just be attributed to the difference in melt flow-rate of the two samples. The current adjustment in melt flow-rate for Lot #486 brought a decrease in extruder gate pressure to approximately 15% greater than the control material.

TABLE 1
Melt Flow-Rate and Die Swell Data for XYZ Co. Polypropylene Resin

Resin Sample	Melt Flow-Rate (G/10 Min)		Die Swell (D1/D0)	
	Before	After	Before	After
XYZ Co. Standard Polypropylene Resin, Lot #312	3.30	4.12	1.38	1.35
XYZ Co. U.S.-made PP Resin, Lot #486	2.93	3.47	1.26	1.34
XYZ Co. European-made, Lot #687	1.92	—	1.52	—

Outline Reports

An expanded outline is a common type of report, set up like a résumé, with distinct headings. This form often accompanies an oral presentation. The speaker follows the outline, explaining detail at the appropriate places. Since members of the audience have copies of the report and do not have to take notes, they comprehend more. Procedural specifications and retail management reports often use this form. The writer assumes that the reader already knows the terminology. Usually the reader does, especially if the form is specifications. The brevity of the form allows the writer to condense material, but of course the reader must be able to comprehend the condensed information. To write this kind of report, follow these guidelines:

- Use heads to indicate sections *and* to function as introductions.
- Present information in phrases or sentences, not paragraphs.
- Indent information, like in an outline, underneath the appropriate head.

SAMPLE OUTLINE REPORT

**CNC PUNCH PRESS PURCHASE:
WARDELL MAGNUM XQP VS. WEBER 150B
DECEMBER 16, 19XX**

Researcher	Karl Jerde	Side heads serve as introductions to each section
Purpose	To compare the Wardell Magnum XQP and Weber 150B CNC punch press to determine which is a better purchase.	

Method	Interviewed tooling engineer and reviewed sales materials. Used these criteria:
	    Machine capacity
	    Machine capabilities
	    Cost
		Essential material presented in phrases

Conclusions	Magnum XQP has the necessary capacity
	Magnum XQP has better capabilities in
	    Accuracy
	    Table speed
	    Tonnage	

Recommendation	I recommend we purchase a Magnum XQP CNC punch press.	

Preprinted Forms

The preprinted form is common. This section explains two commonly used types: a brief message form and a controlled information form.

   Brief Messages   Most firms use printed memo forms for brief hand-written notes. The forms have spaces for entering the recipient's name, the sender's name, and a brief message. Some forms have boxes to check to indicate "please return call" or "for your information" or "file" or whatever. Sometimes these forms are very small — 4 x 4 inches — and sometimes they are the size of standard paper — 8½ x 11 inches.

   A typical memo form, used by Ohio Datagraphics Company, is shown in Figure 11.2. The top of the memo form provides space for the writer's name, the reader's name, the subject, and the date. Company practice or policy will dictate the content of these spaces. To fill in a printed form, use these guidelines:

Ohio Datagraphics

INTERNAL MEMO

To: Date:

From:

Subject:

FIGURE 11.2
Memorandum Form

- Write legibly or type.
- Indicate the contents of the memo on the subject line.
- Write the main point first, then give support.

Controlled Information Form   The controlled information form provides spaces that the reader must fill in. Examples include tax forms, policy statements, and proposal applications. The reader places the required information, as directed, in the required space. A university grant proposal form (Research Initiative Proposal) is shown in Figure 11.3. It has spaces for all pertinent items: name and date, department, proposal title, and total budget.

To fill in controlled information forms, use the following guidelines:

- Write legibly or type.
- Draft several versions before you attempt the final version. (Photocopy the original to use for planning and writing your drafts.)
- In each space, place the main idea first, then give support.

Research Initiative Proposal

Name _____   Date _____

Department _____

Proposal Title _____

Total Budget _____

Chair approves: _____

Answer each question *only* in the space provided.

Name the objective of your proposal:

What methodology will you use to achieve the objective?

Explain how you will evaluate your success:

Explain how this project fits your department's mission:

On a separate sheet, provide budget estimates for salary, fringes, materials, supplies and services, and travel.

FIGURE 11.3
Controlled Information Form

If you design a form like this, follow these guidelines:

- Make all rules long enough for their intended information.
- Allow enough space between lines so that the respondent's words will fit. While 10-point type is easy to read, it is difficult to produce handwriting that small.

- Carefully consider how much space to provide for open-ended questions. (If respondents will type in answers, how many lines can they fit in the space you leave? If respondents handwrite answers, allow enough space to accommodate all the idiosyncrasies of handwriting.)
- Ask only for information you will use.
- Arrange items so that they are clumped by topic.
- Arrange items so that you can compare answers easily.

■ SUMMARY

To provide readers with timely information, informal reports are common in industry. The informal report, usually only one or two pages, often uses a memo format and may use a preprinted introductory page. Always as concise as possible, the informal report often follows this outline: introduction, summary, background, conclusions and recommendations, and discussion. The object of the introduction is to orient the reader quickly; three common types of introductions are (1) to state the objective, (2) to provide context or general background, or (3) to orient the reader to a problem. Many kinds of informal brief reports are used in business and industry, especially short analysis, trip, and lab reports.

■ WORKSHEET FOR INFORMAL REPORTS

☐ *Informal Format*
  Select and write the type of introduction you need:
    To give objective of the report
    To provide context, background
    To alert readers to the problem
  Prepare a style sheet for heads (2 to 4 levels), margins, page numbers, visual aid captions.
  Decide whether to use a memo format or a preprinted introductory sheet (if available).
  List conclusions.
  List recommendations.
  Will each new section start at the top of a new page?
  Prepare the visual aids you need.

☐ *Preprinted Forms*
1. Photocopy the blank form to use for planning and for writing rough draft.
2. Study the questions to pick out key words. What kinds of data will supply the answers?
3. Who will read this form? What kind of information do they need or expect?
4. Brainstorm answers for each question.
5. Write a topic-sentence answer to each question supply additional detail as needed.
6. Type up the final form.

■ MODEL

A professional model of an informal report follows. Derived from research in data bases, the report is an in-house memo describing the actions of competitors. The purpose of this report is to inform its audience. Notice that it presents all the basic information in the introductory sections and then gives the information again, in more complete form, in the discussion section.

GALAXY FOODS

INTRA-COMPANY CORRESPONDENCE

To	Phil Becker	At DCNTC 2319	Copy to	R.H. Fry
From	Rick Martinson	At DCNTC 6-300		R.R. Flint
Subject	COMPETITOR CULINARY SKILLS			R.J. Ostenso
	DEVELOPMENT. DCN 86-027			R.M. Voorhes
				file/3

OBJECTIVE

Determine what competitive food companies are doing to improve the culinary skills of their employees in order to facilitate more rapid formulation of high-quality foods, entrees, and restaurant products.

EXECUTIVE SUMMARY

The following companies actively support development of culinary skills within their R & D organizations:

> Real Delight — sends master chefs to culinary schools and seminars.
>
> Foote International — hires bakers from a French baking school and operates an in-house school for franchisers.

BACKGROUND

This literature search on culinary-skills development was comprehensive, covering the U.S. food-processing industry in general. Although major competitors such as National, Wheeler, and Real Delight were looked at specifically, results were not limited to any predetermined list of food companies. I am confident that the information provided in this search is all that can be found in the publicly available literature discussing this topic.

DISCUSSION

The two consumer food and beverage companies identified have publicly discussed the training of their employees in culinary skills either by professionals outside the organization or through in-house training programs.

Real Delight.   Last September, Pro Food Consultants, an industry-service division of Pro Foods, Inc., conducted a healthy foods workshop for master chefs from the Real Delight Company. The three-day seminar, Pro Food's Cooking, was held at the Pro headquarters in Expensive, PA. The program was designed to share Pro Food's expertise regarding the health-conscious consumer's concerns and interests, and to explore the wide variety of nutritious ingredients and preparation techniques that appeal to this market.

Four Real Delight chefs — V. Wolfe, S. Huftel, E. Teye, and J. Medelman — joined the Pro Food's Cooking staff of food technologists, culinary artists, nutritionists, chemists, and researchers in discussions and hands-on cooking exercises related to consumer

FIGURE 11.4
Informal Report

interests in health, especially salt, sugar, fat, fiber, and protein in the diet. According to [2] the director, Marion Schweisguth, Pro Food's Cooking is the avenue through which Pro Food's information and expertise become available in the food industry.

Earlier this month, Real Delight announced that its Delightvan will tour ten U.S. cities, serving free cups of hot soup to many of the nation's outdoor workers and needy. The ten cities that the Delightvan will visit this month are Chicago, Tulsa, Nashville, Portland, Seattle, Baltimore, Denver, Philadelphia, Pittsburgh, and Indianapolis. Bill Jones and Dave McCordick, chefs with extensive food-service experience and culinary training, will prepare and serve Real Delight's soup from the Delightvan. Between them, the chefs have served such notables as Prince Philip of England and Presidents Ford, Carter, and Nixon. In addition, Jones attended a culinary training seminar at the Cordon-Bleu in Paris.

Foote International.   The production, marketing, and distribution efforts of Foote International have made croissants the dominant item in the corporate product line. While sales of Foote croissants — through product-line and market expansion — have advanced dramatically, the company has just begun to tap the potential market for its bread and related pastry products.

Access to baking technology at the Clare Kelly School of Baking has aided the company in its production process and new product development. Foote International currently employs approximately 15 French-trained bakers. Their training involves a minimum of four years' experience in France, before spending six months at the School of Baking and then being selected to join Foote International.

In 1981 the company began to franchise its retail bakeries to restaurants and other food retailers who bake and sell the complete product line, under the name Foote International, in their own businesses. The franchise manager and baker must successfully complete a two-week course at the Foote International Baking School, which covers such areas as proprietary frozen dough processes, production control, baker management, and product merchandising.

Materials used in preparing this search report are available for perusal.

Bibliographic Resources

1. Dialog Information Retrieval Service. File 15: Abi/Inform, August 1971–January 1986.
2. Dialog Information Retrieval Service. File 192: Arthur D. Little/Online, 1977–November 1985.
3. Dialog Information Retrieval Service. File 196: Find/Svp Reports and Studies Index, 1977–November 1985.
4. Dialog Information Retrieval Service. File 545: Investext (Business Research Corporation), July 1982–January 9, 1986.
5. Dialog Information Retrieval Service. File 47: Magazine Index, 1959–March 1970, 1973 — January 1986.
6. Dialog Information Retrieval Service. File 79: *Foods Adlibra,* 1974–January 1986.

FIGURE 11.4 (continued)

■ EXERCISES

1. In groups of three or four, analyze the results and discussion section of the report on pp. 236–237. Each paragraph has a specific function. Decide what the functions are and evaluate the effectiveness of the paragraphs and of the whole section. Make a brief oral report to the class.

2. In groups of three or four, analyze the sections of the "Galaxy Foods" report on pp. 243–244. How does the summary relate to the discussion? Do you feel that you know everything you need to know after reading the first few paragraphs? What does the discussion section add to the report?

■ WRITING ASSIGNMENTS

1. Assume that you have been assigned by your department manager to explain an important process, concept, or mechanism to the new vice president of your division. Write a memo report using one of the introductory methods explained in this chapter. Consider topics such as just-in-time manufacturing, sterilizing bottles before filling them with milk, the method of depreciation used by your company, the way your hospital analyzes a patient's nutritional intake, or see lists of topics given at the ends of Chapters 2, 3, 8, and 9.

2. Write an informal report in which you use a table or graph to explain a problem to your manager. Use the "alert the reader to the problem" introduction, summary, and discussion sections explained in this chapter. Select a problem from your area of professional interest, for example, a problem you solved (or saw someone else solve) on a job. Consider topics such as pilferage of towels in a hotel, difficulties in manufacturing a machine part, drop in sales in a store in a mall, difficulties with a measuring device in a lab, or problems in the shipping department of a furniture company.

3. Write an outline report in which you summarize the major points of two articles on a topic in your field.

4. Write an outline report in which you summarize a long report that you have written or are writing. Depending on your instructor's requirements, use a report you have already written in another class or one that you are writing in this class.

5. Say that you have to give an oral presentation; write and hand out an outline report that will serve as "notes" for your speech.

6. Form into groups of three or four. Create a form that requires students to fill in the blanks. The form should require essay answers, not just yes or no answers. For the next class, type the form and make enough copies for your classmates. Have the rest of the class fill it out. Then do one of two things (or both depending on your instructor): (1) read all the answers and evaluate the effectiveness of the form, then write a memo report giving the results of your evaluation, or (2) create a table or graph that displays the information that you have collected, then write a memo report explaining the results. Possible topics include opinions on issues in student life, such as the effectiveness of certain courses in a major or the effectiveness of distribution requirements in the total curriculum. Or you could create a mock work situation in which you require workers to express opinions about safety concerns.

◼ WORKS CITED

Riordan, Timothy. *City of Dayton: 1986 Program Strategies.* Dayton, OH: City of Dayton, 1986.

Reimold, Cheryl. "The Trip Report Part 1: Preparation." *The Tappi Journal* 69.10 (1986): 168.

———. "The Trip Report Part 2: Writing the Draft." *The Tappi Journal* 69.11 (1986): 180.

———. "The Trip Report Part 3: The Report Itself." *The Tappi Journal* 69.12 (1986): 168.

12

Formal Format

THE ELEMENTS OF A FORMAL FORMAT

ARRANGEMENT OF FORMAL ELEMENTS

FRONT MATERIAL

FORMAT DEVICES IN THE BODY OF THE FORMAL REPORT

END MATERIAL

Formal reports provide the information that management needs in order to make decisions affecting the future of departments or entire firms. As part of a company's decision-making process, a formal report communicates with many people: executive and management personnel, senior engineers, perhaps legal and financial officers, and others whose areas will be affected by the decision. The technical knowledge of these people varies, but the report serves as the main source of information for all of them.

THE ELEMENTS OF A FORMAL FORMAT

When producing a formal report, the writer employs a number of elements that orient readers to the report's topics and organization. These elements occur at the front, in the body, or at the end of the report. Most of the elements appear before the body of the report. Two of the elements — the conclusion and recommendation sections — can be placed either at the beginning or at the end. Here are the elements of a formal report:

Front	*Body*	*End*
Transmittal correspondence	Headings	References
Title page	Pagination	Appendix[es]
Table of contents		
List of illustrations		
Conclusions		
Recommendations/Rationale		
(or at end)		
Glossary and List of symbols		
Abstract or Summary		
Introduction		

Some of these format elements are traditional and are always handled the same way. Some, however, can vary, depending on the roles of the writer and the readers in the particular situation. This chapter explains and exemplifies all of the elements.

ARRANGEMENT OF FORMAL ELEMENTS

You can arrange these elements into one of two patterns, administrative or traditional. The difference between the two patterns is in the placement of the conclusions and recommendations. In the traditional pattern, conclusions and recommendations appear at the end of the report. In the administrative pattern, they appear towards the beginning of the report. Logically, they belong at the end of the report because they represent the outcome of the investigation described in the body. However, since they are often the part that a reader is most interested in, writers frequently place them first. The theory is that the parts read more frequently should be placed first to avoid the need for cumbersome paging through the report.

Administrative	*Traditional*
Title page	Title page
Table of contents	Table of contents
Summary	Summary
Introduction	Introduction
Conclusions	Body sections
Recommendations/Rationale	Conclusions
Body sections	Recommendations/Rationale

Administrative	*Traditional*
Appendixes	Appendixes
References	References

FRONT MATERIAL

Transmittal Correspondence

Transmittal correspondence is a memo or letter that directs the report to someone. A memo is used to transmit an internal, or in-firm, report. An external, or firm-to-firm, report requires a letter. (See Chapter 19 for a sample letter.) In either form, the information remains the same. The correspondence contains the following:

- the title of the report
- a statement of when it was requested
- a very general statement of the report's purpose and scope
- an explanation of problems encountered, for example, some unavailable data
- acknowledgment of those particularly helpful in assembling the report

SAMPLE MEMO OF TRANSMITTAL

To: Mr. William O'Neill   Date: December 5, 1990
        Vice President
        Industrial Engineering
From: Kevin Harris   Memo head format
        Production/Inventory Department
Subject: Feasibility Report on Industrial Scrubbers

    I am submitting the attached report, entitled "Recommendation on Industrial Scrubbers for Production and Warehouse Areas," in accordance with your request of November 1, 1990.

Title of report

Cause of writing
Purpose of report

    The report examines two industrial scrubbers to determine which one would best alleviate the safety problems that have arisen in the production and warehouse areas. The scrubbers are compared in terms of five criteria established at a joint meeting of the Inventory, Production, and Industrial Engineering Departments. The Tennant 527 scrubber is recommended.

Background for credibility

    This project flowed smoothly from beginning to end. All departments involved cooperated generously with requests for time and data.

Praise to coworkers

Title Page

A well-balanced, attractive title page makes a good first impression on the reader. Some firms have standard title pages just as they have letterhead stationery for business letters. As Figure 12.1 shows, a title page contains

Report title
Name and title of the writer, (and his or her firm's name if the report is external)
Date
Report number (if required by the firm)
Name and title of the person to whom report is addressed, and the name of his or her firm

Here are some guidelines for writing a title page.

- Name the contents of the report in the title.
- Set the left-hand margin for the title and all elements at about 2 inches.
- Use either all caps or initial caps and lower-case letters; use boldface as appropriate.

Documentation Page

A *documentation page* contains boxes for all the appropriate information for identifying a report. This page, used in documents that have a hard cover, replaces the title page and summary. A sample documentation page appears in Chapter 11, p. 228.

Table of Contents

A table of contents lists the sections of the report and the pages on which they start. (See Figure 12.2.) The contents page previews the report's organization, depth, and emphasis. Readers with special interests often glance at the table of contents, examine the abstract or summary, and turn to a particular section of the report. Here are some guidelines for writing a table of contents.

- Present the name of each section in the same wording and format as it appears in the text. If a section title is all caps in the text, place it in all caps in the table of contents.

**RECOMMENDATION OF INDUSTRIAL
SCRUBBERS FOR PRODUCTION
AND WORK AREAS**

By
Kevin Harris
Production/Inventory
December 1, 1990
Final Project Report
PIFR 06-03R

Prepared for:
Mr. William O'Neill
Vice President
Industrial Engineering

Title all caps,
first line longer
than second

Left margin
about 2"

Author
Department
Date
Type of report
Report number

Recipient

FIGURE 12.1
A Title Page for a Formal Report

- Do not underline in the table of contents; the lines are so powerful
  that they overwhelm the words.
- Do not use "page" or "p." before the page numbers.
- Present only two levels of heads.
- If the table of contents and the list of illustrations are both short, put
  them on the same page; do not include the list of illustrations in the
  table of contents.
- Use a *leader*, a series of dots, to connect words to page numbers.

TABLE OF CONTENTS

FIGURE 12.2 Table of Contents for a Formal Report

List of Illustrations (Table and Figures)

The term *illustrations* includes both tables and figures. The list of illustrations gives the number, title, and page of each visual aid in the report. Here are guidelines for preparing a list of illustrations.

- Title it "List of Illustrations" if it contains both figures and tables
- If it contains only figures or tables, call it "List of Figures" or "List of Tables"
- If it contains both types, list all the figures first, then list all the tables
- List the number, title, and page of each visual aid
- Place the lists on the most convenient page. If both it and the table of contents are brief, put them on the same page.

LIST OF ILLUSTRATIONS

Glossary and List of Symbols

Traditionally, reports have included glossaries and lists of symbols. However, such lists tend to be difficult to use. Highly technical terminology and symbols should not appear in the body of a report that is aimed at a general or multiple audience. Rather, they should be reserved for the appendix, whose more knowledgeable readers do not need the definitions. When you must use technical terms in the body of the report, define them immediately. Informed readers can simply skip over the definitions. Still, if you need a glossary, follow these guidelines:

- Place each term at the left margin — start the definition at a tab (2 or 3 spaces) farther to the right. Start all lines of the definition at this tab.
- Alphabetize the terms.

Summary or Abstract

A summary or abstract is a miniature version of a much longer piece of writing, in this case, a formal report. (See Chapter 6 for a full discussion of summaries and abstracts.)

In the summary, sometimes called an executive summary, the writer gives the main points and basic details of the entire report. After reading a summary, the reader should know

- the report's purpose and the problem
- the conclusions
- the major facts on which the conclusions are based
- the recommendations

Follow these guidelines to summarize your formal report:

- Concentrate this information into as few words as possible, a page at most.
- Write the summary *after* you have written the rest of the report. (If you write it first, you might explain background, not summarize the contents.)
- Avoid technical terminology, (because most readers who depend on a summary do not have in-depth technical knowledge).

SUMMARY

This report recommends that the company should purchase the Tennant 527 industrial scrubber for the Production and Inventory Departments. Production increases in precision metal parts have caused safety problems due to the condition of the floors. After two accidents, the departments involved requested that a solution be found. Two scrubbers — the

Recommendation given first

Background

Tennant 527 and the Fujico 200 — are compared in terms of
variety of use, multi-shift capabilities, cost, warranty and
service, and special features.

    The Tennant picks up small litter better, has a larger
cleaning capacity, and will withstand multi-shift use better. The
Fujico costs less — $11,000, including all special features   Basic conclusions
and training, compared to $18,000. Tennant's service center   and data
is much more accessible to us. Only Tennant is able to
provide a squeegee wand, an essential in the view of the
departments involved.

    The Fujico is cheaper, but the Tennant is more flexible   Rationale
to use, is more durable in multi-shift use, has a closer service
center, and is able to provide us with the squeegee wand.

Formal Introduction

Introductory material orients the reader to the report's organization and
contents. After finishing the introductory material, the reader should
largely know the answers to a number of questions:

What is the purpose of the report?	purpose statement
What are the divisions of this report?	scope statement
What procedure was used to investigate the problem?	procedure statement
What is the problem?	problem and
What is the significance of the problem?	background
What or who caused the writer to write about the problem?	statements

Purpose, Scope, and Procedure Statements   Statements of *purpose,*
scope, and *procedure* are used to answer the first three questions above.
These statements are usually fairly brief and straightforward.

    State the *purpose* in one or two sentences. Follow these guidelines:

- State the purpose clearly. "The purpose of this report is . . ." (to
  solve whatever problem necessitated the report, or to make whatever
  recommendations).
- Use the *present* tense.
- Name the alternatives if necessary. (In the purpose statement below,
  the author names the problem [to find a scrubber] and the alternatives
  he investigated.)

    A *scope* statement reveals the topics covered in a report. Follow these
guidelines:

- In feasibility and recommendation reports, name the criteria; include statements explaining rank order and source of the criteria.
- In other kinds of reports, identify the main sections, or topics, of the report.
- Specify the boundaries or limits of your investigation.

The *procedure* statement — also called *methodology* — names the process followed in investigating the topic of the report. This statement establishes a writer's credibility by showing that he or she took all the proper steps. Follow these guidelines:

- Explain all actions you took: the people interviewed, research performed, sources used.
- Write this statement in the *past* tense.

Here are the purpose, scope, and procedure statements of the report on industrial scrubbers:

INTRODUCTION

Purpose

The purpose of this report is to present the results of my investigation of two industrial scrubbers, the Tennant 527 and the Fujico 200, and to recommend the purchase of one of them.

> Two-part purpose: to present and to recommend

Scope

At the request of the Production and Inventory Departments, criteria were developed to decide which scrubber to purchase. The chosen scrubber will fulfill the needs of the two departments because it meets most or all of the criteria. The Production, Inventory, and Industrial Departments set the following criteria, in order of importance:

> Source and rank of criteria

1. The scrubber chosen should be able to handle a variety of cleaning needs, solvents, oils, small litter, etc.
2. The scrubber should be able to handle multi-shift use.
3. Capital investment should not exceed $20,000.
4. The scrubber should have at least 18 months warranty on major parts (engine, transmission).
5. An adequate selection of optional equipment should be available.

This report discusses each criterion and gives details on how the Fujico 200 and the Tennant 527 meet them.

Procedure

All information for this report was gathered from either the | Methodology
manufacturers or present users. All specifications and mechanical
data were obtained from manufacturers' literature. Information
on performance in industry for the Fujico 200 was provided
by Fujico Inc. Information about the Tennant 527's performance
in industry was provided by the U.S. Air Force, a present
user of the 527.

Problem and Background Statements   You must explain the problem. Point out the appropriate data, tell the sources of that data, and explain its significance. Also discuss who or what caused you to write. Your goal is to help the readers understand — and agree with — your solution because they view the problem as you do. To orient the reader, you can use a problem statement and a background statement in the introduction. Depending on the complexity of the problem, you can use a short problem statement in the introduction as well as a lengthy section in the body. These statements should explain the origins of the problem, who initiated action on the problem, and why the writer was chosen. Follow these guidelines:

- If your reader is close, use a brief description of the problem and your involvement.
- If the reader is distant, provide more detail about the problem and your involvement.
- Give basic facts about the problem.
- Specify causes or the origin of the problem.
- Name the significance of the problem (short and long term).
- Name the source of your involvement.

The following problem statement succinctly identifies the basic facts (two accidents), the cause (increases in production), the significance (violating safety standards), and the source (departments recommended).

PROBLEM

Increases in production at KLH have caused a problem with | Background:
the safety conditions of the floors in the production and | cause
warehouse areas. Recently two accidents occurred in one | Basic fact
week in these areas. Both were caused by the condition of | Cause
the floors, which might be violating OSHA standards. Both
departments recommended that an industrial scrubber be | Source of impetus
purchased to solve the problem. | to solve problem

Alternatively, you can explain both the problem and its context in a longer statement called either "Problem" or "Background." A *background statement* provides context for the problem and the report. In it you can often combine background and problem into one statement. To write an effective background statement, follow these guidelines:

- Explain the general problem.
- Explain what has gone wrong.
- Give exact facts.
- Tell the significance of the problem.
- Specify who is involved and in what capacity.
- Tell why you received the assignment.

BACKGROUND

KLH Inc. has been extensively involved with the production of precision metal parts, both small and large. As production has gone up, the production and warehouse floors have become too soiled for maintenance to keep up with the cleaning, causing a safety problem. In one recent week, two accidents on the job were caused by the condition of the floors. Both the Production and Inventory Departments requested that the Industrial Engineering Department investigate and solve the problem.

General problem

Data on what is wrong

Who is involved and in what capacity

IE decided that a large industrial scrubber should be purchased to clean the production and warehouse floors. As the industrial engineer who works in both the production and inventory areas, I was asked to recommend which of two scrubbers should be purchased. Investigation narrowed the choice of scrubbers to either a Tennant 527 or a Fujico 200.

Why the author received the assignment

Conclusions and Recommendations/Rationale

As mentioned previously, writers may place these two sections at the beginning or at the end of the report. Choose the beginning if you want to give readers the main points first, and if you want to give them a perspective from which to read the data in the report. Choose the end if you want to emphasize the logical flow of the report, leading up to the conclusion. In many formal reports you will only present conclusions because you will not make a recommendation.

Conclusions   The conclusions section emphasizes the report's most significant data and ideas. You must base all conclusions on material presented in the body. Follow these guidelines:

- Relate each conclusion to specific data.
- Use concise, numbered conclusions.

- Keep commentary brief.
- Add inclusive page numbers to indicate where to find the discussion of the conclusions.

CONCLUSIONS

This investigation has led to the following conclusions. (The page numbers in parentheses indicate where supporting discussion may be found.)

1. The Tennant 527 is the more versatile in handling a variety of detergents for the cleaning of oils, solvents, and cooling fluids. The Tennant is also capable of handling small litter, glass, and metal chip (6–7).

2. The Tennant is more capable of handling multi-shift use. Investigation of charted maintenance history shows that the Tennant will operate in more hostile environments for longer times between maintenance (8–9).

3. The Fujico 200 costs $11,000, including special features and training. The Tennant costs $18,000, including special features and training (10–11).

4. The warranties available for the Tennant and the Fujico are comparable. The service available for the Tennant is superior because the Tennant service center is more accessible to KLH (12).

5. The Tennant is able to supply all the special features that KLH desires. The Fujico cannot supply the essential squeegee wand attachment (13–14).

Conclusions presented in same order as in the text.

Recommendations/Rationale   If the conclusions are clear, the main recommendation is obvious. The main recommendation usually fulfills the purpose of the report, but do not hesitate to make further recommendations. In the example below, the writer recommends two other items that logically follow from the purchase of the Tennant.

In the *rationale,* you explain your recommendation by showing how the "mix" of the criteria supports your conclusions. In this case, the Tennant "lost" on one major criterion — cost. The sentence that follows the recommendation shows how the results of the other four criteria offset that important failing. Follow these guidelines:

- Number each recommendation.
- Make the solution to the problem the first recommendation.
- If the rationale section is brief, add it to the appropriate recommendation. If it is long, make it a separate section.

RECOMMENDATIONS

1. The Tennant 527 Industrial Scrubber should be purchased.

   Although the Fujico is $7000 cheaper, the Tennant is superior in its ability to clean more kinds of debris, in its reliability over a number of years, in the closeness of its service center, and in its ability to supply the essential squeegee wand.

   Solution to the basic problem

2. All the essential features should be purchased, as outlined in the Special Features section of this report.

   Rationale explains reason for not following rank order of criteria

3. A training session should be contracted for and scheduled at the earliest possible date.

   Other recommendations outline how to implement decision effectively

FORMAT DEVICES IN THE BODY OF THE FORMAL REPORT

In planning the format of the body, you will have to make careful decisions about heads and page numbers.

Heads

Levels of heads are described in Chapter 7. Choose a head system that has enough levels for the subdivisions of your text. In addition, follow these guidelines:

- Generally, put first-level heads at the top of pages, giving each section a "chapter" appearance.

- Use content words, not generic words. ("Total Cost, " not "Criterion Section One".) Sometimes you can combine the two types: "Problem: Parts out of Tolerance."

- Maintain parallel structure for each level. Use all "-ing" words, or "how to" phrases, or whatever. The different levels do not have to be parallel with one another.

- Questions can be very effective. In a proposal, instead of using the head "Cost," say, "How Much Will the Solution Cost?"

- Find other attractive reports to use as a model for formatting your heads.

Pagination

Assign each page in a report a number, whether or not the number actually appears on the page. There are many page numbering systems. The key is to be consistent within your own report. Follow these guidelines:

- Place the numbers in the upper right-hand corner or bottom middle of the page, with no punctuation.
- Use headers and footers (phrases in the top and bottom margins) to identify the topic of a page or section. Most word processing programs include the ability to create headers and footers.
- Consider the title page as page 1. Do not number the title page.
- Give each full-page table or figure a page number.
- In very long reports, use small roman numerals (i, ii, iii) for all the pages before the text of the discussion.
- See below for paging the appendix.

Many word-processing systems automatically place the number centered at the bottom, without any punctuation. As these systems are standardized, this method will probably become more prevalent. Most word-processing systems will number the title page as page 1, unless instructed not to.

END MATERIAL

The end material (references and appendixes) is placed after the body of the report.

References

The list of references, included when the report contains information from other sources, is discussed along with citation methods in Appendix B (pp. 441–459).

Appendix

The appendix contains information of a subordinate, supplementary, or highly technical nature that you do not want to place in the body of the report.

Today the trend is toward greater use of appendixes to shorten the main report. Sometimes the writer treats the entire body as an appendix to the introductory material. In most cases, you should not place so much in the appendix that you fail to present significant data in the main report.

To avoid this, prepare simplified versions of complex or detailed appendix data to use within the main report. Follow these guidelines:

- Refer to each appendix item at the appropriate place in the body of the report.
- Name major subsections, under Appendixes, in the table of contents.
- Number illustrations in the appendix in the sequence begun in the body of the report.
- For short reports, continue page numbers in sequence from the last page of the body.
- For long reports, use a separate pagination system. Since the appendixes are often identified as Appendix A, Appendix B, and so on, number the pages starting with the appropriate letter: A-1, A-2, B-1, B-2.

■ SUMMARY

The format for formal reports uses various standard elements to indicate content. Among these elements, *transmittal correspondence* directs the report to the person who requested the project. The *title page* indicates title, author, date, and other pertinent material. The *table of contents* lists first- and second-level heads, indicating the pages on which each section is found. The *list of illustrations* gives the page on which each visual is found. The *introduction* contains brief statements on (1) the purpose of the report, (2) its scope or boundaries, (3) the procedure or method of solving the problem, (4) background information, and (5) an explanation of the problem and its significance. The *summary*, usually placed near the beginning, is a miniaturization of the body. The *conclusions* and *recommendations/rationale* explain the main results derived in the paper. The first recommendation provides the solution to the problem posed in the paper. A clear heading and pagination system make the paper easier to grasp. A *reference list* documents any information used from other sources. The *appendix* presents highly technical material.

■ WORKSHEET FOR PREPARING A FORMAL REPORT

☐ *Decide how knowledgeable the primary audience is about the topic.*

☐ *List the information you need on the title page.*
   Decide where you will place each piece of information.

☐ *List the levels of heads and the type of leader you will use in the table of contents.*

☐ *Prepare the list of illustrations; present figures first, then tables.*

☐ *Prepare a summary of the problem.*
Write your basic recommendations and list basic facts of each section.

☐ *Prepare a style sheet for up to four levels of heads and for margins, page numbers, and visual aid captions.*

☐ *Decide whether each new section should start at the top of a new page.*

☐ *Decide the order of statements (purpose, scope, procedure, and so forth) in the introduction.*
In particular, where will you place the problem and background statements?

☐ *Prepare a glossary if you use key terms unfamiliar to the audience.*

☐ *List conclusions.*

☐ *List recommendations, most important first.*

☐ *Write the rationale to explain how the mix of conclusions supports the rationale.*

☐ *Prepare appendixes of technical material.*
Do this if the primary audience is nontechnical or if you have extensive tabular or support material.

■ MODEL

The following model illustrates a complete formal report written by a student.

RECOMMENDATION OF A
POWER-ACTUATED HAMMER

By
Gerry Bentzler

February 14, 19XX

Prepared for
Mr. Simon Kelly
Manager

FIGURE 12.3
Formal Report

2

TABLE OF CONTENTS

LIST OF ILLUSTRATIONS

FIGURE 12.3 (continued)

SUMMARY

This report investigates two power-activated hammers, the Hunter DX350 and the Emerson R06. I researched the hammers because our current method of installing acoustic partition wall systems is too slow.

Construction crews used both tools for four weeks. During that time I witnessed demonstrations, interviewed the workers, and talked to distributors.

I evaluated the hammers in terms of four criteria: speed in repeat use, ease and safety of use, availability of services and supplies, and cost.

The Hunter DX350 meets our needs better in each criterion, particularly safety. I recommend that we purchase 10 of them.

INTRODUCTION

Background

Builders Unlimited specializes in erecting acoustic partition wall systems. These systems are secured in place to existing concrete and steel building members by the use of several fastening devices. The present devices are proving too time-consuming to use. For the frequent installation of large quantities of metal wall plate, hat channel, wood furring, and acoustic "z" strips, we require a fastening device that is strong, holds well in steel and masonry, and requires little time and effort to use. I recommended to management that a tempered steel stud, power-driven or shot into concrete or steel surfaces, should be used to secure the wall system in place. The installation time for this fastener is much faster than what is presently being used; it will prove to be a great time saver. To install these tempered steel studs, however, a suitable explosive power-actuated tool is required.

As a result of my earlier suggestion, I was asked to survey the tool market and select the one that is most suited to the needs of our installation procedures. Two makes of such a tool were selected and purchased for comparison.

Purpose

The purpose of this report is to determine the power-actuated tool that is most suitable for driving tempered steel studs into concrete and steel building members for various fastening purposes.

Scope

I selected four criteria that reflect the needs of our installation methods. The four criteria, which are of almost equal value, are speed in repeat use; the ease, simplicity, and safety of the tool during use; the availability of maintenance service and related supplies; and cost.

FIGURE 12.3 (continued)

4

Method

After reviewing the market for explosive power-actuated tools, two were selected as suitable for our installation situation, the Hunter DX350 and the Emerson Ramset RO6. Two of each tool were purchased, and one of each given to two of our installation crews. Each crew used both brands of tools and were asked to note the performance of each over a four-week period. At the end of this time period, I interviewed all operators about the performance of each tool. I also witnessed demonstrations of each tool by its operator and interviewed distributors and suppliers of the tools and their related materials.

CONCLUSIONS

1. In speed of repeat use, the Hunter DX350 tool is faster than the Emerson RO6.
2. The Hunter DX350 has an advantage over the Emerson RO6 in safety characteristics; both tools are easy to use.
3. Hunter service and supply availability are superior.
4. Hunter is slightly cheaper, but the difference is minimal.

RECOMMENDATION AND RATIONALE

On the basis of the conclusions of this study, I recommend the Hunter DX350 power-actuated tool for use by our installation crews. The only criterion in which the Emerson rates close to the Hunter is that of easy and safe operation. Both tools are easy to use, but the Hunter has an advantage over the Emerson with respect to safety during reloading procedures.

DISCUSSION

Speed Of Repeated Use

The speed of repeated use of such a tool or gun is of great importance because of the large quantities of materials our crews install. A few minutes saved on the installation of each piece of wall plate, for example, can easily add up to a considerable amount of time saved during an entire installation project.

Hunter.   The Hunter has a plastic magazine that holds ten charges in the handle of the tool and that allows the tool to be reloaded immediately after firing by cocking the barrel of the gun. It is almost identical in operation to a semiautomatic pistol.

FIGURE 12.3 (continued)

5

Emerson.   The Emerson requires the operator to open the breech of the tool, remove the spent charge by hand, load one charge into it, close the breech, and then fire. (See Figure 1.) During the demonstration I witnessed, the Hunter averaged two to three shots to the Emerson one; this is a clear advantage when such a tool is used to install a large number of fasteners.

Conclusion.   Because of the reloading nature of the gun, the Hunter DX350 is superior to the Emerson model for rapid repeat use.

HUNTER DX350

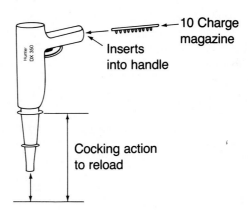

10 Charge magazine

Inserts into handle

Cocking action to reload

EMERSON R06

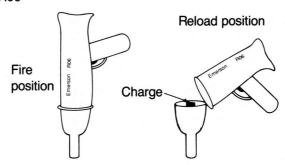

Reload position

Fire position

Charge

FIGURE 1
Reloading Procedures

FIGURE 12.3 (continued)

Easy And Safe To Use

Ease of use and safety are important to the success of any piece of equipment. Unless a tool is easy and safe to use, people resort to using other tools with which they are more comfortable. Especially important is how an unfired charge (a dud) is handled.

Ease of Use.   Both the Hunter and the Emerson guns are simple to use. They have a minimum of moving parts, and the design of each makes their use almost self-explanatory.

Hunter Safety.   The Hunter tool solves the dud problem by having its charges mounted in a plastic magazine contained inside the tool's heavy handle. The Hunter is unloaded by cocking the barrel of the gun, and if a misfired charge should explode, it will be completely contained in the handle of the tool, thus preventing injury to the operator.

Emerson Safety.   With the Emerson tool, the unfired charge had to be removed from the barrel of the gun by hand. The danger of the charge exploding during removal is a real possibility. The operators of the tool seemed to accept this situation to a degree, but I believe it's totally unacceptable if the potential lost time and medical expenses that may result from an injury are considered.

Conclusion.   While both hammers are easy to use, the Hunter is superior in respect to safe loading and unloading of charges.

Availability Of Service And Supplies

The availability of maintenance service and related supplies is critical to our installation crews. Many of our projects are located long distances from the home shop, which can present a problem when a piece of equipment fails or its related supplies unexpectedly run short. Readily available sources of service and supplies over a large area will lessen any lost time due to tool breakdown or lack of supplies.

Hunter.   In talking to Hunter distributors and listing lumberyards, hardware stores, and building-material suppliers that carry studs and charges for the Hunter DX350, it became clear that Hunter is extensively distributed throughout the state. An installation crew can readily find supplies close by when working long distances from the home shop. Maintenance and breakdown service is also well taken care of by Hunter distributors. When needed, they will come out to a project site and service or repair any Hunter product. If a tool cannot be repaired on the site, they will loan a similar tool to the user free of charge until repairs can be made. Response time in this regard is usually the same day or early the next day.

FIGURE 12.3 (continued)

7

Emerson. Emerson supplies are not at all widely distributed. In a one-hundred-mile radius of our office, I was able to locate one Emerson distributor and four merchants that carry studs and charges for the Emerson RO6 gun. Also the Emerson distributor does not perform on-site repairs or service.

Conclusion. The need for service and supplies is overwhelmingly fulfilled by Hunter distributors.

Cost

Cost is the purchase price of the hammers including sales tax. The table below represents the discounted price of each if we buy 10 or more. The cost of the studs is the discounted price for purchases of 100# or more.

TABLE 1
Cost Comparison

	Hunter	Emerson
Purchase cost (incl. tax)	$31.49	$33.08
100# studs	$25.00	$25.00

Conclusion. The price differential between the two is minimal. It is not a factor in the decision.

FIGURE 12.3 (continued)

■ EXERCISES

1. Write a summary of an article you have read in a periodical related to your field. Make it a miniaturization of the original by placing your points in the same sequence as the original. Do not write more than one double-spaced page. Note: A problem with this exercise is to determine the audience you are writing for.

2. Examine the table of contents from several textbooks or manuals to determine their effectiveness in presenting information. Decide which format you like best. Assume that you are an editor for a major publisher. Write a letter to an author explaining which format you have selected and why you want him or her to use it.

3. Ask a professor in your major whether or not your field has a style manual. If it does, examine a copy to see what it recommends for the elements discussed in this chapter. Report your findings to the class in a brief memo or oral report.

4. Write a recommendation/rationale statement based on the following conclusions. They are listed in descending order of importance. Put the statement into a memo to send to your supervisor, who has asked you to make this decision.

    1. A Whizbang computer costs $2,000. A Decisive computer costs $1,500.
    2. Whizbang has two **built-in disk** drives. Decisive has only one drive. A second drive can be purchased for $95.
    3. Whizbang has documentation (instruction manuals) that is very easy to use. Decisive's documentation is difficult to use.
    4. Whizbang gives a free word processing program to a buyer. Decisive does not. The comparable program for Decisive costs $350.
    5. The Whizbang dealer offers five free hours of training at the local store. The Decisive dealer will send a trainer to our office for ten hours for $200.
    6. Whizbang has a one-year warranty. Decisive has a two-year warranty. Both warranties cover parts and labor.

5. Edit the following selection from the discussion section of a formal report. Add at least two levels of heads. Construct an appropriate visual aid. Write a one- or two-sentence summary of the section.

A small Hobart tilting fry pan will replace the electric griddle to help increase the quality of food and conserve labor. The fry pan, which measures 24 in. x 22 in. x 6 in., is used as a fry pan, braising pan, griddle, kettle, thawer, and food warmer-server. The tilting fry pan will increase the quality of the food's color, flavor, and serving temperature. The fry pan improves color and browning of food because it allows for a more even heat distribution than the electric griddle. The fry pan improves the flavor of food because the gear tilting mechanism allows for the removal of extra grease while cooking. Also the serving temperature improves because the fry pan can act as a food warmer-server. Food cooked on the griddle is often served cold. The tilting fry pan will help conserve labor by increasing the speed of food preparation, eliminating heavy lifting, and reducing the amount of pots and pans washed. The fry pan increases the worker's speed in preparing food because it provides for a one-step preparation of menu items. The fry pan eliminates heavy lifting because the gear tilting mechanism enables the workers to simply tilt the pan and pour either excess grease out or various foods into serving bowls. The fry pan reduces the use of pots and pans and, therefore, their washing because many items are prepared directly in the pan.

13

Recommendation and Feasibility Reports

PLANNING THE RECOMMENDATION REPORT

WRITING AND PRESENTING THE
RECOMMENDATION REPORT

Professionals in all areas make recommendations. Someone must choose between alternatives, must say "choose A" or "choose B." The A or B can be anything: which type of investment to make, which machine to purchase, whether to make a part or buy it, whether to have a sale or not, whether to relocate a department or not. The person who makes a choice makes a recommendation based on *criteria*, standards against which the alternatives are judged.

For professionals, these choices often take the form of *recommendation reports*, or *feasibility reports*. While both types present a solution after alternatives are investigated, the two reports are slightly different. Recommendation reports choose between two clear alternatives: this distributor or that distributor, this brand of computer or that brand of computer. Feasibility reports investigate one option and decide yes or no: Should the client start a health club or not? Should the company market this product or not? Recommendation reports tend to be shorter than feasibility reports. While the recommendation of a brand of computer might take eight pages, proving the feasibility of starting a small business might take sixty pages.

Because the two types of report function in the same way — both make a recommendation after considering criteria — they are treated together in this chapter.

PLANNING THE RECOMMENDATION REPORT

In planning such a report, you must consider the audience, devise criteria for making your recommendations, employ visual aids, select an organizational principle, and follow the usual form.

Consider the Audience

Many different people with varying degrees of knowledge (a multiple audience) read these reports. A recommendation almost always travels up the organizational hierarchy to a group — a committee or board — that makes the decision. These people may or may not know much about the topic or the criteria used as the basis for the recommendation. Usually, however, most readers will know a lot about at least one aspect of the report — the part that affects them or their department. They will read the report from their own point of view. The personnel manager will look closely at how the recommendation affects workers; the safety manager will judge the effect on safety, and so on. All readers will be concerned about cost. To satisfy all such concerns, the writer must present a report that allows readers to find and feel comfortable about the information they need.

Devise Criteria

To make data meaningful, you must analyze or evaluate them according to criteria. Selecting logical criteria is crucial to the entire recommendation report because you will make your recommendation based on those criteria.

The Three Elements of a Criterion   A *criterion* has three elements: a name, a standard, and a rank. The *name* of the criterion identifies some area relevant to the situation, for example, "cost." The *standard* is a statement that establishes a limit on a criterion. Consider two very different standards that are possible for cost:

1. The cost of the water heater will not exceed $500.
2. The cheapest water heater will be purchased.

The different standards influence the final decision. In the example above, if the second standard is in effect, the writer cannot recommend a machine that costs more, even if it has more desirable features.

The *rank* of the criterion is its weight in the decision relative to the other criteria. "Cost" is often first, but it might be last, depending on the situation.

As a writer of recommendations, you must identify all three elements. Criteria vary according to the type of problem. There is no formula for selecting them, so you have to use your professional expertise and the information you have about needs and alternatives in the situation. In some situations, a group or person will have set up all the criteria in rank order. In that case, all you have to do is show how the relevant data from the alternatives fit the criteria.

Sample Situations   Suppose you are a social worker who has to recommend a group home in which to place a client. Since this is a routine situation for you, your Human Services Division has compiled a list of criteria and standards. Some of these are

> Cost may not exceed the client's income.
> The ratio of staff to residents should be below five to one.
> Basic independent living skills should be taught.
> Recreational facilities should be available.

As these are the usual or expected criteria in this situation, you use them to evaluate the available group homes before placing the client.

Or say you are asked to choose a machine — a computerized robot for an educational program. If you are familiar with the program and with this kind of machinery, you use your understanding of both to generate a list of criteria that includes

> Features
> Noise level
> Cost

Use Visual Aids

Although you might employ many kinds of visuals — maps of demographic statistics, drawings of key features, flow charts for procedures — you will most commonly use tables and graphs. With these visuals, you can present complicated information easily, for example, cost or comparison of features. For many sections in your report, you will construct the table or figure first, and then write the section. The point of the section will be to explain the data in the table or figure. Visual aids help overcome the problem of multiple audiences. Consider using a visual with each section in your report.

In the following example, the author first collected the data, then made the visual aids, and *then* wrote the section.

COST

Our department has sufficient funds to cover either piece of equipment. Cost is the second most important criterion in choosing between the Bock Grabamatic Robot and the Rough Boy 4000 ZNZ Mill. Both companies also offer 10% educational discounts on the final price of any purchase.

Bock Grabamatic Robot.   The robot comes with all the features needed for operation. The final cost of the Bock is $1,467.90 less than that of the Rough Boy 4000 ZNZ Mill. (See Table 1.)

Point out significance of figures in table

Rough Boy 4000 ZNZ Mill.   The ZNZ Mill has extra costs in addition to the base price of the machine. The extra costs are a result of extra equipment and installation costs required for operation of the machine. (See Table 1.)

Explain reasons for extra cost

Conclusion.   The Bock Grabamatic Robot meets the cost criterion better than the Rough Boy ZNZ Mill because it is $1,467.90 cheaper.

TABLE 1
A Comparison of Costs

	Rough Boy 4000 ZNZ Mill	Bock Grabamatic Robot
Base Price	$ 9,900.00	$10,200.00
Printer	415.00	in base price
Installation	1,000.00	in base price
Additional Tooling	516.00	in base price
Total Cost	$11,831.00	$10,200.00
- 10% discount	1,183.10	1,020.00
Final Cost	**$10,647.90**	**$ 9,180.00**

Select an Organizational Principle and Follow the Usual Form

As you plan your report, you must choose a format, an organizational principle for the entire report, and an organizational principle for each section.

Choose a Format   The format you select will determine how many elements you include. You might use an informal format with one level of heads, or you might choose a formal format, complete with title page, list of illustrations, summary, and chapter divisions. Your decision will depend on the situation: If the audience is a small group that is familiar with the situation, an informal report will probably do. If your audience is more distant from you and the situation, a formal format is preferable. If you use an informal format, you need an introduction and summary, as explained in Chapter 11. The body will remain basically the same. With a formal format, you need to provide the entire range of elements described in detail in Chapter 12. If you select a formal format, the following outline will suffice for almost any recommendation or feasibility report.

Formal	*Informal*
Title page	Introduction
Table of contents	Summary (including recommendations)
List of illustrations	Discussion (same as formal format)
Summary	
Conclusions	
Recommendations/Rationale	
Introduction	
Background	
Purpose	
Method	
Scope	
Discussion	

   A. (If necessary, a section may be devoted to additional background and introductory information.)
   B. Presentation and interpretation of data for the first criterion
   C. Presentation and interpretation of data for the second criterion
   D. Same for any additional criteria . . .

Appendix

Organize the Discussion by Criteria   The discussion of a good feasibility report is organized according to criteria, with each criterion receiving a major heading. Your goal is to present comparable data so readers can evaluate it easily. Such an arrangement, while not easy to write, is much easier to read and understand.

Follow the Usual Form for Writing Recommendation Reports   In the discussion of your report, each section deals with one criterion and evaluates the alternatives in terms of that criterion. Each of these sections should contain three parts: an introduction, a body, and a conclusion. In the introduction, define the criterion and discuss its standard, rank, and source, if necessary. If you discuss the last three somewhere else in the report, do not repeat the information. In the body explain the relevant facts about each alternative in terms of the criterion; and in the conclusion state the judgment you have made as a result of applying this criterion to the facts. You will find sample annotated sections on pp. 278–279.

WRITING AND PRESENTING THE RECOMMENDATION REPORT

As you write the recommendation report, you must carefully develop several sections: the introduction, conclusions, and recommendation/rationale sections.

Writing the Introduction

After you have gathered the data and interpreted it, you must develop an appropriate introduction. The introduction must orient the readers to the problem and to the organization of the report. To do this, you need to offer four statements:

> purpose of the report
> method of investigation
> scope
> explanation of the problem

Writing the Purpose of the Report   Begin a recommendation report with a straightforward statement, such as "The purpose of this report is . . ." You can generally cover the purpose, which is to choose an alternative, in one sentence.

Writing the Method of Gathering Information   State your method of gathering information. As explained in Chapter 5, the four major methods of gathering data are observing, testing, interviewing, and reading. Stating your methodology not only gives credit where it is due but also lends authority to your data and thus to your report.

In the introduction, a general statement of your method of investigation is generally sufficient: "Data for this report were gathered from sales literature supplied by BMI Corporation and Data 100 Corporation."

Writing the Scope Statement   In the scope statement, name the criteria you used to judge the data. You can explain their source or their rankings here, especially if the same reasons apply to all of them. Name the criteria in the order in which they appear in your report. If you have not included a particular criterion because data are unavailable or unreliable,

state this omission in the section on scope so that your readers know you
have not overlooked that criterion.

Explaining the Problem   To explain the problem, you need to define
its nature and discuss its history, its significance, who assigned you to
write the report, and what your position in the corporation or organization
is. There is no formula for writing this important section. Remember that
as the person who has investigated the problem, you know more than your
readers. You may have found that the problem is more complex than it
first seemed. Possibly your investigation has shown that a particular aspect of the problem requires special emphasis. In any case, readers
should know all aspects of the problem so that they will accept your
recommendation.

SAMPLE INFORMAL INTRODUCTION

The following informal introduction effectively orients the reader to the
problem and to the method of investigating it.

Date: April 27, 1987 To: Nicholas Ryan From: Holly Maas Subject: Recommendation for purchase of an ink jet printing system	Memo head of informal format
This memo is in response to your request for a recommendation concerning the purchase of an ink jet printing system. The purpose of this memo is to review two ink jet printing systems and to recommend the best system in terms of the criteria that have been established. The two systems being reviewed are the Thomas system and the Encoder system. I have interviewed several vendors, read their literature, and observed each machine in operation. I recommend the Encoder system.	Cause for writing memo Purpose of the report Method of gathering information Recommendation
As described in my memo of April 10, 1987, the following criteria have been established and are listed in descending order of importance:	Source of criteria
1. Ability to print a readable bar code 2. Printing features 3. Capacity 4. Cost	Scope and ranking
Yost Inc.'s current method of putting information on shipping cartons is with printed labels. This method is time-consuming, costly, and inefficient. Frequently a carton has the wrong	Current situation Explanation of the problem

information on it. Management has decided that an ink jet printing system will solve these problems because it is automated and easier to control.

The remainder of this memo will cover the following: conclusions from research, recommendation of a system, and discussion of the systems and how each met the criteria.

Preview

Writing the Conclusions Section

Your conclusions section should summarize the most significant information about each criterion covered in the report. One or two sentences about each criterion is usually sufficient to prepare the reader for your recommendation. Writers of recommendation and feasibility reports almost always place these sections in the front of the report, right after the introduction (informal) or the summary (formal). Remember, readers want to know the essential information quickly.

CONCLUSIONS FROM RESEARCH

Each system has been compared to the four criteria with the following results:

Conclusions: criteria presented in rank order

1. Both systems have the ability to print a readable bar code; however, the Thomas system requires an extra expenditure for this capability.

2. The Encoder system meets both of the requirements concerning printing speed and automatic adjustment; the Thomas system meets neither of them.

3. Both systems satisfy the capacity requirements.

4. The Thomas system meets the cost requirement; the Encoder system does not.

Writing the Recommendations/Rationale Section

Carefully written reports build steadily toward the recommendation, often reducing it to a formality. For short reports like the samples presented here, one to four sentences should suffice. For complex reports, involving many aspects of a problem, a longer paragraph, or even several paragraphs, may be necessary. As with the conclusions section, place this important section at the beginning of the report.

RECOMMENDATIONS

I recommend that Yost, Inc., purchase the Encoder ink jet printing system. This system meets the three top priority criteria. It exceeds the cost criteria by $1,000. Implementation

Basic recommendation

of the Thomas system would cause a line slowdown, resulting
in a decrease in productivity. This is not feasible, and it
justifies spending an additional $1,000 for the Encoder system.

Rationale to
justify extra cost

Writing the Discussion Section

As was mentioned above, organize the discussion section by criteria, from
most to least important. Each criterion should have an introduction, a body
discussing each alternative, and a conclusion. Here is part of the discussion
section from the recommendation report on ink jet printers.

DISCUSSION

Bar Coding

     The bar code, which is printed on a carton and used for
inventory control, contains information about the product
and is read by a scanning device. In order to be read by the
scanner, the code must be printed in a clear, consistent, and
accurate manner. Any ink jet printing system purchased by
Yost, Inc., must have the capability to print a readable bar
code. Management has given this criterion highest priority
since readability of the code is of utmost importance in
inventory control.

Definition and
background

Source of ranking
and ranking
Significance of
ranking

     Thomas System. The basic Thomas system does not
have the ability to print a bar code. However, an option can
be added to the system, at an additional cost of $700, to print
readable bar codes (see Table 1).

Paragraphs discuss
criterion and
how the systems
meet it

     Encoder System. As shown in Table 1, the basic
Encoder system has the ability to print a readable bar code.
This ability is part of the system, and it requires no additional
investment.

     Conclusion. Both systems have the ability to print a
readable bar code; however, the Thomas system requires an
additional expenditure for this capability.

TABLE 1
Printing Capability of an Ink Jet Printing System

Feature	Thomas	Encoder
Ability to print a bar code	*Yes	Yes
Ability to print a readable bar code	*Yes	Yes

Table constructed
before section
was written

*Note: This ability is available as an option at an extra cost of $700.

Printing Features

Printing features are a system's speed capabilities. The two most important printing features for Yost, Inc., are printing speed and automatic speed adjustment. Printing speed is the rate, measured in feet per minute (fpm), at which the system can clearly print information on a carton. This is important because we have an average line speed of 150 feet per minute. Therefore, we need a system that can print at least 150 fpm. Automatic speed adjustment allows the system to sense the line speed automatically and to print accordingly. This is a desired feature because when a speed variation occurs on the line, an operator does not have to be present to make a manual adjustment. Management has given this criterion second-highest priority.

Definition

Significance

Definition
Significance

Thomas System. As Table 2 shows, the Thomas system has a maximum printing speed of 100 fpm. This speed is less than the acceptable rate of speed that we need. This system does not have an automatic speed adjustment feature (see Table 2). Thus an operator would have to be present at all times to monitor line speeds and make necessary adjustments.

Textual reference
to table

Data
Significance of
data

Encoder System. As Table 2 shows, the Encoder system has a maximum printing speed of 200 fpm. This speed exceeds Yost's current line speed of 150 fpm. The system has an automatic speed adjustment feature (see Table 2). This feature is cost-effective because it saves time and reduces the frequency of illegible printing on cartons.

Textual reference
to table
Data

Significance of
data

Conclusion. The Encoder system meets both of our requirements for printing speed and automatic adjustment; the Thomas system meets neither of them.

Conclusion

TABLE 2
Printing Features of Ink Jet Printing Systems

Feature	Minimum	Thomas	Encoder
Printing speed (fpm)	150 +	100	200
Automatic speed adjustment	Yes	No	Yes

 SUMMARY

Recommendation reports choose between at least two alternatives using some previously established criteria. Feasibility reports investigate one option and decide yes or no. Both types make recommendations after considering criteria. A criterion has three parts: a name, a standard, and a rank. As in most kinds of reports, writers use visual aids, often constructed before they write the section, to convey important information. The informal recommendation report has two parts: The introduction usually includes a summary and the major recommendations. The body discusses each alternative in terms of each criterion, ranked according to importance. The usual organization of the body is from most to least important criterion. The formal recommendation report has many or all the elements of a formal report.

■ WORKSHEET FOR PREPARING A
RECOMMENDATION REPORT

☐ *Who will receive this recommendation report?*

☐ *Who will authorize the recommendation in this report?*

☐ *How much do they know about the topic?*

☐ *Name the two alternatives.*

☐ *How do you know those are the two alternatives?*

☐ *How will you choose the relevant criteria?*
Interview? Read? Inspect the situation?

☐ *Select criteria.*
Each criterion must have a name, a standard, and a rank.

☐ *Prepare background for the report.*
Who requested the recommendation report? Name the purpose of the report. Name the method of investigation. Name the scope. Explain the problem.

☐ *Select a format — formal or informal.*

☐ *Prepare a style sheet of margins, headings, page numbers, and visual aid captions.*

☐ *Select or prepare visual aids that illustrate the basic data for each criterion.*

☐ *Select an organizational pattern for each section*
For example, introduction, alternative A, alternative B, visual aid, conclusion.

☐ *Write each discussion section. Then write the final version of the introductory material.*

■ MODEL

A recommendation report follows. Note the different elements and formats that apply in this situation.

Date: December 17, 1989
To: David Jackson
From: Karl Jerde, Senior Manufacturing Engineer
Subject: Purchase Recommendation for CNC Punch Press

The purpose of this report is to recommend the computer numerical control
(CNC) machine that Tri-State Manufacturing should purchase. After comparing
many different machines, I narrowed my choice down to two: Wardell Magnum
XQF and Williamson Model 150B. To determine which machine I should recommend,
I evaluated them using the following criteria in ranked order:

1. Machine capacity

2. Machine capabilities

3. Cost

Recommendation

I recommend that Tri-State Manufacturing purchase Wardell Magnum XQF,
definitely the best machine for our business. First, this machine has the capacity
necessary to eliminate our high subcontracting labor bill. Second, the capabilities
of this machine are outstanding. Compared to the Williamson Model 150B, the
Magnum XQF has greater accuracy, faster table speed, and more tonnage. Third,
the cost of the Magnum XQF is within our allocated budget.

Scope

As you requested last month, I researched the problems with our current outdated
numerical control (NC) machine. This research clearly pointed out that production
demands have stretched beyond our NC's capacity. The result is a high amount
of subcontracting to outside vendors, which is very expensive. Because this
situation is unacceptable, I have chosen increased capacity as the most important
criterion in recommending a new CNC.

    Our tooling engineer identified the specific capabilities that the new machine
must have. Upon his advice, I ranked capabilities as the second criterion.

    The third criterion is cost. Management has allocated $400,000 to purchase
the machine.

    The remainder of this report will compare both machines to these three
criteria.

FIGURE 13.1
Recommendation Report

Machine Capacity

A new CNC must be able to punch the biggest sheet metal sizes used in the assembly of our final product. To do this, the following subcriteria must be met:

1. Maximum material thickness — punching a 3-in.- diam hole through .250-in.- thick mild steel

2. Maximum workpiece dimensions — punching workpieces up to 48 in. × 60 in.

Magnum XQF.   A big hydraulic mechanism and larger table size allow the Magnum XQF to meet these subcriteria easily. It has the capacity to punch a 4-in.-diam hole through .375 in.-thick mild steel. It also has the capacity to punch workpieces up to 60 in. × 72 in.

Model 150B.   The punching capacity of the Model 150B is restricted by a smaller hydraulic mechanism. It can punch a 3-in.-diam hole, but only through .187-in.-thick mild steel. The table size of the Model 150B is large enough to punch workpieces up to 48 in. × 60 in., just meeting the second subcriterion.

Conclusion.   The Magnum XQF has the capacity needed to punch all of our biggest sheet metal sizes.

Machine Capabilities

Capabilities of the new machine must meet our Tooling Engineer's specified criteria. His criteria are:

1. Accuracy — material movement on the x/y table shall be a minimum of ± .005 in. on hole centers

2. Linear table speed — minimum table speed must be more than 1000 in.-per-minute (IPM)

3. Tonnage — the punching tonnage shall meet or exceed 20 tons

4. Tool loading — tools shall have the ability to be preloaded into a turret that will automatically feed tools into the machine punching head

5. Indexing punch head — the punching head of the machine shall have the ability to index 360° in both directions

6. Laser/plasma capabilities — the machine shall have the designed-incapability to add laser or plasma cutting

The following table compares these criteria in both machines. It is important to notice the outstanding accuracy and speed of the Magnum XQF (see Table 1).

FIGURE 13.1
Recommendation Report (continued)

TABLE 1
Machine Capabilities Comparison

Criteria	Wardell Magnum XQF	Williamson Model 150B
Accuracy	± .002 in.	± .005 in.
Table Speed	2400 IPM	1200 IPM
Tonnage	45 tons	30 tons
Tool Loading	Yes	No
Indexing Punch Head	Yes	No
Laser Capabilities	Yes	Yes
Plasma Capabilities	Yes	Yes

Magnum XQF. The Magnum XQF has the highest accuracy of any CNC punch press on the market, resulting in reduced scrap and better quality parts. The other criteria, such as table speed, tool loading, and indexing punch head, add to the overall speed of the machine. Its faster setup and production speeds will reduce the total production time and increase our sheet metal department's efficiency. The laser or plasma capabilities of this machine will allow us to grow with increasing production demands.

Model 150B. Although the Model 150B meets much of the criteria it falls short in two areas: it lacks automatic tool loading and an indexing punch head. Both are needed to obtain maximum production efficiency.

Conclusion. The Magnum XQF meets or exceeds our tooling engineer's criteria.

Cost

The cost of the machine must not exceed $400,000, including freight and installation, insurance during transit, programming equipment, and tooling.

Magnum XQF. The total cost of the Magnum XQF CNC punch press is $378,450. The following is a breakdown of this price.

Magnum XQF CNC punch press.......	$320,450
Freight and installation	$ 7,000
Transit insurance.....................	$ 2,000
CNC programming equipment	$ 29,000
Tooling.............................	$ 20,000
Total	$378,450

FIGURE 13.1
Recommendation Report (continued)

Model 150B.   The Model 150B CNC punch press has a total cost of $362,900. The following is a breakdown of this price.

Model 150B CNC punch press	$308,900
Freight and installation	$ 6,000
Transit insurance....................	$ 2,000
CNC programming equipment	$ 26,000
Tooling............................	$ 20,000
Total	$362,900

Conclusion.   Both CNC machines meet the third criterion. The Model 150B is $15,550 cheaper, but it does not have the capacity or capabilities of the Magnum XQF.

FIGURE 13.1
Recommendation Report (continued)

■ EXERCISES

1. Edit the following sections from a feasibility report. The problem is whether or not to sell sterling silver jewelry in a boutique. The report is written by a store manager to a district manager. The four criteria are cost, competition, quality of vendor, and profit. If your instructor requires it, write an informal introduction and summary.

REPORT

Survey Used.   To complete the survey I used Uribe's Store in Nisswa, Minnesota, where I had 50 surveys by the register for customers to fill out as they paid for their merchandise. I had help from the other employees in asking customers if they would please take a few minutes to fill out the survey.

Results.   The results of the survey show that 49% of the customers are girls and young women ages 15–20. At least 56% of the customers purchase gold jewelry (14 carat or other). To my surprise, even though the majority of the customers wear gold colored jewelry, 47% feel that Uribe's should carry sterling silver jewelry. I discovered that those wanting the silver jewlery were at least 25 years or older. A frequent comment or suggestion was that customers felt that Uribe's should carry sterling silver. Uribe's even has a list of people to call whenever they start carrying sterling silver. Customers are asking for it all the time.

Cost.   The cost for Uribe's to buy sterling silver is minimal when compared to what we can sell it at. In the trade, the initial markup for sterling silver is 500%. Stores can make 5 times the amount of what they paid for it. The actual cost of sterling silver, per piece of jewelry, is from $2.99 to $7.50. (Refer to Figure 2.)

Competition.   By selling sterling silver, Uribe's could increase its competitive position significantly. In fact, we could take some of the jewelry business away from other higher-priced jewelers such as Larkin Jewelers and Anderson Jewelers. Much of the 14 carat gold jewelry that Uribe's carries right now is similar to that which Larkin and Anderson carry. By selling sterling silver jewelry, we can capture more of the market. Because Uribe's is a low-priced jewelry store, we have the same quality but can sell it at a lower price. We buy it at a lower price and therefore can sell it at a lower price.

Line of Merchandising.   At this time we purchase most of our 14 carat gold jewelry from the Cupid Company. Julie Nelson, who used to be regional Manager for Uribe's, is now a Sales Representative for the Cupid Company. I feel we should purchase our sterling silver from the Cupid Company for two reasons: Julie has had experience with Uribe's and knows what kind of merchandise sells the best, and also, we have had great success with selling Cupid's merchandise. In order to be consistent with the different types of jewelry that we carry in 14 carat gold, I feel we should carry earrings, necklaces, and bracelets in sterling silver.

Profit.   The average markup in Uribe's is only 365%. To give you a better idea of just how much Uribe's can make by selling sterling silver, take a look at

Figure 2 and see the difference in the cost and retail of each of the three lines of sterling silver that we should carry.

	Cost	Mark up (365%)	Retail
Earrings			
4 mm balls	$ 1.90	$ 5.05	$ 6.95
6 mm balls	2.18	5.77	7.95
Hoops	3.00	7.95	10.95
Shapes	3.82	10.13	13.95
Bracelets (one kind)	2.73	7.22	9.95
Necklaces			
16" chain	3.00	7.95	10.95
18" chain	3.55	9.40	12.95
20" chain	4.10	10.85	14.95
24" chain	4.64	12.31	16.95

FIGURE 2
Cost versus Retail Price

You can see the actual dollar markup that Uribe's can make by buying and selling the three lines of sterling silver. For example, 6-mm ball earrings were purchased by Uribe's at a cost of $1.90. The markup dollar amount is $5.05 (365%) so the earrings are sold to the customer at a price of $6.95. It is evident that by carrying sterling silver we can greatly increase our sales.

Display and Promotion. If we do decide to carry sterling silver, we should carefully display and promote it. I feel that the sterling silver should be displayed right alongside the 14 carat gold jewelry. We should keep the higher priced jewelry in the same location. Presently, the 14 carat gold jewelry is displayed in a jewelry case at the register; therefore the sterling silver should also be kept in this case. This way the customer knows that it is sterling silver and not an imitation. When we do carry sterling silver we should put signs up in the store to advertise that sterling silver has arrived at Uribe's. This will draw more customers into the store just to see what Uribe's has to offer.

Sale Price. It is also important to determine a sale price for sterling silver. Keeping in mind Uribe's restrictions on sale merchandise, we could offer 15% off sterling silver jewelry and still make a profit. For example, if we sold a 16" sterling silver chain at 15% off, the price would be $10.95 retail — $1.64 discount = $9.31. We are still making a $6.31 profit on a 16" chain. (Refer to Figure 2.)

2. Decide which display booth to purchase for advertising at a trade show. The budget for the project is $3500. Booths must be at least 8 feet high

but no higher than 10 feet. Redo Table 1, below so that a reader can easily grasp the comparison. The writer is the person who takes the booth to the trade show; the reader is the department supervisor. Write an informal report to justify your decision. You might have to select more criteria than cost.

TABLE 1
Cost Comparison of Two Booths

Portable Display System, Walkart, 8' x 10' unit, with structure and case; price includes cost of seaming with velcro, flameproofing, and art preparation of camera.	$2068.00
One pair of wing panels, 2' x 10' with one pair of scrolls and case. Price includes background color of scroll, but no graphics or prints.	$1000.00
Freight charge	$84.00
Total cost	$3152.00
Portable Display System, Walkart, 8' x 8' unit, with structure and case; price includes cost of seaming with velcro, flameproofing, and art preparation of camera.	$1788.00
One pair of wing panels, 2' x 8' with one pair of scrolls and case. Price includes background color of scroll, but no graphics or prints.	$950.00
Freight charge	$84.00
Total cost	$2822.00

3. Redo Table 2, below. Then write a formal report section in which you conclude that installing the conveyor system better fulfills the cost criterion.

TABLE 2
Various Annual Costs for Using Each Method

Costs	Using Three Forklift Trucks (Dollars)	Costs	Installing Conveyor System (Dollars)
Labor	60,000	Utility	6,000
Floor Space	15,000	Floor Space	3,000
Repair and Maintenance	2,000	Repair and Maintenance	5,000
Forklift Purchase	150,000	Conveyor Purchase	80,000
Damage to Products	4,000	Installation Cost	20,000
Total Annual Cost	231,000	Total Annual Cost	20,000

4. Compare the outline report (chapter 11, p. 238) of the CNC punch press
   to the actual report on pp. 286–287. In which situations would you prefer
   to receive the outline; in which would you prefer the report?

WRITING ASSIGNMENT

1. Assume that you are working for a local firm and have been asked to
   evaluate two kinds, brands, or models of equipment. Select a limited
   topic, (for instance, two specific models of ten-inch table saws, the Black
   and Decker model 123 vs. the Craftsman model ABC) and evaluate the
   alternatives in detail. Write a report recommending that one of the
   alternatives be purchased to solve a problem. Both alternatives should be
   workable; your report must recommend the one that will work better.

   Gather data about the alternatives just as you will when working in
   industry — from sales literature, dealers, your own experience, and the
   experience of others who have worked with the equipment. Select a
   maximum of four criteria by which to judge the alternatives, and use a
   minimum of one table in the report. Aim your report at someone not
   familiar with the equipment.

14

Proposals

A proposal persuades its readers to accept the writer's idea. There are two kinds of proposals: external and internal. In an *external* proposal, one firm responds to a request — from another firm or the government — for a solution to a problem. Ranging from lengthy (100 pages or more) to short (4 to 5 pages), these documents secure contracts for firms. In an *internal* proposal, an employee or department urges someone else in the company to accept an idea or to fund equipment or research. Internal proposals may be unsolicited or assigned.

THE EXTERNAL PROPOSAL

A firm writes external proposals to win contracts for work. Government agencies and large and small corporations issue a *Request for Proposal* (RFP), which explains the project and lists its specifications precisely. For example, a major aircraft company, such as British Airlines, often sends RFPs to several large firms to solicit proposals for a specific type of equipment, say, a guidance system. The RFP contains extremely detailed and comprehensive specifications, stating standards for minute technical items and specifying the content, format, and deadline for the proposals.

Companies who receive the RFP write proposals to show how they will develop the project. A team assembles a document that shows that the company has the managerial expertise, technical know-how, and appropriate budget to develop the project.

After receiving all the proposals, the firm that requested them turns them over to a team of evaluators, some of whom helped write the original specifications. The evaluators rate the proposals, judging the technical, management, and cost sections in order to select the best overall proposal.

Not all proposals are written to obtain commercial contracts. Proposals are also commonly written by state and local governments, public agencies, education, and industry. University professors often write proposals, bringing millions of dollars to campuses to support research in fields as varied as food spoilage and genetic research.

Discussion of a lengthy, 50- to 200-page proposal is beyond the scope of this book; it is a subject for an entire course. But brief external proposals are very common. They require the same planning and contain the same elements as a lengthy proposal. This section will illustrate the planning and elements of an external proposal.

PLANNING THE EXTERNAL PROPOSAL

To write an external proposal, you must consider your audience, research the situation, use visual aids, and follow the usual form for this type of document.

Consider the Audience

The audience for a proposal consist of potential customers. These customers know that they have a need, and they have a general idea of how to fill that need. Usually they will have expressed their problem to you in a written statement (an RFP), or in an interview. You must assess their technical awareness and write accordingly. Generally a committee will make the decision of whether or not to accept your proposal. You must write so that they all understand your proposal. To write to them effectively, you should

- Address each need they have expressed.
- Explain in clear terms how your proposal fills their needs.
- Explain the relevance of technical data.

For instance, if you wish to sell a computer system to a nonprofit arts organization, you cannot just drop code names for microprocessors — say, an 8020 chip — and expect them to know what that means. You need to explain the data so that the people who make the decision to commit their money will feel comfortable.

Research the Situation

To write the proposal effectively you must clearly understand your customer's needs as well as your own product or service. Your goal is to show how your product's features will fill the customer's needs. You must research their needs by means of interviewing them or by reading their printed material. Make sure you understand exactly what they want.

Writers devise different ways to develop their research on the client's needs and the features their product offers. To relate needs and features, many writers compile a two-column table. The following table shows how one firm decided how to fill the needs of a client.

Need	Feature that Meets Need
Director must be able to access latest financial data and public relations data	available on Apple talk network, hard disk
Director must be able to access data at any time	Director needs work station in her office
Secretary enters data, but not continuously	Secretary needs access to work station
Secretary does accounting	Secretary can use Accountant Inc. 2.1c
Artist enters data	Artist needs access to work station
Artist does desktop publishing	Artist needs Aldus Pagemaker and laser printer
$15,000 maximum	2 lower-end computers, 1 laser printer; software for artwork, word processing, accounting software, desktop publishing

Once you establish the needs, you can easily point out a reasonable method of fulfilling them. This research requires careful preparation.

Use Visual Aids

Many types of visual aids may be appropriate to your proposal. Tables might summarize costs and technical features. Maps (or layouts) of the situation, for instance, might show where you will install the work station and the electrical lines in the office complex. Illustrations of the product with callouts can point out special features. Your goal, remember, is to convince the decision makers that your way is the best; good visuals are direct and dramatic, drawing your client into the document.

WRITING THE EXTERNAL PROPOSAL

To write an external proposal, follow the usual form for writing proposals. The four main parts of a proposal are an executive summary, and the technical, managerial, and financial sections.

Writing the Executive Summary

The *executive summary* contains information designed to convince executives that the proposers should receive the contract. In a lengthy external proposal, this summary might be 10 to 15 pages long, summarizing all the sections in lay, not technical, terms. Often these summaries read like magazine articles, artfully designed to sell the proposal. In short external proposals, this section should be a proportional reduction of the body (see Chapter 6). It should present the contents of the technical, managerial, and financial sections in clear terms. This section is often designed to make nontechnical people feel comfortable with the proposal.

Writing the Technical Section

A proposal's technical section begins by stating the problem to be solved. The proposers must clearly demonstrate that they understand what the customer expects. The proposal should describe its approach to solving the problem and present a preliminary design for the product if one is needed. Sometimes the firm offers alternative methods for solving the problem and invites the proposal writer to select one. For instance, in the computer network example above, the proposal might explain three different configurations that fulfill needs slightly differently but still stay within the $15,000 maximum cost.

Writing the Management Section

The management section describes the personnel who will work directly on the project. The customer wants to know that the project will receive top priority, and the proposal writer must respond by explaining what technical personnel and levels of management will be responsible for the project's success. In a large external proposal, this section often contains organization charts and résumés. In a short proposal, this section usually explains qualifications of personnel; the firm's success with other similar projects; and its willingness to service the product, provide technical assistance, and train employees. This section also includes a schedule for the project sometimes with deadlines for phases of the project.

Writing the Financial Section

The financial section provides a breakdown of the costs for every item in the proposal. Often this section is not just a table of costs. At times a brief introduction and the table may be all you need, but if you need to explain the source or significance of certain figures, then do so. Below is the financial section from a brief external proposal. A complete external proposal appears on pp. 314–317.

COST AND TERMS

The cost of the total Go Smooth system is $559.00. The breakdown of component cost is shown in Table 1. The price of this Go Smooth system is $200.00 less than Fillmore's Ultracon SCR Electronic Drive and $175.00 less than Anderson's Variable System.

TABLE 1
Component Cost Breakdown of Go Smooth Speed Control

Go Smooth Adjustable Speed Drive (Y1)	$276.00
General Electric 1/2 HP 3 Phase Motor	$183.00
C-Flange Adapter	$ 75.00
Jones Flexible Coupling	$ 15.00
Misc. Mounting Hardware	$10.00
Total:	$559.00

The system package can be shipped by April 24, 1988. Our representative in Torrance can install it at a minimum charge, or your plant technicians can do it with assembly instructions.

THE INTERNAL PROPOSAL

The internal proposal is a common writing assignment in business and industry. Like the external proposal, it is written to persuade someone to accept an idea — usually to change something or to fund something, or both. Covering a wide range of subjects, internal proposals may request new pieces of lab equipment, defend major capital expenditures, suggest adding more departments to a store, or recommend revised production control standards. The rest of this chapter explains the internal proposal's audiences, visual aids, and design, and presents three different forms of informal proposals.

PLANNING THE INTERNAL PROPOSAL

The goal of a proposal is to convince the person or group in authority to allow the writer to implement his or her idea. To achieve this goal, the writer must consider the audience, use visual aids, understand organizational principles, and design a format.

Consider the Audience

Writers consider the audience of a proposal in at least three ways; according to their involvement, their knowledge, and their authority.

How Involved Is the Audience?   In most cases, readers of a proposal either have assigned the proposal and are aware of the problem, or they have not assigned the proposal and are unaware of the problem. For example, suppose a problem develops with a particular assembly line. The production engineer in charge might assign a subordinate to investigate the situation and recommend a solution. In this assigned proposal, the writer does not have to establish that the problem is a problem; but he or she does have to show how the proposal will solve the problem.

More often, however, the audience does not assign the proposal. For instance, a manager could become aware that a new arrangement of her floor space could create better sales potential. If she decides to propose a rearrangement, she must, in a situation like this, first convince her audience — her supervisor — that the problem is a problem, and then offer a convincing solution.

How Knowledgeable Is the Audience?   The audience may or may not know the concepts and facts involved in either the problem or the solution. You must learn to *estimate* the level of knowledge that your audience possesses. If the audience is less knowledgeable, take care to define terms, give background, and use common examples or analogies.

How Much Authority Does the Audience Have?   The audience may or may not be able to order implementation of your proposed solution. A manager might assign the writer to investigate problems with the material flow of a particular product line, but most likely the manager will have to take the proposal to a higher authority before it is approved. So the writer must bear in mind the multiple readers who may see and approve (or reject) the proposal.

Consider Your Own Position

Your position mirrors the audience's position. You either will have been assigned to write the proposal or will have discovered the problem or need yourself. In the first case, you don't have to establish that the problem is a problem, but you do have to show how your proposed solution matches the dimensions of the problem. In the second case, you have to establish that the problem is a problem and then explain your solution.

Use Visual Aids

Since the proposal probably will have multiple audiences, visual aids can enhance its impact. Visuals can support any part of the proposal — the problem, the solution, the implementation, the benefits. In addition to the tables and graphs described in Chapter 7, Gantt charts and diagrams are two other types of visual aid that can help your proposal.

Gantt Charts   Gantt charts visually depict a schedule of implementation. A Gantt chart has an x- and y-axis. The horizontal axis displays time periods; the vertical axis displays individual processes. Lines inside the chart illustrate when a process will start and stop. By glancing at the chart,

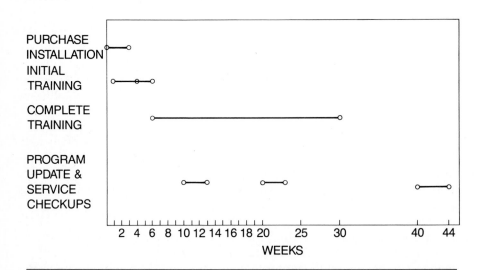

FIGURE 14.1
Gantt Chart

the reader can see the project's entire schedule. See Figure 14.1 for an example of a Gantt chart.

Diagrams Many kinds of diagrams, such as flow charts, block diagrams, organizational charts, and decision trees, will enhance a proposal. Layouts, for instance, are effective for proposals that suggest rearranging space. A good example of a layout is Figure 1 on p. 311. In that proposal, the layout simplifies the writer's job by communicating complicated details.

Organizing the Proposal

The writer should organize the proposal around four questions:

1. What is the problem?
2. What is the solution?
3. Can the solution be implemented?
4. Should the solution be implemented?

What Is the Problem? Describing the problem is a key part of many proposals. You must establish three things about the problem:

the data
the significance
the cause

The *data* are the actual facts, that a person can perceive. The *significance* implies the falling off from a standard you hope to maintain. To explain the significance of the problem, you show that the current situation has negative effects on productivity or puts you in an undesirable position. The *cause* is the problem itself. If you can eliminate the cause, you will eliminate the negative effects. To be credible, you must show that you have investigated the problem thoroughly by talking to the right people, looking at the right records, making the right inspection, showing the appropriate data, or whatever. In the following section from a proposal, the writer describes a problem:

Problem

The current problem in Junior Sportswear is a 20% decrease
in sales over the past 1989 fiscal year. In 1988, sales averaged   Data
$30,000 per month and contributed 10% to total store sales.
In 1989, sales averaged $24,000 per month, and contributed
only 7% to total store sales.

The cause of the decrease in sales can be traced to the   Cause of problem
display fixtures. The current wall fixtures and 5-foot bar
racks do not let us call attention to special merchandise at the   Significance of
department entrance. The wall fixtures line the back and side   each aspect of the
walls of the department; their position makes it difficult to   problem
capture the attention of customers walking past. The 5-foot
bar racks hold the merchandise sideways in mass quantities,
and make it impossible to show the special features and
colors of the new merchandise.

The current placement and setup of the fixtures do not
allow the customer traffic pattern to flow through the merchandise
from the front to the back of the department. Nor do customers
have enough room to step back and view the merchandise.

The fixtures that are now in the department are made of
wood and molded plastic. These materials do not blend in
with the sleek and modern "hi-tech" image of the department.
The fixtures also do not utilize the available 2300 square
feet of selling space to its fullest sales potential. Currently
300 square feet are wasted by the placement of the 5-foot bar
racks in straight lines.

What Is the Solution?   To explain the solution, explain each of its parts. Explain how your solution will eliminate the cause, thus eliminating whatever is out of step with the standard you hope to maintain. If the problem is causing an undesirable condition, say a high percentage of damaged parts, then the solution must show how the condition can be eliminated. If the old layout of a department in a store is not attracting customers, then the writer must show how the new layout will.

Solution

The solution that I propose is to replace the current fixtures   Solution explained
with black and silver chrome T-stands, spirals, and waterfall
fixtures. We could still make use of the wall fixtures and
transfer the 5-foot bar racks to the Athletic Sportswear
department. With the new fixtures, we should be able to
attract more customers into the department and display the
merchandise to optimum advantage.

Can the Solution Be Implemented?   The writer must show that all the systems involved in the proposal can be put into effect. To make this clear to the audience, you would explain

the cost
the effect on personnel
the schedule for implementing the changes

This section may be difficult to write because it is difficult to be certain about what the audience needs to know.

Cost of Implementation

The purchase cost of the new fixtures is $1128.00, as shown in Table 1. Total cost of the project, including $1080 for installation, is $2208.00.

Cost explained

TABLE 1
Cost of Fixtures and Installation

Fixture	Cost	Number Needed	Total
T-stands	$49.00	12	$ 588.00
Spirals	$42.00	10	$ 420.00
Waterfalls	$ 6.00	20	$ 120.00
Total Purchase			$1128.00
Installation			$1080.00
			$2208.00

Implementation

We can purchase the fixtures from our vendor, Walseth Inc. Walseth can fill our order within two weeks. The reconfiguration will take three days and will cost $1080.

Schedule for implementation

Day 1 — Empty wall racks; place merchandise in storeroom.
— Maintenance crew removes wall fixtures and fills holes in the walls.

Day 2 — Painters paint all walls. (The color, sandstone, will match the carpet used throughout the Sportswear department.)

Day 3 — Install new fixtures; place clothing on them.

During these three days, the department will operate at reduced capacity for a day and a half. Salespeople will help empty and fill the racks. One salesperson will take three vacation days, and the part-time help will not be scheduled.

Effect on personnel

Should the Solution Be Implemented?   Just because you *can* implement the solution does not mean that you *should*. To convince someone that you should be allowed to implement your solution, you must demonstrate either that the solution has benefits that make it desirable or that it fits the established criteria in the situation, or both. In an unsolicited proposal, you will find it easier to demonstrate benefits after you have carefully defined the problem. When you are assigned a proposal, concentrate on showing how the solution fits the given problem. In the following section, the writer pinpoints the benefits of the proposed solution, stressing how it will eliminate the negative aspects of the situation.

Benefits of the Solution

By replacing the fixtures, we can place the new merchandise near the department entrance to draw in customers. They will be able to see special styles, details, and colors of the merchandise. The customers will also be led through a flow pattern by the proper placement of the new fixtures.

    The new fixtures will tie together the overall hi-tech theme of the Junior Sportswear department. Using this theme is important: it helps make the target customer more excited about shopping and more inclined to make purchases.

    The new fixtures also will benefit the department by increasing the amount of merchandise carried. We can carry a wider and more current selection to attract a broader customer base. With the addition of new lines, customer sales will increase.

Significance: each aspect of solution eliminates each aspect of the problem

Designing the Proposal

To design a proposal, select an appropriate format, either *formal* or *informal.* A formal proposal will have a title page, table of contents, and summary (see Chapter 12). The format for an informal proposal can be a memo report or some kind of preprinted form (see Chapter 11). The format depends on company policy and on the distance that the proposal must travel in the hierarchy. Usually the shorter the distance, the more informal the format. Also the less significant the proposal, the more informal the format. For instance, you would not send an elaborately formatted proposal to your immediate superior to suggest a $50 solution to a layout problem in a work space.

WRITING THE INTERNAL PROPOSAL

Use the Introduction to Orient the Reader

The introduction to a proposal needs careful thought. The introduction must orient the reader to the writer, the problem, and the solution. The introduction can contain one paragraph or several. The following are important points to clarify:

- Why is the writer writing? Is the proposal assigned or unsolicited?
- Why is the writer credible?

- What is the problem?
- What is the background of the problem?
- What is the significance of the problem?
- What is the solution?
- What are the parts of the report?

Introductory sections often contain a separate summary paragraph that restates the main points of the body. The summary is a one-to-one miniaturization of the body. (Be careful not to make the summary a background; background belongs in a separate section.) If the body contains sections on the solution, benefits, cost, implementation, and rejected alternatives, the summary should cover the same points. This paragraph usually appears after the context-setting paragraph that answers the questions listed above.

Remember, a proposal is not a murder mystery. Readers should know the basic facts after reading the introduction and the summary. The audience does not want to wait until the end to discover your solution and reasons.

Sample introductions to an assigned proposal and an unsolicited proposal follow.

INTRODUCTION TO AN ASSIGNED PROPOSAL

On April 24, you requested that I review our present manufacturing capabilities in an attempt to identify inefficient production	Reason for writing
facilities. In doing so, I have found that we manufacture a product that is available from a more efficient source. I am	Methodology
proposing that we purchase the #A-100 wiring harness from a supplier rather than manufacture it in our plant. The	Proposal
information in this report will enable you to make this purchasing decision.	Purpose

INTRODUCTION TO AN UNSOLICITED PROPOSAL

A new technology — desktop publishing — has made it possible to create pages of graphics and text directly on a computer. The staff of the *Clarion* is interested in setting up	New development
an electronic publishing system that would allow us to lay out the paper on our computers and to produce a master copy ready for mass production. This memo will describe the	Significance of fact to local situation
hardware and software we need, how it works, and how it can save Hastings Publishers, Inc., a lot of money.	Proposal based on significance

Use the Discussion to Convince Your Audience

The discussion section contains all the detailed information that you must present to convince the audience. A common approach functions this way:

The problem

- Explanation of the problem
- Causes of the problem

The solution

- Details of the solution
- Benefits of the solution
- Ways in which the solution satisfy criteria

The context

- Schedule for implementing the solution
- Personnel involved
- Solutions rejected

In each section, present the material clearly; introduce visual aids whenever possible; and use headings and subheadings to enhance page layout.

Which sections to use depends on the situation. Sometimes you will need an elaborate implementation section; sometimes you won't. Sometimes you should discuss causes, sometimes not. If the audience *needs* the information in the section, include it; otherwise, don't.

TYPES OF INTERNAL PROPOSALS

This section presents several types of internal proposals. The examples that follow demonstrate ways to handle the assigned proposal and the unsolicited proposal.

The Assigned Proposal

The model that begins on page 305 reflects an assignment that a student completed during his internship at a company. It has three parts: an *introduction* including a summary, a *description* of the actual system, and an *explanation* of how the proposed solution meets preset criteria. Two visual aids — a table and a layout — reinforce the writer's points.

In the introduction the author explains that he has been assigned the project. He also explains what he did to become credible in the situation and briefly discusses his proposal. In the criteria and summary sections, he notes that his solution will fulfill the five criteria that have been agreed upon for the situation.

In the discussion section, or the body, the author describes the new machine and explains how it meets all the criteria. Notice how the author uses visual aids throughout to communicate his ideas.

INTRODUCTION

This report is a response to your request for a proposal from the production department on ways to increase productivity at work station 128. After speaking with a number of staff members, I recommend adding another numerical control (NC) lathe to work station 128. The machine we have chosen is the Stamford MR 7000. This report covers *two* main areas: a description of the new lathe and an explanation of how it meets the criteria.

Criteria

The following criteria were set at a meeting of the management group in March:

- Total cost not to exceed $200,000
- Payback period of one year
- Productivity increased by 20% at work station 128
- Vulnerability at work station 128 reduced
- Implementation time of eight weeks

Summary

This report shows that the machine we propose to purchase will meet all the criteria.

- Total cost of adding another machine is $182,000
    New lathe — $150,000.
    Additional hardware — $15,000
    Cost of labor to install — $17,000
- Payback period of one year
- Increase in productivity of 20% at work station 128
- Reduced vulnerability
- Implementation time of six weeks

DISCUSSION

The discussion segment of this report shows how the proposed solution will increase productivity at work station 128. This section describes the new machine and its advantages, discusses how the machine meets all five criteria, and also explains why we rejected the other alternatives. The five criteria are spending allowance, payback period, increase in productivity, reduced vulnerability, and implementation time.

The New NC Lathe

Description of Machine.   The new NC lathe recommended by the production department is the Stamford MR 7000. The machine is capable of computerized numerical control (CNC); the operator can program an operation right at the machine location via a keyboard. It is a five-axis machine that will handle the complex tooling required at work station 128. It has a turret-type tool mount on either side of the bed ways. Each turret is capable of holding four tools, which gives the machine a total holding capacity of eight tools.

Advantages of New Lathe.   The existing NC lathe does not have the capabilities of this new one, which might have more features than we need. But owing to the present technology and pressure from our competitors, we feel we are forced to buy a slightly overqualified machine. However, if future demand requires increased capability at work station 128, we will be able to handle it. If there is a need at any time for the R & D department to experiment with the tooling capabilities of this machine, they may do so after the second shift when the demand for time is not so great.

How the Solution Meets the Criteria

Spending Allowance.   The spending allowance allocated for this project is $200,000. This figure covers everything from acquisition to implementation. The total cost to purchase and install a new NC lathe is $182,000 broken down as follows:

- cost of new machine - $150,000
- cost of additional hardware - $15,000
- cost of labor to install - $17,000

Explanation of Cost

The cost of additional hardware needed for the installation of the lathe is $15,000. The hardware is needed to "marry" the new lathe to the existing one. The proposed setup will allow one person to operate both lathes. Thus we will not need any additional labor.

The additional hardware needed consists of 20 feet of conveyor track, all new tooling required for the new lathe, and a new switching unit to alternate the dispersement of machinable parts coming into the work station (see Figure 1).

The labor cost required to install the additional hardware plus the new lathe is $17,000. This figure is based on a total of 720 hours of labor at $24 per hour (includes benefits). We feel it would take a three-person crew a total of six weeks to install the new system. Tasks include hooking up the hydraulic and electrical systems, plus installing the actual hardware.

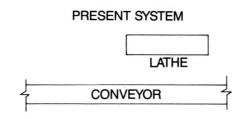

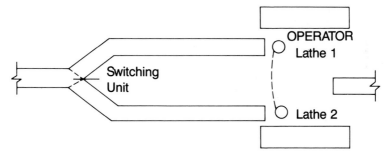

FIGURE 1
Work Station Layout

Payback Period

The payback period required by the criteria was one year. Because the new lathe will increase production at no additional labor cost, we feel we can meet this one-year deadline.

The $182,000 cost of the new machine is a fixed cost. Any additional costs will be from maintenance and from the energy required to operate it. The Stamford sales rep estimates these costs at around $150 per month. The machine has a one-year guarantee on parts and labor.

If we achieve a 20% increase in productivity, and if all other costs presently incurred by production increase at this same rate (which is unlikely), we can expect a 20% increase in net earnings from sales. Our present net earnings from this product line are $920,000 per year. The new machine, therefore, has the potential to increase our net earnings from $920,000 to $1,104,000 the first year, an increase of $184,000. Thus the payback period for the new machine would be 52 weeks (see Figure 2). This 52-week figure does not take into account our tax deduction from interest and depreciation on capital goods. If these figures are taken into account, our payback would be even sooner.

Increase in Productivity

The 20% increase in productivity due to the addition of a lathe at work station 128 justifies the cost of the new machine. The increase in productivity parallels

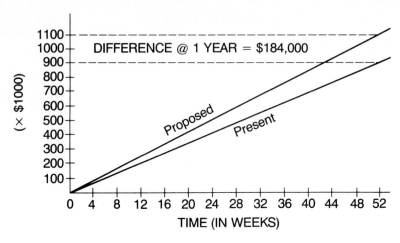

FIGURE 2
Payback Period

an increase in net earnings of at least 20%. Our profit from the increased productivity would not be seen until after the machine has been paid for, as shown by Figure 2. After that point, however, we will make a profit. The increase in productivity can be seen by the man/machine chart shown in Figure 3. Note that the station still needs only one operator.

OPERATOR	Unload 1	Clean 1	Insect 1	Load 1	Unload 2	Clean 2	Inspect 2	Load 2	Unload 1	Clean 1	Inspect 1	Load 1	Unload 2
LATHE 1					Machine		20%→		←—100%—→		←—80%—→		Machine
LATHE 2		20%→			←—80%—→ ←—100%—→			Machine					

FIGURE 3
Man/Machine Chart

Reduced Vulnerability

Because of its increase in productivity, work station 128 will be less vulnerable to slowing down production. Furthermore, having two machines is like having a backup system. If one lathe breaks down, the other can still be producing parts, although production would decrease.

Implementation Time

Implementation time, as set in the criteria, is eight weeks or less. We are confident that the total time required to purchase and install the lathe is six weeks. The vendor states a two-week delivery time. After the lathe is here, we would require an additional four weeks to install the machine and put it into operation (see Figure 4).

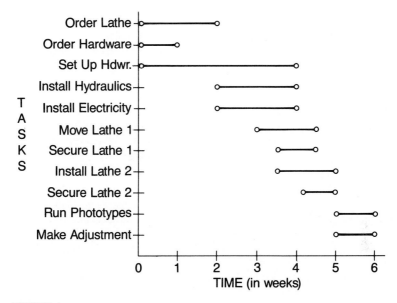

FIGURE 4
Schedule for Project

The Unsolicited Proposal

In the following proposal, a student is bringing a problem that he has experienced to the attention of the lab supervisors. He assumes that the readers are unaware of the problem. The proposal has an introductory section, a discussion, and two visual aids. Notice that the reader knows the proposed solution by the end of the second paragraph.

The introduction has two sections. First, the writer explains the source of his credibility: he has worked in the lab in the role of the person it was designed for. Second, he names the problem, tells immediately how much the project will cost, and shows how his solution will answer the problem.

In the body of the proposal, the writer outlines the problem in one section, his solution in the next, and the method of implementation in the last. He accompanies his proposal with two visual aids, one that depicts the new layout he proposes and the other a table of costs. Throughout the report, the writer anticipates questions and objections. He has obviously done his homework, as demonstrated by the layout and cost table. Unless a major difficulty arises from some other source, the solution could be implemented easily.

Introduction

As a student who uses the software development computers in Room 214, Fryklund Hall, I see a problem with the arrangement of the computer terminals in the existing work area. This problem will be further magnified with the arrival of the new computer terminals expected for January 1990. The cost to implement a more efficient layout would be less than $60.

Suggestion

I suggest that the software development computers and their clustered computer terminals be arranged as shown in the following layout (Figure 1). This layout is designed to eliminate the existing problems of the inefficient arrangement of the computer terminals and blocked access to the printer.

Problem

The existing layout of the software development computers is very inefficient. During peak use of the development computers, only five out of the six terminals can be used. The computer terminal located in the corner of the work area is blocked off whenever people use the computer terminals on each side of it. If this computer terminal were used, people would be on top of each other — an uncomfortable situation that discourages people from using it. I have heard fellow classmates and instructors complain about all the terminals being in use when in fact the terminal in the corner of the work area was free.

　　With the existing layout, the printer location also creates a problem. The position of the printer now is just above one of the computer terminals. This location is very inconvenient for both the person using the computer terminal and the one using the printer. This inconvenience happens quite regularly because there is only one printer to service all the computer terminals. The problem with overcrowding will only worsen if the same arrangement is kept when the new computer terminals arrive.

Solution

To eliminate the problem of inefficient use of the computer area, the enclosed plan for a new layout should be implemented. This layout includes the four new terminals that will be acquired by the university for January 1990. The area just to the right of the doorway is the proposed new work area, an area chosen because few people sit there during class periods. This expanded area, along with the old area, will give ample room for the computer terminals. This layout will also

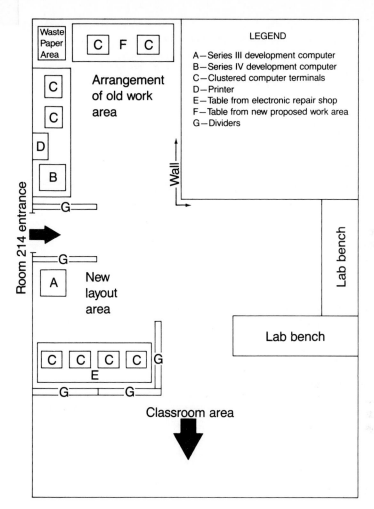

FIGURE 1
Proposed Layout

free space for the printer, making it more accessible and increasing the amount of paper that it can hold. This solution will also make the repair of main development computers easier.

Implementation

Implementation concerns focus on cost, material, and labor.

Cost. The new program costs less than $60. The only cost involved would be for female and male computer cable connectors. A total of four female connectors and four male connectors would be needed to complete the proposed layout. Table 1 shows a comparison of computer connector costs from venders used by the university. The other materials needed, as described in the next section, will not cause any additional expense to the university since they come from materials the university already has on hand.

TABLE 1
Cost Comparison

	Radio Shack		Digi-Key		JDR Corp.	
Connector	Price each	Total	Price each	Total	Price each	Total
Male	6.99	27.96	6.49[a]	25.96	7.49[a]	29.96
Female	6.99	27.96	6.70[a]	26.80	6.95[a]	27.80
Total cost		55.92		52.76		57.76

[a]Includes shipping costs

Material.   The existing materials already in the work area would be used — except for the work bench. The work bench would have to be removed and used elsewhere, perhaps to the electronic repair shop (Fryklund 210). The long table for the new work area is presently stored in the electronic repair shop; the five dividers can be obtained from the Electronics Department or from the university's warehouse. The needed cable for the computer terminals is available in the electronic repair shop.

Labor.   Student employees from the electronic repair shop can supply the labor. These students are very knowledgeable about how the software development computers function and would have no problems in implementing the layout plans suggested in Figure 1. These students would also assemble the different lengths of cables with the connectors that would be purchased.

■ SUMMARY

Proposals convince the reader to allow the writer to solve a problem or perform some activity. External proposals are used to convince one company to buy another company's services or products. Written to respond to a client's needs, external proposals explain the service or product offered in the technical section, review the firm's credentials and abilities in the management section and list the costs in the financial section. Internal proposals report on assigned problems or on unsolicited solutions for problems. Usually such proposals attempt to answer four questions: What is the problem? What is the solution? Should we implement the solution (justification)? Can we implement the solution (implementation)? Writers of proposals must carefully consider the audience's knowledge level and authority. A visual aid commonly found in proposals is the Gantt chart, which illustrates the components of a schedule.

■ WORKSHEET FOR PREPARING A PROPOSAL

Who will receive this proposal?
Who will decide on the recommendations in this proposal?
How much do they know about the topic?
Prepare background. Why did the proposal project come into existence?
Select a format — formal or informal.
Prepare a style sheet of margins, headings, page numbers, and visual aid captions.

☐ *External Proposal*
Write a statement of the customer's needs.
Prepare a matrix (2-column listing — see p. 294) of the customer's needs and the features of your proposal.
In the written sections, do you present your features in terms of the customer's needs — using the customer's terminology?
Are the financial details clearly explained?
Do you explain in detail why your company has the expertise to do the job?

☐ *Internal Proposal*
Was the report assigned, or is it an unsolicited proposal?
Outline your methodology for investigating the problem
Decide the level of detail you will need to explain the problem.
Prepare a list of the dimensions of the problem and show how your proposed solution eliminates each item (this list is the basis for your benefits section).
Prepare a schedule for implementation.
List rejected alternatives, and in one sentence tell why you rejected them.

■ MODEL

A solicited proposal in an informal format follows.

DATE: December 5, 1988
TO: Harold Crane, President
             Company Store
FROM: Curt Everson, Sales Engineer
             Wieland Controls Inc.
SUBJECT: Proposal for a 1301 Halon System for Company Store Data-Processing
             Room.

Summary

This is the proposal you requested for a halon system for your data-processing
room. Wieland Controls can offer you several benefits if you choose our firm
to install and maintain your system. I have described in this proposal how the
system will work in the event of an alarm. I have also outlined how we will meet
your insurance company's requirements. Wieland will complete your job in eight
weeks at a total cost of $47,250. An optional $1,300 annually renewable maintenance
agreement is explained as well.

About Wieland Controls

We have experience. We have installed 16 halon systems in the past five years.
Our recent jobs include Bogus Engineering, Rochester; Franklin Research,
Phoenix; and Midwest Foods, Des Moines. We have used a Fyrcontrol Always
There system in all of these installations and have had extremely good results.
Fyrcontrol is the leader in the field of halon fire extinguishing systems. Our
proposed system is also expandable. If you need to add to your system later on,
you will only need to purchase the necessary boards, not a whole new system.
We also provide the needed support to our clients using halon systems. Our
Chicago branch office has the facilities to fill and test halon tanks. In emergency
situations, the Chicago office can have our tanks filled within 24 hours. This is
important because not everyone has the expertise to fill halon tanks, especially
on short notice. We can provide all the support you will need now and in the
future in maintaining and adding to your system.

System Operation

In simple terms, a halon fire extinguishing system is a large automatic fire
extinguisher. Since halon gas is very expensive, a false alarm can be extremely
costly. We overcome this problem by using what is called cross zone release.
False alarms can be caused by dust, humidity, heat, and faulty detectors. Cross
zone release means that both an ionization and a photoelectric smoke detector
must be sent an alarm at the same time before the halon will dump. This configuration
prevents the problem of one detector dumping the halon. The system will dump
the halon gas for one of two reasons: the owner pulls a manual pull station, or

FIGURE 14.2
External Proposal

2

smoke activates a cross zone alarm of two dissimilar smoke detectors. When a dump happens, the following sequence of events will occur:

1. The alarm horns will sound.
2. All doors will shut.
3. All smoke dampers will shut.
4. The air handler supplying the room will stop.
5. The main fire alarm panel will be signaled.
6. The fire department will be signaled.
7. After 30 seconds, the strobes will flash.
8. After a further 30 seconds, the halon gas will dump.

If only one smoke detector is triggered, the same sequence of events will occur, except for 7 and 8.

How Our System Meets Your Requirements

Your insurance company has the following requirements for the halon system you are thinking of installing.

1. Must meet the minimum standards required by national and local codes.
2. Must use halon 1301 gas.
3. Must be of the total flooding type.
4. Must maintain at least a 7% concentration for ten minutes.
5. Discharge test after installation must prove that concentration level requirements are met.
6. Must automatically trigger the main fire alarm panel, which in turn will signal the fire department.

Here is how our system meets your requirements:

1. The proposed system meets or exceeds the requirements outlined by the National Fire Protection Association (NFPA) and local codes. This system will be installed in accordance with NFPA 12A "Standard on Halon 1301 Fire Extinguishing Systems."
2. Our proposed Always There 1301 system uses 1301 halon gas. The reason for requiring 1301 halon gas is safety. In all the years that halon 1301 gas has been used, there has not been a single death or injury associated with it.
3. Our proposed system is of the total flooding type. A total flooding system is used when an entire enclosed area needs to be protected. As required by

FIGURE 14.2 (continued)

NFPA 12A for a total flooding system, when fire or smoke is detected: all doors will automatically close, all air ducts will be sealed via dampers, and the air handling unit for your data-processing room will shut down.

4. In order to meet your insurance company's requirements for a 7% concentration for ten minutes, you will need to have 177 pounds of halon. This figure was derived from a table in NFPA 12A. With all air exchange between rooms cut off as required by NFPA 12A, this concentration will be maintained for 10 minutes.

5. A concentration meter will monitor the results of an actual test and give a printout of the results. This printout will be used as proof for the insurance company that the concentration requirements have been met. We will perform a complete test of the system after completion. A concentration meter will be present in the room, which will monitor the concentration level for 10 minutes. We will rectify a failed test at no additional cost to you.

6. We will have a provision in our panel to signal your main fire alarm system. Since you have indicated you already have a tie-in with the fire department, no further modifications to your existing panel will be required. We will only need to run wires from the data-processing room to the main panel to signal it.

Performing the Job

The time we take from receiving your order to signing off the job will be eight weeks. The job will be performed with minimal inconvenience to your data-processing department. The smoke detectors will be mounted on the ceiling, and this may cause some inconvenience when workers need to get above desks or pieces of equipment. The piping of the halon and mounting of the tanks will all be done above the false ceiling and along the wall. This shouldn't cause any problems. The biggest inconvenience will come with the actual test of the system and the dumping of the halon gas. This can be expected to last about two hours. We will give two days' advance notice so your employees can properly prepare.

Cost

We will perform the complete job for a cost of $47,250. This will include the following.

1. Fyrcontrol System 3 flush mount control panel with provisions to complete the job.

2. Provision in the panel for up to 6 additional auxiliary devices or zones for later expansion.

3. Four Fyrcontrol SI-X6 ionization smoke detectors.

4. Four Fyrcontrol SP-X6 photoelectric smoke detectors.

5. One Fyrcontrol MP-60 manual pull station.

FIGURE 14.2 (continued)

4

6. Three smoke dampers and electric motors.

7. Two 24-hour standby gel cell batteries.

8. Two Weldon horns.

9. Two Weldon strobes.

10. Two magnetic door holders.

11. All necessary conduit and wire to install the system.

12. Two 80-pound halon tanks and 1301 gas.

13. All necessary piping, brackets, and hardware to install the halon tanks.

14. Complete test of the system with actual halon dump.

15. All labor required from design to sign off.

16. Two sets of blueprints.

17. Train all data-processing room employees.

Maintenance Agreement

For an additional $1,300 a year, we will test and maintain the system. This will include the following.

1. Semiannual check out of all equipment and components in accordance with NFPA 12A, encompassing the following:

   a. Sensitivity test of all smoke detectors.

   b. Operational check of panel and all associated devices.

   c. Check weight and pressure of both halon tanks.

   d. Confirm tie to fire department.

   e. Clean all smoke detectors.

   f. Battery load test.

   g. Certification form after check out.

2. All parts required to repair and maintain the system in normal operating order.

3. Up to three emergency trouble calls per year.

4. This will not include the cost to replace the halon in the event of a false alarm.

FIGURE 14.2 (continued)

■ EXERCISES

1. Rewrite the following paragraph. The writer is a product designer; the reader is the head of manufacturing. The proposal suggests the use of plastic as the casing material for a flashlight. The purpose of this section is to show that plastic meets the criterion of "availability." Eliminate all unnecessary sentences.

> One of the first things that many designers ask when initially considering the use of a particular material is whether the material is on hand. It is not wise to use a proprietary material that is available only from one supplier if it is likely that the product in question will be made again in five years. If you are a captive buyer, you are at the mercy of the supplier on cost and delivery. Sometimes a raw material fabrication technique may limit availability. As an example, it is not common to get delivery times of 60 weeks on special forgings. In lean times castings may be obtained in 4 weeks; other times delivery times can be 12 weeks. These are all important factors under the category of availability. I have checked with several potential plastic suppliers (AVW Plastics Inc. and Jasper Plastic Supply) within our area and they seem to be able to handle what our production demands, and they can deliver it when we desire.

2. Rewrite the following paragraphs. Use the two figures and the table as a basis for your discussion. The writer is an area supervisor who has discovered the problem; the reader is the layout supervisor of the printing company. The project is to produce guest checks for a restaurant by cutting the checks from a large sheet of paper and then printing them. Your instructor might also assign you to write an introduction and summary to make this a brief informal report.

Problem

The problem with the present process is that it wastes paper. The cutoff on the printing press being used is permanently set at 17 inches. The problem gets worse when the printed pages must be cut, perforated, padded, and then cut again. The first cuts are irregular (Figure 1), which means the paper cutter must be reset after each cut. After the cut you are left with four stacks of paper, which must be further processed as mentioned above.

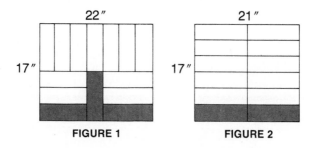

FIGURE 1   FIGURE 2

Solution

Positioning the forms as indicated in Figure 2 and using a 21-inch roll of paper instead of 22 inches will waste less paper (Table 1). Even though only 10 instead of 11 checks can be positioned on a page, this method wastes less paper and makes processing easier. Only one initial cut would be needed (Figure 2) before perforating, padding, and final cutting. There would also be only two stacks to handle instead of the current four stacks.

Benefits

The proposed process would maximize profits by reducing paper waste (Table 1). The changed layout would also reduce the time of further processing on the guest checks (Table 1). There would also be less wear and tear on the paper cutter since only one initial cut instead of four is needed.

TABLE 1						
No. of Checks per Sheet	Roll Size (width)	Area of Waste per Sheet	Est. Time of Processing per Skid	Price of Paper per Roll	No. of Sheets per Roll	Amount of Waste per Roll
11	22 in.	28.25 sq. in.	5.3 hrs.	$220	2500	$16.61
10	21 in.	21.00 sq. in.	3.2 hrs.	$210	2500	$12.94
10	22 in.	38.00 sq. in.	4.1 hrs	$220	2500	$22.34
7	17 in.	51.50 sq. in.	4.0 hrs.	$170	2500	$30.28

3. Rewrite the following two sections. Use the two figures. The writer is the department manager; the reader is the merchandise manager. The writer has discovered a need: the department should expand because its sales have exceeded all previous peaks. The background is that six months earlier the department added several brand-name clothing lines. The new lines have drawn in more customers than expected.

Discussion

Profits.   Profits will increase dramatically with the new expansion of the department. Because of the additional brand-name clothing we implemented, our sales have shown a noticeable increase in the past year. With the added space, our sales volume could reach an additional profit. A new expansion of the department would allow twice the extra space to display more of the merchandise that is selling. For example, the brand TUTA is a growing category. If, by chance, we can expand our line of TUTA merchandise, we can create what is known as multiple purchases. Multiple purchases are defined as a customer buying an additional item besides the intended purchase. For example, a customer may purchase a TUTA shirt; however, on the way to the register, she notices a TUTA belt and decides to purchase it also. Multiple purchases lead to profits because of the unexpected purchase.

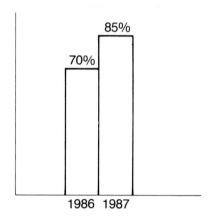

EXPECTED PROFIT

FIGURE 1

Increased Customer Market. The increased customer market will be evident with the expansion of the department. This means we will reach other customers who would normally not shop the store. With the expansion of the store, we could display an extended line of merchandise we now carry, along with

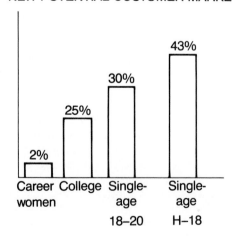

NEW POTENTIAL CUSTOMER MARKET

FIGURE 2

the new merchandise. Our department will carry, for example, TUTA pants, shirts, socks, belts, and hats. We will also carry such brand names as Tex, Arrowhead, Zach's, and Boe. We will attract many customers who may be in the same age bracket, yet have different tastes (see Figure 2). Each customer has her own unique taste; with the extended department of various vendors, we can please each one of them. By reaching other customers, we will also help to increase profits. Increased profits are a result of the additional new customers and their purchases.

WRITING ASSIGNMENT

1. Write a proposal in which you suggest a solution to a problem. Explain the problem and solution. Show how the solution meets established criteria or how it is beneficial in terms of the dimensions of the problem. Explain cost and implementation. If necessary, describe the personnel who will carry out the proposal. Explain why you rejected other solutions. Use at least two visual aids in your text. Base the proposal on factual data and make it convincing. Your instructor will assign either an informal or a formal format. Topics for the assignment could include a problem that you have worked on (and perhaps solved) at a job or a problem that has arisen on campus, perhaps involving a student organization.

15

Progress Reports

PLANNING THE PROGRESS REPORT

INITIAL PROGRESS REPORTS

SUBSEQUENT PROGRESS REPORTS

P rogress reports inform management about the status of a project. Submitted regularly throughout the life of the project, they let the readers know whether work is progressing satisfactorily, that is, within the project's budget and time limitations. Firms who win contracts through external proposals must provide regular reports on progress to their clients. On the internal level, departments often furnish reports to managers on ongoing projects involving research, construction, and installation.

As with all writing projects, you must plan the progress report. You must consider your audience, research the situation, use visual aids, and follow the usual form for this type of report.

PLANNING THE PROGRESS REPORT

Consider the Audience

Most readers are not fully informed about all aspects of the project. They may have walked through the laboratory or visited the construction site, but they probably did not see the whole project or understand everything they did see. To understand the progress made to date and the problems that are anticipated, readers must fully grasp what the project involves. If the report goes only to your immediate supervisor, you can assume that he or she probably knows technical terms related to the project. The supervisor, however, does not know the details — of what has been done and what still needs to be done — so you will have to supply those. If the report goes a greater distance from you, you should assume that the readers do not know the technical details and perhaps not the technical concepts. In other words, you must go into as much detail as necessary to inform your specific audience.

Research the Situation

To plan the project, you must select the categories that you need to use to discuss the project. Usually two major categories are budget and schedule, but many other categories are possible. Sometimes those categories can reflect criteria explained in the original proposal or report. Sometimes you will have to devise additional categories, depending on your audience's knowledge and interest in the project.

Use Visual Aids

Visual aids are as effective for progress reports as for any other type. If you need to use a table, graph, or illustration, do so.

Follow the Usual Form for Progress Reports

Progress reports usually follow the form shown in the outline below. You need to provide three sections: Introduction, Work Completed, and Work Scheduled. Sometimes you will have to add special sections.

     I. Introduction
        A. Purpose of report
        B. Purpose of project
    II. Work Completed (July 1–July 31)
        A. Installation of cut-to-length line
            1. Conveyor system
            2. Crane
        B. Installation of banding line
            1. Hydraulic system
            2. Expansion of banding area
   III. Work Scheduled (August 1–August 31)
        Same second- and third-level heads as in Work Completed section.

INITIAL PROGRESS REPORTS

An *initial progress report* contains a brief introduction and a body that describes the work. In the introduction, you can help your reader, who might be reading several such reports, by supplying statements on the purpose of the report and the purpose of the project.

Writing the Introduction

Begin by stating the purpose of the report. A single sentence can name the project, define the time period covered by the report, and tell the purpose: to inform readers about the current status of the project. Mention the project's objectives and scope, and name the major work areas. This statement gives the readers an overview of the project and is, in effect, a preview list for the body of the report. If subsequent progress reports are written, state the report's number in the series.

Writing the Work Completed Section

In the Work Completed section, specify the time period, and divide the project into major tasks. Second-level heads will identify these tasks in this and subsequent reports. The naming of these major tasks should be comprehensive, allowing these same heads to appear consistently in subsequent reports. On the other hand, third-level headings ("Conveyor system," and so on, in the outline above) should be sequential, and these will change as the project moves forward.

Writing the Work Scheduled Section

The Work Scheduled section again specifies the time period and repeats second-level and third-level heads from the Work Completed section. The heads help readers grasp the continuity of work in major project areas. If readers require a more detailed chronology of future work, divide this section into two parts:

> Work Scheduled for Next Report Period
> Work Proposed for Future (Dates)

SUBSEQUENT PROGRESS REPORTS

Second and succeeding progress reports maintain continuity and refresh the reader's memory by adding one new section, a summary of work completed prior to the present reporting period:

I. Introduction
II. Summary of Work Previously Completed (Dates)
III. Work Completed (Dates)
IV. Work Scheduled (Dates)

The new section contains the same heads for major tasks (second-level) that have appeared in earlier reports but condenses the information. To prepare this section, you examine the Work Completed sections of your previous progress reports and write a capsule version of them.

Adding Special Sections

Sometimes other sections need to be added. If readers want specific information on some aspect of the project that they particularly want to control, you should provide that information. Budget, or any other item of special concern, can easily be given a main heading of its own and a thorough accounting.

■ SUMMARY

Progress reports update readers on the status of projects during a specific time period. These reports explain the project's goals, then describe the work already completed and future work scheduled. If necessary, writers explain special problems encountered in the course of the project.

■ WORKSHEET FOR PROGRESS REPORTS

☐ *Who will receive the progress report?*

☐ *How much do they know about the project?*

☐ *What time period does the progress report cover?*

☐ *Are you required to use a specific format, especially for the introduction?*

☐ *List all the categories of information you need to report.*

☐ *Explain the work completed in this time period for each category.*

☐ *Explain work projected for the next time period for the same categories.*
    If there is a new category, make special note of it.

☐ *Explain special problems or requests.*

☐ *What visual aids will help readers grasp information quickly?*

☐ *Select a style sheet for margins, headings, and page numbers.*

■ MODEL

A progress report follows. Notice the sections it contains and how it explains material to the supervisor who knows in general what is going on.

To: Ray Schweisguth   November 20, 19XX
From: John F. Furlano
Subject: Progress Report on Recommendation Assignment

Purpose

The purpose of this memo is to inform you of the progress I have made on my recommendation report during the weeks November 6–20. This memo will cover the work I have completed to date, the work I have yet to complete, and problems I have encountered with this assignment.

Work Completed To Date

    Writing.   I have developed and organized criteria that would be required in a real-life decision on whether to rebuild or replace an extruder. My extruder data include information on maintenance and energy costs, output capacities, and life expectancies. I have also written the introduction and several short paragraphs of discussion for my final recommendation.

    Visual Aids.   I have finished rough drafts of both of my tables. The first table summarizes extruder specifications, such as output capacity and power requirements. The second table shows ten-year total costs and total profit estimates on which to base a rebuild-versus-replace decision.

Work Scheduled

I have yet to decide on subdivisions for my final report; therefore, I have not yet completed an organized rough draft. I also need to develop a good format for my two tables so they will be clear and easy to read. I expect to complete my final report by December 1.

Problems

My first problem was that I could not find an actual example of a decision to rebuild or replace a piece of equipment. However, after looking at several books, I was able to design my own example of such a decision. Here are my remaining questions: (1) I know I need to explain the figures in my tables, but do I have to explain how these figures were obtained? (2) I have not decided on a table format for the Extruder Specifications Chart because many different specifications are being compared in the same chart, for example, costs, power usage, output capacity, and life expectancy. Do I need to label the columns and rows, or can I put the labels right after the units?

FIGURE 15.1
Progress Report

■ WRITING ASSIGNMENTS

1. Your instructor may require a progress report approximately midway through your writing of a longer project, such as a manual or a recommendation report. The progress report should be 350 words in length, aimed at a reader who is less knowledgeable than you. You can use as an example the model in this chapter (p. 327), which describes progress on a recommendation report. Notice the informal format and the writer's use of heads to clarify the report's organization.

2. Write a progress report on one of the following hypothetical situations. You may have to supply some data.

   a. The Upper Room Computer Consultants are installing a computerized inventory system in a branch of Grover's Fashions for Men, a retail store. Send a progress report to James Tanner, a vice-president of Grover's. The facts follow. The electrical work is finished in the store: several new lines and outlets have been installed. The Superneat computerized cash registers are back-ordered from Boe Supply; they will not be in for two weeks. The programmer has finished the inventory survey and has assigned codes to all brands and items of clothing. The training booklet for the clerks has been written; the manual for the system program will not be completed until after the programmer is finished. Training sessions will begin next week: Upper Room will install a computerized cash register for training that is operated in the same way as those that will be installed. The new bar code tags are in. Upper Room will hold a one-hour training session with clerks to explain how to use the tags.

   b. Bain Americana will sponsor the American Folk Craft Exposition in Dallas, Texas, April 1–4. Write a progress report from the Exposition Manager, Jerome Cokeby, to the Bain Board of Directors. Explain the work completed and the work remaining: selecting artists, creating publicity, and organizing equipment for the display.
      Of 27 weavers and 12 potters who applied to exhibit in the exposition, 15 weavers and 8 potters were selected. Each submitted three pieces of original work. Other applicants included 97 watercolorists, 20 lawn decorators, 191 trinket makers, and 76 food suppliers. Of these, 21 watercolorists, 6 lawn decorators, 11 trinket makers, and 15 food suppliers were accepted. A public relations firm — Mjork, Bjork, and Johnson — was hired in January. The firm has set up an entire campaign, including handbills (done in a Grandma Moses motif), newspaper ads, TV and radio spots, feature stories, and posters. The posters and handbills are distributed. The radio and TV ads will begin two weeks before the event and run three times a day on KTTZ TV and KOOO FM 101. The feature stories will focus on the methods used to create the pots and weavings, and on the authentic designs the artists are using.
      Each display artist must furnish a booth. Each artist must supply a drawing of the booth, complete with requirements for electricity and water. The exposition has received 26 descriptions so far. Each artist is assigned a space based on electricity/power needs. The committee rejected the designs for three booths because of inadequate safety or small size. These artists must resubmit a design or lose their $500 entrance fee.

16

Operator's Manuals

PLANNING THE MANUAL
WRITING THE MANUAL

Manuals are a common and extremely important type of technical writing. Companies sell not only the mechanisms they manufacture but also the knowledge to use those products properly. That knowledge takes the form of a manual. Both the manufacturer and the buyer want a manual that will ensure safe and successful assembly, operation, maintenance, and repair of the mechanism.

Extremely complex mechanisms have separate manuals for assembly, check-out, operation, and so on — and sometimes several volumes for each. The most common kind of manual, however, is the operator's manual which must be written for virtually all mechanisms.

Operator's manuals have two basic sections: descriptions of the functions of the parts, and sets of instructions for performing the machine's various processes. In addition, the manual also gives information concerning theory of operation, warranty, specifications, parts lists, and locations of dealers for advice or parts. This chapter will explain how to plan and write an operator's manual.

PLANNING THE MANUAL

Writing a manual requires careful planning so that you communicate information clearly and accurately, using clear prose, effective page design, and visual aids. To plan effectively, you must consider the audience, analyze procedures and parts, select visual aids, format the pages, arrange the sections, and prepare for the review process.

Consider the Audience

The audience for an operator's manual varies widely in knowledge and background. Some readers will be beginners with little or no technical knowledge; others will be experienced. The person learning his first word-processing program knows nothing about "save," "cut," "paste," "open," "close," and "print." Readers learning their fifth program, however, already understand basic word-processing concepts.

Often readers will use the manual in an emotionally charged situation: perhaps they have a deadline to meet and must fix the machine quickly. To help the audience, you must design the pages to allow easy access to the contents. Manual readers do not read the manual like a story — first page to last page. They select the section they need. As a result, you must use devices — such as heads and tables of contents — to make it easy to find information.

Analyze the Procedures and Parts

Analyzing the procedures includes two activities: discovering all the procedures and analyzing the steps in each procedure. Usually engineers (who have designed and built the machine) assist writers in this stage. While the following discussion focuses on machines, most of what is said transfers easily to software, a common topic for manuals. (I am indebted to Cohen and Cunningham for this section).

Discover Procedures   Discovering all the procedures requires that you learn the machine so thoroughly that you are expert enough to talk to an engineer about it. You must learn to operate it, disassemble it, and fix it. You must learn the name and function of each part. You must learn all the procedures the machine can perform and all the ways it performs them. These procedures may include

> how to assemble it
> how to start it
> how to stop it
> how to load it
> how it produces its end product
> how each part contributes to producing the end product
> how to adjust parts for effective performance
> how to change it to perform slightly different tasks

For example, the writer of a manual for a piston-filler, a machine that inserts liquids into bottles, must not only learn how to start and stop the machine but also grasp how the machine injects the liquid into the bottle. Gaining this knowledge requires observing the machine in action, interviewing engineers, and assembling and disassembling sections.

Analyze the Steps   To analyze the steps in each procedure requires the same methodology as writing a set of instructions (see Chapter 10). The writer determines both the end goal and the starting point of the procedure, and then provides the in-between steps to guide the users from start to finish. A helpful device for this analysis is a combination flow chart/ decision tree. Make a flow chart for the entire procedure, each step represented by a box, and then convert the chart to a decision tree.

Here is part of a larger sequence of steps taken from the piston-filler manual. The object of the sequence is to explain how to insert a specified amount of liquid into a bottle. Figure 16.1 below shows the flow chart; Figure 16.2 on page 332 shows the decision tree based on the flow chart.

Here is the text developed from the two figures:

1. Set the fill distance for the proper volume.
   a. Check specifications, p. 10, for bottle volumes.
   b. To determine this distance, find out the diameter of your piston.
   c. Go to volume chart on p. 11.
   d. Find the piston diameter in the left column.
   e. Read across to the volume you need.
   f. Read up to determine the length you need.
   g. Adjust the distance from *A* to *B* (Figure 6 [not shown]) to the length you need.
2. Add the product to the hopper.
   If you are unsure of the product type, see specifications, p. 11.
3. Press the left button (*A* on Figure 6 [not shown]) for single cycle.

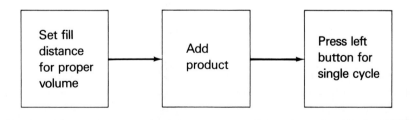

FIGURE 16.1
Flow Chart

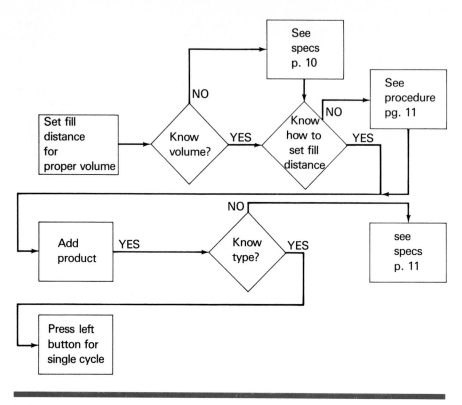

FIGURE 16.2
Decision Tree Based on Flow Chart

Analyze the Parts   To analyze the parts, list each important part and explain what it does. In general, you will set up a specific section where you explain each part. A VCR manual, for instance, will have a section that points out each part in the machine and describes its function. Here is a sample of a *part analysis*; in this case the part is a stop button:

The red emergency stop button immediately stops all functions of the machine.

Select Visual Aids

Manuals depend heavily on visual aids. Photographs, drawings, flow charts, decision trees, schematics, and troubleshooting charts are all used to help out the reader. As you write the instructions, you must decide whether or not each step needs a visual aid. If you write a software manual, use a screen dump to illustrate your text. (A *screen dump* is a picture of exactly what appears on the screen.) Many manuals have at least one visual aid per page. Recently, manual writers have started to use one visual aid per step.

For illustrating steps, many manual writers use photographs and drawings. When writers employ large numbers of photographs and drawings, they usually require assistance from photographers and artists. The manual writer must carefully plan the visual image needed to illustrate the step. He or she must decide which image to include and which angle to view it from. If the user will see the part from the front, then a picture of it from the rear would be useless. Manual writers often employ a storyboard, such as the one shown in Figure 16.3, to plan the visual aids.

Format the Pages

The manual writer must design the manual's pages to make them easy to read. Two basic concerns are laying out the page and choosing type sizes.

Laying Out the Page   Lay out the page so that it is easy to read, following the elements of formatting explained in Chapter 7. Two additional concepts will help here:

- Make a template.
- Use a grid.

A *template* is an arrangement of all the elements that will appear on each page, including page numbers, headers, footers, rules, and blocks of text. The page number should appear in an easy-to-find spot — usually the upper right corner. *Headers* and *footers* are the lines of type that appear on each page, top and bottom. For example, you may want to put the name of the process in the header and the name of the chapter in the footer. If

| 1. Pull out the stop switch (C). | Photo of top switch
Pulled out
3/4 view from front top |

FIGURE 16.3
Storyboard

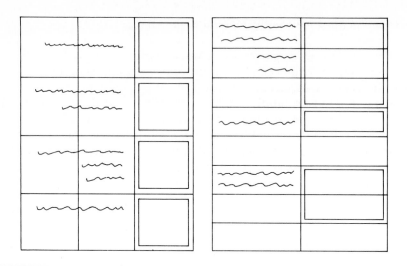

FIGURE 16.4
Page Grids

you use rules (thin lines drawn across the page), place a rule below a header and above a footer. Attention to these details will make a pleasing page.

A *grid* is a group of imaginary lines that divide a page into rectangles. Designers use a grid to ensure that similar elements appear on pages in the same relative position and proportion. One common grid divides the page into three imaginary columns and four rows. Writers place text to the left and visuals to the right. Figure 16.4 shows two sample page grids.

Other grids — and arrangements — are possible, as you will see if you compare several manuals. Two very good producers of manuals are Apple Computer and John Deere, Inc. A sample Apple page appears in Chapter 1 (p. 12); selections from a John Deere page appear in Chapter 10 (p. 208).

Choosing Type Size   Manuals generally use 10-point type in the text. Typeset or laser-printed manuals have head systems of larger point sizes, usually boldfaced. Boldfacing for heads is available on many word-processing systems; heads in typed manuals usually are underlined.

Designating Levels of Heads   As discussed in Chapter 7, you can designate head levels in two ways, numbered and open. The numbered organization uses headings and a rigid section-numbering system to iden-tify each section clearly:

```
1.0 SECTION
   1.1 COMPONENT
      1.1.1 Subpart
      1.1.2 Subpart
   1.2 COMPONENT
      1.2.1 Subpart
      1.2.2 Subpart
```

Many long manuals, especially those written for the military, use this method of organization.

The open organization uses headings of various sizes and various locations on the page to indicate divisions. In this method, common in consumer user manuals and in many computer manuals, the first level might be 18 point bold centered, the second, 14 point bold at the left margin. A boldface system might use the following levels:

FIRST-LEVEL HEAD   18 point bold

SECOND-LEVEL HEAD   14 point bold

THIRD-LEVEL HEAD   12 point bold

Fourth-Level Head   10 point bold

Arrange the Sections

You must decide on the order of the sections. The basic principles are these.

- Always describe a part, then tell how to use it.
- Always cross-reference; never assume the reader will have read an earlier section.

Basically, a manual has two major sections:

- description of the parts (for example, the function of all the buttons on the control panel)
- instructions for all the procedures (how to start the machine from the control panel)

The instruction section is usually much longer than the description section. The best order for the instruction section is chronological: use the order in which readers will encounter the machine. First, tell how to assemble, then how to check out, then how to start, perform various operations, and maintain.

Manuals traditionally have a number of other sections, although not all of them appear in all manuals or in the exact arrangement specified here. These sections are the front matter, the body, and the concluding section. The front matter and introduction may include

> Title page
> Table of contents
> Safety warnings
> A general description of the mechanism
> General information, based on estimated knowledge level of the audience
> Installation instructions

The body may include

> A theory of operation section
> The actual operation of the processes associated with the machine

The concluding section may include

> Maintenance procedures
> Troubleshooting suggestions
> A parts list
> The machine's specifications

Outline the Review Process

A *review process* is a plan for having other knowledgeable people check the accuracy of the contents and the acceptability of the design. The usual method is to write the text, then have another person or a group review it. In industrial situations, this person might be the engineer who designed the machine. If you write for a client, it will be the client or some group designated by the client. Usually this process produces numerous changes and revisions, which you must incorporate into the text. After the text review, you usually have a design review so that everyone concerned agrees on the layout of the pages. You should show several possible layouts so your reviewers can select one. At the outset of the project, set dates for each of these reviews, and decide who will be part of the review team.

WRITING THE MANUAL

The following student sample shows parts of an operator's manual for software designed to prepare payroll. The sections included here — the introduction and the operation instructions — are essential ones used in all manuals. The full manual included a title page and a table of contents. This manual uses an open-head system.

Introduction

An introduction explains how to use the manual, gives appropriate background, and defines the subject of the manual. If appropriate, it states the level of training needed to use the mechanism, for example, as pointed out below, what basic concepts users need to know. Introductions also provide any special information that will help familiarize readers with the equipment.

INTRODUCTION

How to Use This Manual

This manual has three parts:

- The Introduction gives an overview of the program
- "Using the Payroll Program" explains how to use the program's 12 procedures
- The Appendices show sample reports and related references.

Preview of sections used in manual

The following conventions are used in this manual:

- **Bold** type means to enter something
- Square brackets [ ] are used to identify a key that cannot be represented by a single character.

Explanation of symbols used in text

Example: Press **[enter]** to continue. This means you should press the [enter] key.

About the Payroll Program

The payroll program will help you manage your student payroll. Information about each student employee is stored in a data base. From this data base you can print payroll labels, time cards, a budget report, and an employee list.

Definition

How will the payroll program help you? It will allow you to keep an accurate and up-to-date record of your student employees. By entering the hours an employee works, you will be able to keep track of how much you are spending on student payroll. In short, you will have control over your student payroll.

Significance

The Support Services Division has been using an older version of this program since 1982. It has been rewritten so that each campus operation can use it to manage its own student payroll. Similar programs are also being used in the Maintenance Office and Leisure Center.

The following conventions are used in the program:

- The date is displayed in the upper left corner of the screen.
- Your location in the program is displayed in the upper right corner of the screen. For example, when you are at the Main Menu screen, "Main Menu" will be displayed.

Background

Explanation of display screens

What You Need to Know About the [enter] Key

The [enter] key has two functions:

- It completes an entry of data into a field. When an instruction tells you to enter data, you then need to complete the entry by striking the [enter] key. For example, "Enter the last name" means you should type in the last name and then strike the [enter] key.
- It tells the program your answer to a question. When an instruction tells you to press a key, you only need to strike that key. For example, "Press [enter] to continue" means you should strike the [enter] key if you want to continue.

Specific information about a key aspect of the program

Before You Use the Program

Before you use the program, you should

a. have a basic understanding of data bases, i.e., know what a record and field are
b. be familiar with an IBM PC, its keyboard, and how to use diskettes
c. know how to set up, start, and use your printer

Basic concepts that the user must understand

What You Need to Start the Program

To start and use the program you will need:

- an IBM PC or IBM-compatible with a printer
- the Startup Diskette
- the Files Diskette
- the Backup Diskette

Materials needed to use the program

Operation

The operation section explains each process that the user must follow. Directions for operation should be divided into short, numbered, easy-to-follow sequences. Within the directions, refer to components by their callout numbers to help the readers locate controls and gauges.

All manual writers must be aware of the dangers associated with running a machine. Keep in mind that if you leave out a step, the operator will probably not catch the error, and the result might be serious. Also, you must warn readers about potentially dangerous operations by inserting the word *warning* in capital letters and by providing a short explanation of the danger. These warnings should always appear *before* the actual instruction.

The following pages present one section from a small software program manual.

USING THE PAYROLL PROGRAM

This section will enable you to start the Payroll Program and use it from the Main Menu. To start the program:

Objective of the section for the user

1. Insert the Startup Diskette in drive A (the left drive).

2. Insert the Files Diskette in drive B (the right drive).

Actual instructions

3. Turn on the computer, monitor, and printer if the system is off. Press **[ctrl] [al]** simultaneously as soon as the system is on.

4. Wait for the computer to boot up and for the files to be loaded into memory.
   Note: The name of each file will be displayed on the screen. Do not press any keys until you are prompted at the bottom of the screen.

Special note for the user

5. Press **[K]** to assent to the License Agreement.

6. If you do not want to continue with the program, press **[ctrl] c**. You will end the program and return to the A: prompt.

7. Press **[K]** and you will be taken to the Main Menu (see Figure 1).

The Report Submenu

Pressing **G** on the Main Menu will take you to the Report Submenu (see Figure 2). From this menu, you can generate and print the budget report and the employee report, print the employee list, and enter the budget data.

From the Main Menu, you will be able to generate and print the budget report. Use 8 1/2-inch by 11-inch paper in the printer.

To generate and print the budget report:

1. Press **G** from the Main Menu for Generate and Print Reports.

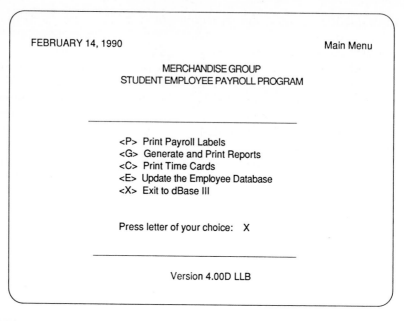

FIGURE 1
The Main Menu Screen

FIGURE 2
The Report Submenu: Generate and Print the Budget Report

2. Press **B** from the Report Submenu for Generate and Print the Budget Report.

3. Press **[enter]** to return to the Report Submenu if you do not want to continue. Note: After you start step 4, you cannot exit until the Budget Report has been printed.

4. Put 8 1/2-inch by 11-inch paper in the printer and align the print head with a page perforation.

5. Put the printer in 80-character draft mode by
   a. turning the printer off
   b. waiting ten seconds
   c. turning the printer back on
   Note: You can skip this step if you have already done it in another procedure.

6. Press **R** to print just the Budget Report. Press **R** to generate and print the Budget Report. Note: When you generate the Budget Report, data already entered in the Enter Hours for the Pay Period procedure will be compiled. You should only generate the report once per pay period.

7. Return to the Report Submenu after the budget report has been printed. Then press **X** to exit to the Main Menu from the Report Submenu.

Entering the Budget Data

From the Main Menu, you will be able to enter the budget data. The data are taken from the control sheet for student payroll.

To enter the budget data:

1. Press **G** from the Main Menu for Generate and Print Reports.

2. Press **E** from the Report Submenu for Enter the Budget Data.

3. Press **Y** to keep the year-to-date totals. Press **X** to reset the year-to-date totals to zero.

4. Press **[enter]** if the data are correct. You can correct an error by pressing **Z** and reentering the data.

5. You will return to the Report Submenu after pressing **[enter]**. Then press **X** to exit to the Main Menu from the Report Submenu.

▪ SUMMARY

This chapter explains how to plan and write a manual. The writer must choose his or her audience level, stating in the introduction any assumptions about the knowledge level of the user. When planning a manual, it is essential is that the writer thoroughly learn the processes and parts of the machine to be described. Helpful visual aids and pages designed for easy reading also must be planned early in the writing process. The general plan of a manual is (1) to describe the parts and their function, and (2) to explain the steps in the procedures. The writer also should provide an introduction (with background, warnings, and any special information a user must know) and a conclusion (with maintenance procedures, a troubleshooting list, and so forth).

■ WORKSHEET FOR PREPARING A MANUAL

☐ *Name the audience for this manual.*

☐ *How much do readers know about the general terms and concepts?*

☐ *What should they be able to do after reading the manual?*
Know basic terms and where to find help? Have a clear grasp of processes and be able to work independently?

☐ *On what date is the last version of the manual due?*

☐ *List the people who have to review each stage.*
How long will each review cycle take?

☐ *List where you can obtain the knowledge you need to write the manual.*
A person? Reading? Working with the mechanism?

☐ *List the major parts and their functions.*

☐ *List the processes a user must know to make the machine work.*

☐ *List the sequence for presenting the processes.*
Choose an organizational pattern for the sequence — chronological or most-to-least important.

☐ *Prepare a style sheet of up to four levels of heads, captions for visual aids, margins, and page numbers.*
Also select rules, headers, and footers as needed to help make information easy to find on pages.

☐ *Select heading and body typefaces if a word processor is available.*

☐ *Select an open or numbered heading system.*

☐ *Select a visual aids strategy.*
  Will you use drawings or photographs?
  Will you use a visual aid for each instruction?
  Will you use a visual aid on each page?
  Will you use callouts?

■ MODEL

The professional model that follows includes many sections, though not all, of an operator's manual for a piston filler.

TABLE OF CONTENTS

FIGURE 16.5
Operator's Manual

2

PISTON FILLING PRINCIPLE

In the pneumatic piston filling system, the amount of product discharged from the hopper into the container is controlled by the length of the piston stroke. Using the capacity chart in the appendix, you may establish the proper piston stroke length to discharge the appropriate amount of product.

Once the stroke length is set, each cycle of the piston will discharge the same amount of product into the container placed under the nozzle. On the intake stroke, the valve plug opens to the hopper and the proper amount of product is pulled into the cylinder. As the feed stroke begins, the valve plug rotates and closes off the hopper. The valve plug is then open to the fill tube. On the discharge stroke, the piston pushes the product through the cylinder and discharges it into the container.

HOW TO INSTALL YOUR PISTON FILLER

1. Remove the upper structure of the crate, leaving the unit bolted to the skids.
2. Examine the unit for any possible shipping damage, such as loose parts.
3. Check to be sure your kit includes a spanner wrench for cylinder removal, and a piston wrench (Fig. 1).

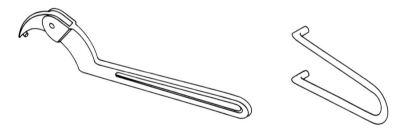

FIGURE 1
Spanner Wrench and Piston Wrench

4. If there is no damage, move the unit on the skids to your production area.
5. Connect the unit to your plant air supply.
6. Set the piston filler regulator at 40–60 psi (Fig. 2). Most products will use 40 psi. Set the regulator closer to 60 psi for products that have a heavier consistency.

FIGURE 16.5 (continued)

3

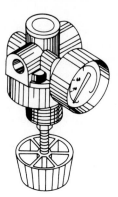

FIGURE 2
Piston Filler Regulator

SETTING THE CONTROLS

Your pneumatic piston filler will accurately discharge a fixed volume of flowable product into a container. You must adjust the valves for product discharge air flow control and the air cylinder retract stroke to set the speeds of the forward and retract strokes. You must also adjust the length of the piston stroke to control the amount of product discharged into the container.

Adjusting the Product Discharge Air Flow

1. Rotate the yellow valve plug (A on Fig. 3) to control the product discharge flow.

   a. For a faster product discharge, rotate the valve counterclockwise to a higher number, lining up the number on the valve with the red line.

   b. For a slower product discharge, rotate the valve clockwise to a lower number, lining up the number on the valve with the red line.

   c. This valve should be set at approximately the same value as the retract stroke (B on Fig. 3).

Adjusting the Air Cylinder Retract Stroke

1. Rotate the yellow valve plug (B on Fig. 3) to control the speed of the retract stroke.

   a. For a faster retract stroke, rotate the valve counterclockwise to a higher number, lining up the number on the valve with the red line.

FIGURE 16.5 (continued)

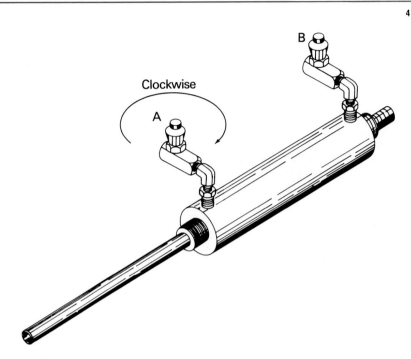

FIGURE 3
Setting the Controls

    b. For a slower retract stroke, rotate the valve clockwise to a lower number, lining up the number on the valve with the red line.

    c. This valve should be set at approximately the same value as the product discharge stroke valve (A on Fig. 3).

2. The other two yellow valves located closer to the hopper are factory set and do not need to be adjusted.

Setting the Piston Stroke Length

1. Using the capacity chart in the appendix, look down the left-hand column for the diameter of your cylinder.

2. Follow across the row to locate the number of ounces you will be using.

3. When you have found the correct number of ounces, look at the top of the column to find the stroke length.

4. Measure the stroke length from point A to point B as shown on Figure 4.

FIGURE 16.5 (continued)

5

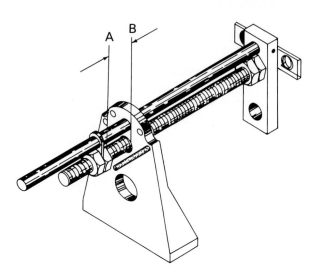

FIGURE 4
Measuring the Stroke Length

OPERATING PROCEDURES

Your pneumatic piston filler has two different cycle settings:

1. A single cycle, which will complete one cycle and then stop.
2. An automatic cycle, which will provide a continuous cycle until you stop it. In order to familiarize yourself with the two cycles, make a dry run with the two types of cycles before beginning production.

Single Cycle Operation

To make a dry run:

1. Set the selector switch (A on Fig. 5) in the downward position.
2. Pull out the red emergency stop button (B on Fig. 5).
3. Push in the green start button (C on Fig. 5). The machine will complete one cycle, filling one container, and then will stop automatically.
4. Observe the rate of the product discharge stroke and the retract stroke. Make any adjustments necessary to increase or decrease the speeds of the strokes.
5. Add your product to the hopper.

FIGURE 16.5 (continued)

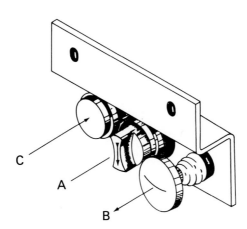

FIGURE 5
Single Cycle Operation

6. Place the container under the nozzle.
7. Push the green button and the machine will fill the container.

Automatic Cycle

To make a dry run:

1. Push in the red emergency stop button (A on Fig. 6).
2. Set the selector switch (B on Fig. 6) in the upward position.
3. Pull out the red emergency stop button. The machine will begin a continuous automatic cycle.
4. Observe the rate of the product discharge stroke and retract stroke. Make any necessary adjustments to increase or decrease the speed of the strokes.
5. To stop the machine immediately, push the red stop button in. You can also stop the machine by turning the selector switch to the downward position. This will cause the machine to complete the cycle and then stop.
6. Add your product to the hopper.
7. Place the container under the nozzle.
8. Pull out the red emergency stop button to begin the continuous cycle.
9. To stop the machine, push in the red emergency stop button or turn the selector switch to the downward position.

FIGURE 16.5 (continued)

7

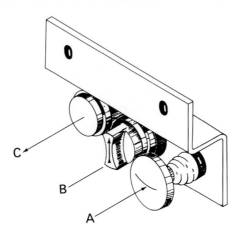

FIGURE 6
Automatic Cycle Operation

CLEANING

Your piston filler should be cleaned after production is completed, or if you are changing to a different type of product. The cleaning procedure is done in two phases:

1. Flushing out the system.
2. Cleaning the valve.

Flushing Out the System

1. Run the machine until the hopper is empty.
2. Fill the hopper with soapy water or appropriate cleaning solution.
3. Place a large container or bucket under the nozzle.
4. Set the machine for automatic cycle.
5. Continue to cycle the machine until the hopper is empty.
6. Stop the machine by pushing in the red emergency stop button.
7. Add clear rinse water to the hopper.
8. Start the automatic cycle and continue to cycle the machine until the water runs clear, adding more water if necessary.

FIGURE 16.5 (continued)

8

Cleaning the Valve

1. Disconnect the piston filler from the plant air supply.
2. Remove the piston:
   a. Remove the two hand knobs.
   b. Using the spanner wrench provided, loosen the piston cylinder.
   c. Grasp the air cylinder and slide it out to the right.
   d. Grasp the piston cylinder and slide it out to the right.
3. Remove the two hex nuts that hold the nozzle to the valve.
4. Remove the nozzle.
5. Clean the nozzle with soapy water. Rinse and wipe dry.
6. Disassemble the valve plug.
   a. Remove the pin arm. (A on Fig. 7)
   b. Remove the spring clip. (B on Fig. 7)
   c. Remove the four bolts that are holding the end plate. (C on Fig. 7)
7. Flush out all the parts with soapy water or appropriate cleaning solution. Rinse and wipe dry.

TABLE 1
Troubleshooting

Trouble	Probable Cause	Possible Solution
Uneven fill	Piston stroke changing	Positive stop nuts loose
		Piston and linkage worn
	Air leaking into cylinder	Piston seals worn out
		Packing gland worn out
		End cap loose
		Cylinder not screwed tight to valve block

FIGURE 16.5 (continued)

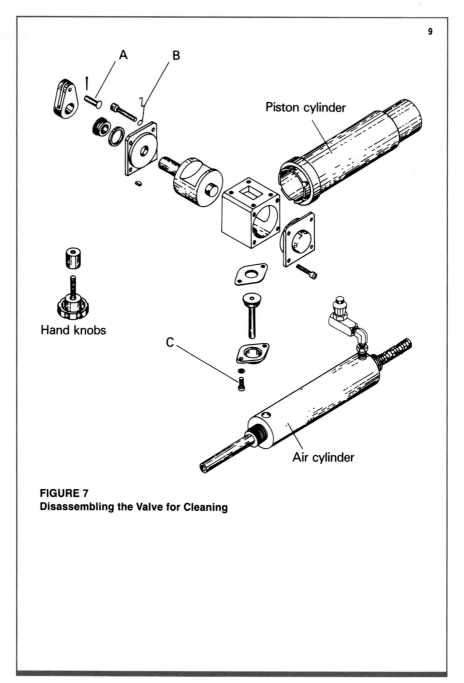

9

FIGURE 7
Disassembling the Valve for Cleaning

FIGURE 16.5 (continued)

■ EXERCISES

1. Examine two or three operator's manuals. Use all the same kind if possible
— sewing machines, VCRs, electrical generators, computers, oscilloscopes,
electronic microscopes, whatever. If you cannot find several of the same
type, use any three manuals. Analyze the manuals for page layout, use
of visual aids, number and arrangement of sections, and style. Depending
on your instructor's assignment, perform one of the following: (1) Write
a brief persuasive memo to convince your manager at work to accept
one of the formats. (2) Prepare a five-minute speech in which you explain
why one format among the two or three is best. Use visual aids to explain
your position.

2. Prepare a troubleshooting chart for a machine or process that you know
well.

■ WRITING ASSIGNMENT

1. Write an operator's manual. Choose any machine you know well — or
would like to learn about. The possibilities are numerous — a bicycle, a
sewing machine, part of a software program such as Wordstar or Lotus
1-2-3, a computer system, any laboratory device, a welding machine.
Because of the difficulty of the assignment, your instructor will advise
you on the type of visual aids to use. If you need to use high-quality
photographs or drawings, you may need assistance from another student
who has the necessary skills. Your manual must include at least an
introduction, a table of contents, a description of the parts, and the
instructions for procedures. You might also include a troubleshooting
section. Do not forget to give warnings when appropriate.

■ WORK CITED

Cohen, Gerald, and Donald H. Cunningham. *Creating Technical Manuals: A Step-by-Step Approach to Writing User-Friendly Manuals.* New York: McGraw, 1984.

17

Oral Reports

PLANNING THE ORAL REPORT
ORGANIZING THE ORAL REPORT
PRESENTING THE ORAL REPORT EFFECTIVELY

A s you advance in your career, your speaking skills become increasingly important. At meetings, you may be called on to explain the results of investigations, propose solutions to problems, report on the progress of projects, or justify your department's requests for more employees and equipment. Every kind of written report has its oral counterpart; sometimes an oral report supplements a written one, and often an oral presentation takes the place of a written report.

This chapter explains the audience, the visual aids, and the organization of an oral report, and concludes with guidelines for delivering the speech effectively.

PLANNING THE ORAL REPORT

Consider the Audience

Your oral report must engage your audience, who will have different levels of knowledge and emotional involvement with your topic. Your speeches will reach listeners more effectively if you understand a few essential differences — both positive and negative — between a reading and a listening audience.

Speakers Use Personal Contact   One of the advantages of a speech is that you have personal contact with your listeners. You can make use of personality, voice, and gestures, as well as first-person pronouns, visuals, and feedback from listeners. Use this personal contact to your advantage. Be a person speaking to people. Your audience will react positively.

Listeners Are Present for the Entire Oral Report   That listeners are present for the entire report may seem advantageous, but it also may make communication more difficult. Many listeners only want to hear selected parts of a report — the parts that apply directly to them. Let's assume that your listeners are the plant manager and her staff. The plant manager would probably prefer a capsule version of the report, which the abstract of a written report would provide, leaving the details for staff members to examine. An oral report, however, gives the manager no choice but to listen to all your detailed information — a situation that might put her in a negative frame of mind.

Even the manager's staff members might prefer a written report. With a written report, they can read the abstract and then use the table of contents to locate the financial, technical, personnel, or other sections critical to their work. However, all the staff members must listen to the entire oral report, and they might become restive as they hear all about sections that they have little interest in.

Listeners Have Only One Chance to Grasp Information   Even though a question-and-answer period may follow the report, listeners cannot study the information as a reader would study a formal report.

Listeners Cannot React to Formatted Pages   The oral report does not provide headings to identify sections of particular interest to the listeners and to indicate parallel and subordinate ideas. Instead, you have to provide oral cues or use visual aids to help an audience understand when one section ends and another begins.

Use Visual Aids

Visual aids can reinforce major points and clarify complex ideas in an oral report. As you construct an outline for your report, ask yourself whether a visual aid will help listeners grasp the point or the section, and then organize the report with all your visual aids in mind. Good visual aids are often the difference between an effective and an ineffective presentation. Research shows that visuals cause audiences to perceive the speaker as better prepared and more professional. Color graphics enhance the speaker's effectiveness even more (Meng 137–143). To use them advantageously, you should learn the kinds of visual aids available and how to select, use, create, and display them.

Kinds of Visual Aids   Appropriate visual aids for a speech include outlines; slides or drawings; tables, graphs, and charts; and handouts.

The basic *outline* shows listeners the sections and subsections of the report. This device orients the audience to the relationship among sections in the speech — what is a major section, what is a subsection — as well as to the sequence of sections. Remember, though, that outlines are boring to look at for any length of time.

Slides and *drawings* can introduce listeners to important images. Using good-quality slides, a speaker can present exact representations. With drawings, a speaker can illustrate procedures, such as the path of products through a sterilizing machine.

Tables, graphs, and *charts* can present data in a way that allows listeners to grasp relationships immediately. An oral explanation of the relationship among the percentages that affect a pay increase is hard to follow, but a table or graph will clarify the point. (See Chapter 7 for more on tables and graphs.)

A handout can replace or supplement projected visual aids. Often a handout of the outline is effective (use an outline report as explained in Chapter 11). You could also pass out copies of a key image, perhaps a table. Listeners can make notes on it as you speak.

Uses of Visual Aids   Decide how you will use visual aids. There are two basic options:

- to illustrate a point
- to begin a lengthy explanation

For instance, if the writer of the power scrubber report discussed earlier in Chapter 12 was giving an oral presentation and wanted to dramatize the effect of the scrubber, he might show before-and-after slides — a picture of a dirty floor and another of a clean floor. He would not discuss the details of the image but would just let the contrast make the point. However, if he wanted to familiarize the audience with the machine, he might project a photograph or drawing of it and then discuss each part in detail.

If a section of your speech contains a complicated explanation — of a process or a mechanism or an abstract relationship — a visual aid will always help listeners. Project the image first; then explain it in detail. This strategy is more effective than reading a long explanatory section from a paper and then showing the image.

Use a Storyboard to Choose Visual Aids Experienced speakers use storyboards to determine which visual aids they will use. A *storyboard* is simply a list of topics opposite a list of visual aids. To make a storyboard, follow these guidelines:

- Determine the major points of your presentation and list them down the left side of a sheet of paper.
- List the visual aids you plan to illustrate each point down the right side.

Here is an example of a storyboard.

Point	*Visual Aid*
Introduction	
Source of assignment	Outline of main topics
Preview	
Section 1	
Method of researching	List of main methods
Section 2	
Process of laminating	Flow chart of process
Machines used	Drawing of laminator
Section 3	
3 types of laminates	For each type:
Advantages of each	Cross-sectional view
	List of advantages (both on same page)
Section 4	
Cost	Table of costs
Section 5	
Recommendations	List of recommendations

Creating Computer Visuals Several computer programs, such as Powerpoint, Cricket Presents, and More, allow you to design visuals on a computer screen. Most of them allow you to start from an outline and use a storyboard. After you design your visuals, you can print them and then make overhead transparencies from them. Or you can duplicate them and hand them out as notes to your speech. You can also have slides made from them; more sophisticated computers will allow you to use color. These options are discussed on pages 358–359.

The following guidelines will help you design effective computer visuals (Scoville; Tessler). These guidelines apply equally well to handmade visual aids used in a presentation.

1. Know the parts of the visual — title, text or graphics, and border (see Figure 17.1).
   The *title* appears at the top, usually in the largest type size. Use it to name the contents of the visual clearly.
   The *text* makes the points you wish to highlight. Use phrases that convey specific content rather than generic topics.
   The *graphic* consists of a table, chart, or drawing.
   The *border* is a line that provides a frame around the visual.
2. Create a template or "master." Make all the visuals consistent, with the same elements in the same place and in the same color. (For instance, make all titles 24 point, black, centered at the top.)
3. Use only one main idea per visual.
4. For text visuals (visuals that use only words):
   Use no more than seven lines of text.
   Restrict each line to seven words or less.
   Use initial capitals followed by lower-case letters.
   Use 18-point type for body text, 24-point type for titles.
5. For graphic visuals (tables, charts, pictures):
   Simplify the chart so that it makes only one point.
   Use charts for dramatic effect. A line graph that plunges sharply at one point calls attention to the drop. (Your job is to interpret it.)

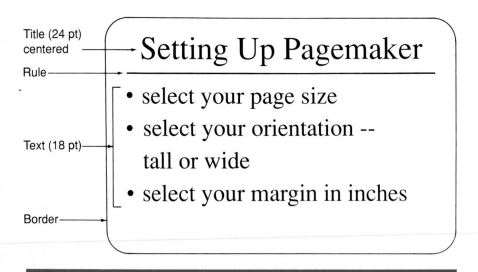

FIGURE 17.1
Parts of a Visual

Use tables for presenting numbers. (Be prepared to point out the numbers you want the audience to notice.)

Use pictures to illustrate an object that you want to discuss (for instance, the control panel of a new machine).

6. Use color intelligently.

Give each item in the template its own color.

Use a background color; blue is commonly used.

Use contrasting colors — white or yellow text on green or blue background.

Use red sparingly; it focuses attention on itself. Long passages in red are hard to read.

Avoid hard-to-read color combinations, such as yellow on white, or black on blue; violet can be very hard to read.

Methods of Displaying Visual Aids Whether you use a computer or make visuals by hand, you must have some method to display them. Three common methods are overhead transparencies, slides, and flip charts.

An *overhead transparency* is a clear sheet of plastic that carries an image. An overhead projector transmits the image to a screen. These transparencies are simple to make and easy to handle during a presentation. They can be framed with a commercially available cardboard holder, much like a giant slide mount. The cardboard frame makes them easy to handle; you can also use the frame to record key information, such as the topic and the sequence number. Overhead projections can be used in normally lighted rooms and thus allow you to maintain eye contact with your audience.

Slides are an effective medium, allowing you to add color to your presentation, but they are difficult to make. Either you send your camera-ready copy to a special agency that will convert them, or you make them yourself using a copy stand. The computer programs mentioned above have made slide presentations more common. Two problems with using slides are that you lose eye contact because you darken the room, and you increase the formality of the presentation, thus reducing personal contact.

Flip charts are large pads of paper on an easel. The speaker discusses the image on each sheet, then flips it over when it is no longer needed. The speaker draws the image (or prints the words) during or before the speech. This method eliminates electronic equipment with its possible failures, but the act of drawing can distract the audience.

Make sure before your presentation that the room has the equipment you need. Many experienced speakers bring a backup visual aid to use if the first one should fail. They will, for example, have handouts to supply if a projector bulb burns out during a key visual.

ORGANIZING THE ORAL REPORT

Effective oral reports contain an introduction, body, and conclusion. The audience should recognize each of these sections.

Develop the Introduction

The introduction establishes both the tone and the topic of the speech. Your *tone* is your attitude toward the listeners and the subject matter. You need to be serious but not deadly dull. And avoid being so intense that no one can laugh or so flip that the topic seems insignificant.

To establish a topic, you need to introduce it succinctly. Be explicit about your purpose. Follow these guidelines:

- In industry or business, you do not need to begin your report with a humorous story, a quotation by an authority, or an anecdote.
- Capitalize on your listeners' initial attention by saying something closely related to your topic in terms of time, space, money, equipment, personnel, or policy.
- Explain how your report is important to your audience.
- Present your conclusions or recommendations right away. Then the audience will have a viewpoint from which to interpret your data as you present it.
- Link yourself to the audience by explaining how you assembled your report.
- Indicate your special knowledge of or concern with the subject.
- Identify the situation that required you to prepare the report (or the person who requested it).
- Preview the main points so your listeners can understand the order in which you will present your ideas.

The order of your ideas should be appropriate to the subject. Most subjects can be structured into chronological order, problem-to-solution order, least-to-most-important order, or spatial order, such as from outside to inside or from north to south. If your preview is clear, the listeners can follow and understand the major points in the report.

Develop the Body

Time is to an oral report as space is to a written one. An oral report, however, does not lend itself to the concise presentation good writers achieve in a written report. To communicate an idea verbally, a speaker must state a generalization, provide details to support it, and reinforce it with a summary. Numerous studies have shown that listeners simply do not hear everything the speaker says, and if they miss an idea or an important detail, they have no recourse. Therefore, the verbal communication process should consume several minutes per main idea — long enough for the speaker to get each main point across.

Use Transitions Liberally   Clear transitions are very helpful to audience members. By your transitions, you can remind them of the report's structure, which you have established in the preview. Indicate how the next main idea fits into the overall report and why it is important to know about it. For instance, a proposal may seem very costly until the shortness of the payback period is emphasized.

Select Important Details   Although providing extensive details to support main ideas is not possible in the time permitted for an oral report, you should select enough significant details to make the point valid. Choose details that are especially meaningful to the audience. Explain any anticipated changes in equipment, staff, or policy, and how these changes will be beneficial.

Impose a Time Limit   Always impose your own time limit on the report, and narrow your number of main ideas accordingly. It is much better to present two or three main ideas carefully than to attempt to communicate more information than your listeners can grasp. If you select only the most important ideas, your speech will be concise enough to please the plant manager and detailed enough to satisfy her staff members.

Develop a Conclusion

The conclusion section restates the main ideas presented in the body of the report. Follow these guidelines:

- As you conclude your report, you should actually say "In conclusion . . ." to capture your listeners' interest.
- For a proposal, stress the main advantages of your ideas and urge your listeners to take specific action.
- For a recommendation report, emphasize the most significant data presented for each criterion and clearly present your recommendations.
- Use a visual to summarize the important data.
- End the report by asking if your listeners have any questions.

PRESENTING THE ORAL REPORT EFFECTIVELY

The best oral report is extemporaneous rather than read or memorized. An extemporaneous report, however, is not a spontaneous, off-the-top-of-your-head presentation. Rather, an extemporaneous report follows a prepared, clear outline, with the speaker supplying appropriate detail and explanation as needed. In fact, an extemporaneous report is carefully rehearsed and delivered.

Rehearse Your Presentation

To be successful, you should rehearse your extemporaneous report. During rehearsals, go straight through the speech, using note cards. At least once, wear the same clothes you will use in the actual presentation. Use the outline you have prepared on the note cards (5- by 8-inch cards are preferred) only as a reference and a reminder. Do not clutch the note cards with both hands, but hold them graciously with one hand. This will allow you to be more free and expressive with your hands during the presentation.

At each trial run, attempt to give the presentation a conversational quality, and practice using your voice and gestures to emphasize important points. You have rehearsed enough when you feel secure with your report, but always stop short of memorization. If you do not, you will ultimately grope for memorized words rather than concentrating on the listeners and letting the words flow.

For reports to large groups, final rehearsals should simulate conditions under which you will make the speech. Use a room of approximately the same size, with the same type of equipment for projecting your voice and your visuals. Rehearsals of this type not only guard against technical problems but allow you to become comfortable in an environment similar to the one for the final report.

Carefully arrange visual aids in the correct order and decide what you will do with them as you finish with them. If a listener asks you to return to a visual, you want to be able to find it easily. If you are using handouts, decide whether to distribute them before or during the presentation. Distributing them before the presentation eliminates the need to interrupt your flow of thought later, but since the listeners will flip through the handouts, they may be distracted as you start. Distributing them during the presentation causes an interruption, but listeners will focus immediately on the visual.

After you have rehearsed privately several times, ask a colleague to attend at least one rehearsal, to comment on how well he or she can hear you and see the visuals, and to offer a critique of the speech, including any possibly distracting mannerisms. Finally, record the report on an audio cassette recorder and listen yourself for verbal mannerisms or lack of clarity.

Deliver Your Presentation

A well-prepared presentation is pleasant to give and pleasant to attend. After all your preparation, you should give a strong report. You will increase your effectiveness, however, if you use notes, adopt a comfortable extemporaneous style, and overcome stage fright.

Use Notes Experienced speakers have found that outlines prepared on a few large note cards (one side only) are easier to handle than outlines on many small note cards. Some speakers even prefer outlines on one or two sheets of standard paper, mounted on light cardboard for easier handling.

The outline should contain clear main headings and subheadings. Make sure your outline has plenty of white space so you can keep track of your place.

Adopt a Comfortable Style The extemporaneous method results in natural, conversational delivery and concentration on the audience. Using this method, you can direct your attention to the listeners, referring to the outline only to jog your memory and to ensure that ideas are presented in the proper order. Smile; take time to actually look at individual people and to collect your thoughts. Instead of rushing to your next main point, check to see if members of the audience understood your last point. Your word choice may occasionally suffer when you speak extemporaneously, but reports delivered in this way still communicate better than those memorized or read.

Overcome Stage Fright

Stage fright is natural in a speaking situation. The best way to conquer it is to prepare carefully and to know your subject. If you are the best-informed person in the room, you can stop worrying about your subject and concentrate on communicating.

The following suggestions will help you as you face your listeners and deliver the speech.

1. Begin positively. Smile and mention that you are glad to be present. An audience is usually more willing to accept your information if you speak positively.

2. Make sure you can be heard, but try to speak conversationally. You should be able to feel a sense of round, full voice in your rib cage. You should also feel that your voice fills the space of the room, with the sound of your voice bouncing back slightly to your own ears. If you are enthused, your voice will have variety and emphasis that will add to clarity. The listeners should get the impression that you are just talking to them rather than that you are presenting a report. Inexperienced speakers very often talk too rapidly.

3. If the situation calls for you to use a microphone, practice with one beforehand. Fumbling with the microphone will distract your audience. Just speak naturally into the microphone; it will do the rest for you.

4. Some projection equipment runs loudly. Be certain that you can speak over the hum of the motor.

5. Look directly at each listener at least once during the report. With experience, you will be able to tell by your listeners' faces whether you are communicating. If they seem puzzled or inattentive, be prepared to adapt by repeating the main idea, by giving additional examples for clarity, or by asking for questions.

6. Fight the tendency to use your outline when you do not need it. When collecting your thoughts, do not say "uh"; instead, pause and remain silent. Remaining silent requires fortitude. Smile. Look at your audience. Think over the last idea presented. Chances are that the transition to your next idea will occur to you easily and quickly. What seems like a very long pause may actually be only a few seconds.

7. Try to learn — and stop — your distracting mannerisms. No one wants to see speakers brush their hair, scratch their arms, rock back and forth on the balls of their feet, or smack their lips. If the mannerism is pronounced enough, it may be all the audience will remember. Stand firmly on both feet without slumping or swaying. During transitions, take a moment to make sure that your upper body, shoulders, neck, and face are relaxed and comfortable before you continue with your report.

8. Learn how to use visual aid equipment. Speakers who fumble over the equipment and apologize about it lose their credibility.

9. When you are finished with a visual, remove it so that it does not compete with you. If you are using a pointer, set it down to avoid tapping with it. If you are using an overhead projector, cover the lighted glass with a piece of paper.

10. To point out some aspect of a visual projected by an overhead projector, lay a pencil or an arrow made of paper on the appropriate spot of the transparency. Do not point with your finger. Some speakers are

very effective using a pointer such as a yardstick directly on the screen.

11. When answering questions, make sure everyone understands the question before you begin to answer. Have reference materials ready for instant access if required. If you cannot answer a question during the question-and-answer session, say so and assure the questioner that you will find the answer. Thank the audience for their questions and interest in the report.

■ SUMMARY

An effective speech begins with a careful analysis of the audience; keep in mind important points such as audiences only hear the speech once. Speakers should use visual aids to help listeners grasp the main points of the speech. An effective device for preparing a speech is a storyboard, on which the speaker lists main points and the visual that will be used to support each point. Every speech has an introduction, body, and conclusion. The introduction sets the tone and previews the speech. The body has clear transitions and important details, and is presented in a set time frame.

To give a good speech, rehearse several times until you can speak comfortably from note cards. Try to speak in a conversational manner as if talking extemporaneously to friends. The secret to conquering stage fright is to be well informed and well prepared.

■ WORKSHEET FOR PREPARING AN ORAL REPORT

☐ *Identify your audience.*

☐ *What is your listeners' knowledge level of the topic?*

☐ *What is their interest level in the entire speech?*

☐ *What is your main point?*

☐ *What are your main subpoints?*

☐ *What visual aid will illustrate each main subpoint most effectively?*

☐ *Will you need any kind of equipment?*

☐ *If so, do you know how to use it?*

☐ *If you have visual aids, will you have enough room at the speaking location to have at least two piles: "to use" and "already used?"*

☐ *How will you point to the screen?*

☐ *Have you prepared clearly written note cards — with just a few points on each?*

☐ *Have you rehearsed the speech several times, including how you will actually handle the visual aids?*

■ EXERCISE

1. Examine at least two indexes or abstract services that list periodicals in your major or area of interest. Prepare a brief oral report on them. Use at least two visual aids (make one of them an overhead transparency — you should learn how to make and use transparencies). Discuss topics such as the number of periodicals the index or abstract presents and the ease and method of using it. You might want to bring in a photocopy of a page from a periodical and demonstrate how you found it, starting with the index and then explaining how you found the journal itself.

■ SPEAKING ASSIGNMENT

Your instructor may require an oral presentation of a formal report you have written during the term. The speech should be extemporaneous and approximately ten minutes long. To prepare for your oral presentation, follow the suggestions in this chapter for converting a written report into a successful speech. Make your visuals, outline the speech, and most important, rehearse. A question-and-answer session with other members of the class should follow your presentation.

■ WORKS CITED

Holmes, Nigel. "Get Smart about Charts." *Publish!* 4.3 (1989): 42–45.

Meng, Brita. "Get to the Point." *Macworld* 5.4 (1988): 136–143.

Scoville, Richard. "Slide Rules." *Publish!* 4.3 (1989): 51–53.

Tessler, Franklin. "Step-by-Step Slides." *Macworld* 5.12 (1988): 148–153.

Professional Communication

18

Letters

THE THREE BASIC LETTER FORMATS

ELEMENTS OF A LETTER

PLANNING BUSINESS LETTERS

TYPES OF BUSINESS LETTERS

B usiness letters are an important, even critical, part of a professional's job. They are written for many reasons, in many situations, to many audiences. They may be written to an expert to request information, to a client to transmit report, or to a supplier to discuss the specifications of a project. Letters represent the firm, and the quality of the letter signifies the quality of the firm. Since they are so important, you must handle them well. This chapter introduces you to effective, professional letter writing by explaining the common formats for business letters, the standard elements of letters, and several common types of business letters.

A good professional letter has a crisp, precise format that impresses the reader.

THE THREE BASIC LETTER FORMATS

The three most common formats are block, modified block, and simplified.

Block Format

In the *block* format, place all the letter's elements flush against the left-hand margin. Do not indent the first word of each paragraph. The full block

1427 Seventh Avenue
Grand Forks, ND 58201
  (skip 2 or 3 lines)

February 14, 19XX
  (skip 2 or 3 lines; varies with letter length)

Paul Schmidt
President XYZ Corporation
1427 Magnolia Street
Bloomsville, WI 54751
  (double-space)
Dear Mr. Schmidt:
  (double-space)
SUBJECT: ABC CONTRACT *(optional)*
  (double-space)
  XXXXXXXXXXXXXXXXXXXXXXXXXXXXXXXXXXXXXXXXXXXXXXXXXXXX
  XXXXXXXXXXXXXXXXXXXXXXXXXXXXXXXXXXXXXXXXXXXXXXXXXXXX
  XXXXXXXXXXXXXXXXXXXXXXXXXXXXXXXXXXXXXXXXXXXXXXXXXXXX
  (double-space between paragraphs)
  XXXXXXXXXXXXXXXXXXXXXXXXXXXXXXXXXXXXXXXXXXXXXXXXXXXX
  XXXXXXXXXXXXXXXXXXXXXXXXXXXXXXXXXXXXXXXXXXXXXXXXXXXX
  XXXXXXXXXXXXXXXXXXXXXXXXXXXXXXXXXXXXXXXXXXXXXXXXXXXX
  (double-space)
Sincerely yours,
  (skip 4 lines)

John K. Palmer

John K. Palmer
Treasurer
  (double-space)
abv *(typist's initials)*
enc: (2)
  (skip 1 or 2 lines; depends on letter length)
cc: Ms. Louise Black

FIGURE 18.1
Full Block Format

format, shown in Figure 18.1, may seem unbalanced, but it is widely used because it can be typed quickly.

Modified Block Format

The *modified block* format, shown in Figure 18.2, is the same as the full block with two exceptions: the date line and closing signature are placed on the right side of the page. The best position for both is five spaces to the right of the center line, but flush right is acceptable. A variation of this format is the *modified semiblock.* It is exactly the same as the modified block except that the first line of each paragraph is indented five spaces.

Simplified Format

The *simplified* format (Figure 18.3) is a recent development that departs from conventional letter formats. Its streamlined form contains no salutation and no complimentary close, but it almost always includes a subject line. It is extremely useful for impersonal situations, or for situations where the identity of the recipient is not known. Because it requires less typing and so is faster to produce, it is gaining popularity. In personal situations, writers start the first paragraph with the recipient's name.

(placement of date: 5 spaces to right of center; skip 3 lines below letterhead)

                                                      February 14, 19XX

(skip 2 to 6 lines; depends on length of letter)
L. Hanson
Personnel Director
On-Line Systems, Inc.
4444 Computer Drive
Saginaw, MI 48600
    (double-space)
Dear L. Hanson:
    (double-space)
Subject: Application for Printing Engineer Position (optional)
    (double-space)
XXXXXXXXXXXXXXXXXXXXXXXXXXXXXXXXXXXXXXXXXXXXXXXXXXXXXXXXX
XXXXXXXXXXXXXXXXXXXXXXXXXXXXXXXXXXXXXXXXXXXXXXXXXXXXXXXXX
XXXXXXXXXXXXXXXXXXXXXXXXXXXXXXXXXXXXXXXXXX
    (double-space)
XXXXXXXXXXXXXXXXXXXXXXXXXXXXXXXXXXXXXXXXXXXXXXXXXXXXXXXXX
XXXXXXXXXXXXXXXXXXXXXXXXXXXXXXXXXXXXXXXXXXXXXXXXXXXXXXXXX
XXXXXXXXXXXXXXXXXXXXXXXXXXXXXXXXXXXXXXXXXX
    (double-space)

                                          Sincerely,
                                          (skip 4 lines)

                                          Robert Schuler

(signature placed 5 spaces to right of   Robert Schuler
center)   *(double-space)*
  Enclosures: (1)
      (double-space)
  cc: M. Lang

FIGURE 18.2
Modified Block Format

(skip 2 to 6 lines)

February 2, 19XX
   (skip 2 to 3 lines)

Personnel Department
Western Fruit & Produce Co.
P.O. Box 14236
St. Paul, MN 55164
   (double-space)
Subject: Application for Managerial Opening
   (double-space)
(Recipients's name if known), XXXXXXXXXXXXXXXXXXXXXXXXXXXXXXXXXXXX
XXXXXXXXXXXXXXXXXXXXXXXXXXXXXXXXXXXXXXXXXXXXXXXXXXXXXXXXXXXXXX
   (double-space)
XXXXXXXXXXXXXXXXXXXXXXXXXXXXXXXXXXXXXXXXXXXXXXXXXXXXXXXXXXXXXX
XXXXXXXXXXXXXXXXXXXXXXXXXXXXXXXXXXXXXXXXXXXXXXXXXXXXXXXXXXXXXX
XXXXXXXXXXXXXXXXXXXXXXXXXXXXXXXXXXXXXXXXXXXXXXXXXXXXXXXXXXXXXX
XXXXXXXXXXXX.
   (skip 5 lines)

Mary Hyman
Mary Hyman
   (double-space)
enc: résumé

FIGURE 18.3
Simplified Format

ELEMENTS OF A LETTER

As a professional, you should know the standard ways to handle each element of a letter. This section describes the elements from the top to the bottom of a letter.

Heading   In personal letters, include both your address and the date, positioned according to the requirements of the format you have chosen. Use these guidelines:

- Spell out words such as *Avenue, Street, East, North,* and *Apartment* (but, use *Apt.* if the line is too long).
- Put an apartment number to the right of the street address. If, however, the street address is too long, put an apartment number on the next line.
- Spell out numbered street names up to Twelfth.
- To avoid confusion, put a hyphen between the house and street number (1021–14th Street).
- Either spell out the full name of the state or use the U.S. Postal Service zip code abbreviation. If you use the zip code abbreviation, note that the state abbreviation has two capital letters and no periods, and the zip code number follows two spaces after the state. For example:

4217 East Eleventh Avenue
Apartment 3
Austin, TX 78701

Date   Dates can have one of two forms: May 1, 19XX, or 1 May 19XX. In American correspondence, the former prevails. In Europe and the American military, the latter is used more frequently. Use these guidelines:

- Spell out the month.
- Do not use ordinal indicators, such as *1st* or *3rd*.

Inside Address   Readers are sensitive about their names, titles, and firms, so the inside address requires special care. Use these guidelines:

- Make sure that you use the correct personal title (Mr., Ms., Dr., Professor) and business title (Director, Manager, Treasurer).
- Write the firm's name exactly, adhering to its practice of abbreviating or spelling out such words as *Company* and *Corporation*.

- Place the reader's business title after his or her name or on a line by itself, whichever best balances the inside address.
- Use the title *Ms.* for a woman, unless you know that she prefers to be addressed in another way.

Ms. Susan Wardell
Director of Planning
Acme Bolt and Fastener
23201 Johnson Avenue
Arlington, AZ 85322

Attention Line   Attention lines are generally used only when you cannot name the reader ("Attention Personnel Manager"; "Attention Payroll Department"). Use these guidelines:

- Place the line two spaces below the inside address.
- Place the word *Attention* against the left margin. Do *not* follow it by a colon.

Salutation   The salutation always agrees with the first line of the inside address. A colon always follows the salutation. Use these guidelines:

- If the first line names an individual (Ms. Ann Burdick), say "Dear Ms. Burdick:"
- If the first line names a company (Dougherty Contracting), repeat the name of the company ("Dear Dougherty Contracting:" or just "Dougherty Contracting:"), or use the simplified format with a subject line.
- If the first line names an office, address the office, use an attention line, or use a subject line.

Personnel Director
Firari & Firari, Accountants
1535 Goodrich Avenue
Lewiston, ME 04240

Dear Personnel Director: (*or*)
Attention Personnel Director (*or*)
Subject: Application for Finance Analyst

- If you know only the first initial of the recipient, write "Dear B. Smith" or else use an attention line.

- If a job advertisement, for example, lists only a post office box, use a subject line.

Box 4721 ML
The Daily Planet
Gillette, WY 82716

Subject: APPLICATION FOR OIL RIG MANAGER

Subject Line   Subject lines are common in business letters. Use these guidelines:

- Follow the word *Subject* with a colon.
- For emphasis, you may either completely capitalize or underline the subject.

Body   Single-space the body. Generally try to balance the body on the page. It should cover the page's imaginary middle line (located 5 1/2 inches from the top and bottom of the page). Use several short paragraphs rather than one long one. Use 1-inch margins in the right and left.

Complimentary Close and Signature   Use simple closings, such as "Sincerely" or "Sincerely yours," to end business letters. Use these guidelines:

- Capitalize only the first word of the line.
- Place a comma after the close (or whatever company policy specifies).
- Place the company's name immediately below the complimentary close (if necessary).
- Allow space for the handwritten signature.
- Place the writer's title or department, or both, below his or her typed name.

Examples:

Sincerely yours,   Sincerely,
DAVIS MANUFACTURING CO.

Ronald C. Purvis   *William S. Yale*

Ronald C. Purvis   William S. Yale
Personnel Director   Manager, Drafting Division

Optional Lines   A number of optional lines provide notations below the typed signature.

- Place the typist's initials in lower-case letters, flush left.
- Add an enclosure line if the envelope contains additional material. The line may start either with " Enclosure:" or the abbreviation "enc:" Place the name of the enclosure (résumé, bid contract) after the colon, or put the number of enclosures in parentheses — enc: (2).
- If copies are sent to other people, place "cc:" (for carbon copy) at the left margin and place the names to the right. Note: some authors use "c" or "copy to" when they do not actually make a carbon copy.

bvm	typist's initials
enc: résumé	enclosure noted
cc: Joanne Koehler	copies noted

Succeeding Pages   For succeeding pages of a letter, place the name of the reader, the page number, and the date in a heading:

Ms. Burdick	-2-	May 15, 19XX

Envelopes

The standard business envelope is 9 1/2 by 4 3/16 inches. Place the stamp in the upper right-hand corner. Place your address, the same one that you used in the letter, in the upper left-hand corner.

Traditionally, the recipient's address has been placed at the top left of the lower right quadrant of the envelope, as shown in Figure 18.4. However, the U.S. Postal Service now uses optical character recognition (OCR) machines, which have a "read area" (see Fig. 18. 5). As a result, you may place the address anywhere in the read area. The Postal Service recommends that you place apartment numbers to the right of the street address or, if it is too long, on the line below. If for some reason the zip code number cannot be placed to the right of the state, place it on the next line at the left margin. The Postal Service prefers but does not require that you type the address in all capital letters.

Bar coding is now used by many businesses. If an envelope has a bar code, no other printing should appear in the bar code read area. Figure 18.5 shows the Postal Service's recommended method for formatting an envelope for OCR.

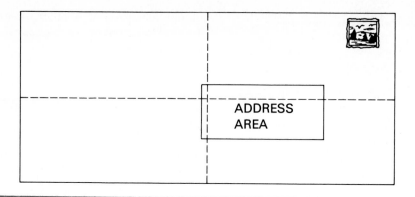

FIGURE 18.4
Traditional Placement of Recipient's Address

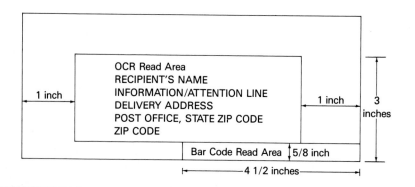

FIGURE 18.5
OCR Area of Envelope

PLANNING BUSINESS LETTERS

Business letters require clarity and tact. You must make a good impression, writing clearly without being pushy. To do so, you must approach your audience with a personal tone — treating them as people, not machines.

Approaching Your Audience

Style   In business letters, you want to sound natural; you are, after all, one human being addressing another. You can use several stylistic strategies to achieve a natural sound. First, use the first-person *I* and *we*, and refer to the reader in the second-person *you*. These pronouns will give your letters the natural, conversational style needed for personal communication. Second, write in plain English. The "businessese" style of writing sounds like a poorly programmed robot rather than a person. Consider this brief passage:

> Pursuant to our discussion of February 3 in reference to the L-19 transistor, please be advised that we are not presently in receipt of the above-mentioned item but expect to have it in stock within one week. Enclosed herewith please find a brochure regarding said transistor as per your request.

Such stilted, awkward prose is so common in business correspondence that many young writers think they are supposed to write that way. Here is the paragraph rewritten in a more direct style:

> I've enclosed a brochure on the L-19 transistor we talked about on February 3. Our shipment of L-19's should arrive within a week.

The new phrasing is more conversational, and it makes the contents much easier to grasp.

The "You" Approach   The "you" approach is a matter of tone — your attitude toward your subject and your reader. When you write a letter, the personality you reflect is often as much the message as the contents. The wrong personality can obscure the message. The basic way to set a tone that puts the reader at ease is to adopt the "you" approach.

The "you" approach requires only a common-sense awareness of human nature. Be professional, not emotional. Talk to the other person as you would like to be talked to. Talk person to person, using "I," "we," and "you."

Consider the difference in the following two examples, both of which start a letter intended to explain why a manual did not accompany a machine.

> There was a question asked by you in regard to the complete fulfillment of contract 108XB (Manual Effector Arm Robot A). Complete documentation of same has not been fulfilled. The specifications are interpreted by this office to mean that no such documentation was required.

> I'm writing in response to your phone call about the technical manual, which you understood was to accompany your new robot. We did not feel that the contract called for a manual. Since this is an important concern, we would like to explain our actions.

Which one would you rather receive? Probably the second. It sounds like one human being talking to another.

To look at it another way, the "you" approach requires an analysis of each reader's point of view, which is invariably different from yours. If you have to respond to a complaint, for instance, that some equipment you sold was defective, try to show an understanding of the customer's position. Here is the body of a "you" approach letter to a person with a complaint. Notice how the writer deals with the emotions in the situation.

Dear Mr. Franklin:

Let me start by apologizing for the shredded conveyor belt. I realize from our earlier (and happier) conversation how important that conveyor system is to your operation. I have already taken steps to fix the problem.

First, I have shipped you a new belt. It left here today, so I hope it arrives before this letter does. Since the defective belt is under warranty, the new belt is free.

Second, I reviewed your problem with our design engineer. She feels that the belt exactly fills your specifications, and so the fault must be in the metal limbs that join the rubber lengths. As your employees install the new belt, would you have them check the pins that attach the limbs to the rubber? The limbs should not "wobble" on the pin. If they do, please call me immediately.

Third, our sales representative will inspect the part on Friday, June 19. If you have other concerns, he will be glad to answer them.

Again, let me apologize for the inconvenience. Your business is important to us.

Sincerely,

Clement Williams

Clement Williams
District Sales Manager

TYPES OF BUSINESS LETTERS

The rest of this chapter explains several types of business letters and suggests how to structure their contents.

Letters of Inquiry

Letters of *inquiry* request information from another company. Simply identify the information you need in a one- or two-paragraph letter.

As a student, you may occasionally write a letter of inquiry requesting that a company send you information for use in a class project or report.

Since firms receive many requests of this type, they appreciate a courteous but concise letter. State specifically the information you need, why you need it, and why you selected the particular firm as a source of information. Figure 18.6 is a letter of inquiry from a student requesting information for a research project.

Transmittal Letters

A *transmittal* letter conveys a report from one firm to another. Begin the letter by identifying the report enclosed and stating the date it was requested. Then provide a brief paragraph or two explaining the report's purpose and scope. Close the letter by indicating your availability if the reader has any questions about the report. Figure 18.7 is a transmittal letter. (A transmittal memo is explained in Chapter 11.)

Specification Change Letters

During the course of many projects, the original specifications must change. *Specification change* letters indicate whether or not a company accepts the change that the other company proposed. These letters are short and to the point. Notice, however, in the following example, Figure 18.8, that the writer uses the informal context-setting introduction to orient his reader before he states his points. These letters are common in many industries and, since much money rides on them, they must be absolutely clear.

221 West Third Street
Menomonie, WI 54751

February 16, 1990

Mr. Harold Stonewall
Director of New Technologies
Mead Packaging
1040 West Marietta Street
Atlanta, GA 30302

Dear Mr. Stonewall:

I'm a student at the University of Wisconsin-Stout, majoring in Packaging Engineering. In the process of researching aseptic packaging for my senior research class, I discovered your name highlighted in an article, "Cross Check Aseptic Systems" in *Food Processing,* June 1988, page 56.

Could you please take a few minutes to supply more information on the following questions?
    — How does the hot water and citric acid solution compare to other aseptic procedures?
    — How did hot water and citric acid originate as an aseptic procedure at Mead Packaging?

Your cooperation would be greatly appreciated and would help me to further my education in the area of aseptic packaging. You can write to me at the above address, or call (715) 235–8391 on Monday and Wednesday (9 a.m.-12 p.m.), or Friday after 10 a.m. If you can't help me with these questions, could you suggest other sources of information? Also, if you have any additional information that may be helpful in my research, please send it along. Thank you.

Sincerely,

Todd Hainer

Todd Hainer

FIGURE 18.6
Letter of Inquiry

MANUALS UNLIMITED
"Instructing the World"
1021 Coho Drive
Kalispell, MT 59901
(406) 456-2727

February 14, 19XX

Mr. Robert Horan, Project Manager
KLH Robots, Inc.
3209 Golden Gate Avenue
San Francisco, CA 94100

Dear Mr. Horan:

As required by contract 1–1–06, October 1, 19XX, we are sending you a final copy of the master copy of the User's Manual for the P–2000 Dexter Robot. The robot that we used to develop the manual is already on its way to you via UPS.

This manual is written at an eighth-grade level, and it covers 16 functions, as specified in the contract. In addition, it has 9 exploded drawings of various subassemblies, a troubleshooting chart, and a change-parts section. It is in the required three-ring binder cover.

We have enjoyed our relationship with your company. Your engineers, especially Robert Meier, have been extremely helpful in filling our requests for information.

We are available to work on other projects of a similar nature. If you have questions about the manual, please direct them to me.

Sincerely,

Pinckney Hall

Pinckney Hall
Project Manager

FIGURE 18.7
Letter of Transmittal

July 14, 19XX

Ms. Michelle Smith
Goldenglo Investment Company
54160 John Muir Avenue, Suite 471
Los Gatos, CA 95032

Dear Michelle:

REFERENCE: Sun Bld. EB-10, CELTIC/Northern Letter Dated 6/29/XX

As a result of yesterday's construction meeting and discussions with Cokeby &
Hankedo and CELTIC/Northern, we have elected to reverse our decision on two
items listed on the attached letter:

ITEM #5: We will NOT accept the GE HID lowbay fixture. We want to use the
          Daybrite fixture as specified.

ITEM #6: We will NOT accept the MC cable as suggested by Valley for the
          installation of the warehouse fixtures. We will use wire and conduit as
          specified originally.

In addition to these items, we discussed a slight modification to the cable tray
along the exterior wall near the warehouse offices. This will be coming, in writing,
as a clarification from Wallace Cokeby.

Sincerely,
AVANT MICROSYSTEMS, INC.

Bill Coldwell
Bill Coldwell
Project Manager

mes

Enclosure (1)

cc: Arthur Friest (Avant)
     Wallace Cokeby (Cokeby & Hankedo)
     Myrtle Wert (SMWM)

FIGURE 18.8
Specification Change Letter

■ SUMMARY

Letters have three formats — block, modified block, and simplified. The last is becoming much more common, particularly in impersonal situations. Good letter writing requires careful attention to all the elements of the letter, including the inside address, the salutation, and the complimentary close. Good letter writers approach their readers with the "you" attitude, and attempt to treat them as they would like to be treated. Other special types of letter are letters of inquiry, letters of transmittal, and specification change letters.

■ WORKSHEET FOR WRITING A BUSINESS LETTER

Draft the body of the letter.

☐ *Use an informal introduction, state your main points succinctly, and conclude briefly.*

☐ *Write out the date.*

☐ *Write out your address. Do not use abbreviations.*

☐ *Write out the recipient's address. Pay attention to the following:*
-Correct spelling of name
-Correct title
-Correct corporation title

☐ *Repeat the recipient's name, followed by a colon, in the salutation.* If you have no name, use a subject line.

☐ *Write the complimentary closing, and place a comma after it.*

☐ *Type your name four spaces below the closing.*

☐ *Sign your name between the closing and typed signature.*

☐ *Add enclosure and copy lines if needed.*

☐ *Reread the letter slowly, word for word, for spelling and grammar errors.*

☐ *Reread the letter to check for a "you" attitude.*

☐ *Reread the letter to make sure your facts are accurate.*

■ EXERCISES

1. Find two or three business letters — sales letters are the easiest to find. Analyze them for format. Write a brief memo to your manager to suggest using one of the formats for all correspondence from your division. Explain why you chose that format. Easy to read? Looks good on the page? Fits the image of the company?

2. Write the same business letter in two different formats. Make a transparency of each. Present a brief (2 to 3 minutes) speech recommending one of the formats as the preferred one for your company. Indicate why certain details of the format make it preferable in terms of easier reading.

■ WRITING ASSIGNMENT

1. As part of a research project, write a letter of inquiry to a professional. Ask him or her for information about your topic. Your questions should be as specific as you can make them. Ask, "How does Wheeler Amalgamated extrude the plastic used in the cans for Morning Bright orange juice?" Avoid questions such as, "Can you send me all the information you have on the extruding process and any other processes of interest?"

■ WORKS CITED

United States Postal Service. *Addressing for Optical Character Recognition.* Notice 165. N.p. n.d., June 1981.

19

Job Application Materials

Résumés and letters of application are essential types of writing for college students and for anyone changing positions. This chapter will explain the process that will help you produce an effective résumé and letter. You need to understand your goals and your audience, to plan the contents of the résumé and letter, to present each in an effective form, and to perform effectively at an interview.

ANALYZING YOUR GOALS AND YOUR AUDIENCE

Before writing a résumé and letter of application, you must understand both your goals and your audience for these critical documents.

Understanding Your Goals

Your résumé and letter of application have two goals: to get an interview and to indicate the skills you will bring to the company. The letter and résumé open the way to an interview, and if you can present your strengths and experiences convincingly in these key documents, the employers will ask you for an interview. To be convincing, you must explain in writing what you can do for the reader; you must show how your strengths meet the firm's needs.

The letter and résumé also provide topics for discussion at an interview. It is not uncommon for an interviewer to say something like, "You say in your résumé that you worked with material requirements planning. Would you explain to us what you did?"

Understanding Your Audience

The audience for your résumé and letter could be any of a number of people in an organization, from the personnel manager to a division manager. The audience could be just one person or a committee. Whoever they are, they will approach the letter with a limited amount of time and with expectations about writing skills and professional presentation.

The Reader's Time   Employers read letters and résumés quickly. A manager might have one hundred résumés and letters to review. If the manager spent an hour on each one, it would take 2 1/2 weeeks to read them all. Managers do not have that kind of time. It is much more likely that, on the initial reading, the manager will spend 30 seconds to 3 minutes on each application, quickly sorting the applications into "yes" and "no" piles.

Skill Expectations   Managers look for data that show how the applicant will satisfy the company's needs. The data, according to two researchers, should include "the college graduate's previous achievements; special aptitudes/skills; and work-related learning, contributions, and achievements" (Harcourt and Krizar 183). Suppose, for instance, that applicants are responding to an ad specifying "need experience in materials resource planning." The manager will reject any applications that do not give evidence of such a skill.

Professional Expectations   Managers read to discover your ability to write clearly, to handle detail, and to act professionally. Clean, neat documents written in clear, correct English demonstrate all three of these. Bad grammar, unclear sentences, spelling mistakes, typographical errors, erasures, and poor-quality paper will probably offend a manager.

PLANNING THE RÉSUMÉ AND LETTER

To plan your résumé and cover letter, you need to assess your field, your own strengths, and the needs of your prospective employers.

Assess Your Field

You should spend some time discovering what workers and professionals actually do in your field of interest. Make lists that answer the following questions:

> What are the basic activities in this field?
> What skills do I need to perform them?
> What are the basic working conditions, salary ranges, and long-range outlooks for the areas in which I am interested?

These answers will enable you to assess the various strengths of your own background and to discover how you may be useful to an employer.

To find this information, talk to professionals, visit your college placement office, and use your library. To meet professionals, you can ask to interview them about their field. You can also attend career conferences, join a student chapter of a professional organization, or become a student member of an organization. Most professional organizations are happy to help student members find jobs. Your college's placement service probably has much career and employer information available. Most placement services give advice on all kinds of topics — from whom to contact to what to wear to an interview.

In your library you can find books that describe career areas. Two helpful books are the *Dictionary of Occupational Titles* (*DOT*) and the *Occupational Outlook Handbook* (*OOH*), both issued by the U.S. Department of Labor. The *Dictionary of Occupational Titles* presents brief but comprehensive discussions of positions in industry, listing the necessary job skills for these positions. You can use this information to judge the relevance of your own experience and course work when considering a specific job. Here, for instance, is the entry for manufacturing engineer:

> 012.167–038 MANUFACTURING ENGINEER (profess. & kin.)
> Directs and coordinates manufacturing processes in industrial plant: Determines space requirements for various functions and plans or improves production methods including layout, production flow, tooling and production equipment, material, fabrication, assembly methods, and manpower requirements. Communicates with planning and design staffs concerning product design and tooling to assure efficient production methods. Estimates production times and determines optimum staffing for production schedules. Applies

statistical methods to estimate future manufacturing requirements and potential. Approves or arranges approval for expenditures. Reports to management on manufacturing capacities, production schedules, and problems to facilitate decision-making.

The *Occupational Outlook Handbook* presents essays on career areas. Besides summarizing necessary job skills, these essays contain information on salary ranges, working conditions, and employment outlook. This type of essay can help you in an interview. For instance, you may be asked, "What is your salary range?" or "What do you think you're worth?" If you have the appropriate figures in your head, you can confidently name a range in line with industry standards.

Assess Your Strengths

After you have analyzed the field, you need to analyze yourself. Review all your work experience — summer, part-time, internship, full-time — your college courses, and your extracurricular activities to determine what might fill specific needs in your field.

Take this analysis seriously. Spend time at it. Talk to other people about yourself. Make long lists. Put down every skill and strength you can think of; at this stage, don't exclude any experiences because they seem trivial. Many applicants have essentially the same educational background so you must try to think of qualifications that distinguish you from your competitors. Here are some questions to help you analyze yourself (based in part on Harcourt and Krizar 181):

1. What work experience have you had that relates to your field? What were your job responsibilities? What projects were you involved in? What machinery or evaluation procedures did you work with?
2. What special aptitudes and skills do you have? Do you know advanced testing methods? Complicated software?
3. What special projects have you completed in your major field? List processes, machines, systems that you dealt with.
4. What honors and awards have you received? Do you have any special college achievements?
5. What is your grade point average?
6. Have you earned your college expenses?
7. What was your minor? What sequence of useful courses have you completed? A sequence of three or more courses in, for example, management, writing, psychology, or communication might have given you knowledge or skills that your competitors do not possess.

8. Are you willing to relocate?
9. Are you a member of a professional organization? Are you an officer? What projects have you participated in as a member?
10. Can you communicate in a second language? Many of today's firms do business internationally.
11. Do you have military experience? While in the military, did you attend a school that applies to your major field? If so, identify the school.

Assess the Needs of Employers

After deciding what skills to emphasize, you can promote them much more effectively if you are aware of the market. You cannot use the "you" approach to its full potential unless you know how your skills might benefit each prospective employer. In addition, if you can show that you have taken the trouble to investigate a company, you distinguish yourself from your competition.

You can find out about firms from school placement offices (where company literature and the *College Placement Annual* are available), from professional journals, and from professors. Most libraries have collections of annual reports; reading them will give you information that you can apply to your job search. Books such as *Dun's Employment Opportunities Directory: The Career Guide* list employers alphabetically, geographically, and by industry classification. This convenient book lists the name of the person to contact for employment information and gives an overview of the company as well as its career opportunities, training and development programs, locations, and benefits.

WRITING THE RÉSUMÉ

After you have analyzed your field, your strengths, and prospective companies, write your résumé. Since the résumé contains all relevant information, you should write it before the letter. Your letter can then highlight or expand on this information. If you have the résumé, you will find it easier to adapt the letter to the needs of a specific employer.

Your résumé is a one-page document that summarizes your skills, experiences, and qualifications for a position in your field. To be a strong applicant, you must write your résumé with great care, selecting the most pertinent information and a readable format. You can present your résumé in one of two formats: traditional or functional.

The Traditional Résumé

The traditional résumé has the following sections:

- career objective
- personal data
- educational history
- work history

Objective The *career objective* states the type of position you are seeking — usually an entry-level position. If you wish to add anything more, name a position you would like to have in four or five years. To word the objective effectively, ask professionals in your job area. Avoid clichés such as "Energetic accountant wishes to employ fine-tuned skills in challenging position with determined, aggressive growth company." The following are well-written, basic objectives:

Entry-level managerial position in large retail chain.

Systems analyst with opportunity for advancement.

Position in research and development in microchip electronics.

Personal Data The personal data consist of name, address, place to contact for credentials, willingness to relocate, hobbies, and interests. The first four are essential in a résumé, but the last two are optional.

List your current address and phone number. Tell employers how to acquire credentials and letters of reference. If you have letters in a placement file at your college placement service, list the appropriate address and phone numbers. If you do not have a file, indicate that you can provide names upon request. Check with all the people beforehand to make sure they will agree to act as references.

Because of laws passed in recent years, you do not need to reveal your birth date, height, weight, health, or marital status. You may do so if you wish — for instance, many college students give their birth date since age discrimination is rarely practiced on people in their twenties. You can give information on hobbies and interests. They reveal something about you as a person, and they are topics at a surprising number of interviews.

Education The education section includes pertinent information about your degree. List your college or university, major, minor, concentration, years attended, and grade point average (if good). If you attended more than one school, present them in reverse chronological order, the most recent at the top. You can also list relevant courses (many employers like to see technical writing in the list), honors and awards, extracurricular activities, and descriptions of practicums, coops, internships, and special "professional encounters," such as extended field trips. You do not need to include your high school on the list. Two sample sections follow.

SAMPLE 1 (Education Section)

EDUCATION	**University of Wisconsin-Stout,** Menomonie, WI 54751	Bold head
	Degree: Bachelor of Science, May 1989 Major: **Industrial Technology** Concentration: **Manufacturing** **Engineering**	Relevant data
	G.P.A.: 3.35/4.0 scale	Strong G.P.A.

RELATED
COURSES
	Production
	  Processing
	Quality
	  Assurance
	Time and Motion
	  Study
	Manufacturing
	  Cost Analysis
	Production
	  and Inventory
	  Control
	Production and
	  Inventory
	  Control
	  Practicum
		Computer-Aided
		  Manufacturing
		Robotics
		Material Handling
		Production
		  Management
		Mechanics and
		  Electricity
		Organizational
		  Leadership
			Courses in left
			column emphasize
			technical skills;
			courses in right
			column emphasize
			career skills

SAMPLE 2 (Education Section)

EDUCATION	**University of Wisconsin-Stout,** Menomonie, WI 54751 Bachelor of Science, May 1989 Major: **Fashion Merchandising** Minor: **Business Administration** Cum Laude G.P.A.: 3.4/4.0	Most recent university listed first
	University of Wisconsin-Green Bay, Green Bay, WI 54302 Two semesters and one summer session of general studies	

(sample 2 continued)

**EDUCATIONAL
EXPERIENCE**

Intern	**Dayton-Hudson Department Store,** Edina, MN	Job title, location, dates
	September 1988 to December 1988	
	While working full-time, responsibilities entailed managing and supervising the Christmas Trim-the-Home department.	Description of duties emphasizes managerial skills
	Duties included assisting with the setup and organization of the Christmas shop and interviewing, hiring, training, and scheduling of employees. Gained confidence in my professional abilities by conducting department meetings, forecasting sales, and corresponding with upper management.	
Show director	**Promotions Fashion Show,** University of Wisconsin-Stout	Job title, location, dates
	Fall 1987	
	Worked with our Promotions class in directing a fashion show. Responsible for overseeing the duties of the students.	Description emphasizes managerial skills
	Skills necessary for this job included leading, motivating, and meeting deadlines. Specific duties involved selecting appropriate models, finding merchandise, arranging the location and setup of the show, and publicizing the event.	

Work Experience   The work experience section includes the positions you have held that are relevant to your field of interest. List your jobs in reverse chronological order — most recent first. In some cases you might alter the arrangement based on the importance of the experience: for example, if you first held a relevant eight-month internship and then took a job as a dishwasher when you returned to school, list the internship first. List all full-time jobs and relevant part-time ones — as far back as the summer after your senior year in high school. You do not need to include every part-time job, just significant ones (but be prepared to give complete names and dates).

Your work experience should be presented in "entries" — one for each job. Each entry should have four items: job title, job description, name of company, dates of employment. These four items can be arranged in a number of ways, as the examples below show. However, *the job description is the most important part* of the entry. Here you describe what your duties have been, the projects you have worked on, and the machines and processes you have used. Write the job description in the past tense, using words such as *managed, directed,* or *developed.* Arrange the items in the description in order of importance. Put the important skills first, even if you didn't perform them as often as the others.

Emphasize what you think are the most important parts of the job description: To emphasize the time you worked, place the dates to the left. To emphasize your title, place that to the left. The description itself is always to the right. The examples below illustrate how to arrange the four elements to achieve different emphases.

SAMPLE 1 (Dates Placed at the Left)

WORK EXPERIENCE

August 1985–January 1986	Technician. Miller Electric Manufacturing Co., Appleton, WI	Date, job title, location
	Served as a technician exposed to a broad range of manufacturing processes, methods, and manufacturing engineering responsibilities. Involved in part design drawings, equipment installation drawings, office and warehouse relocation layouts, and line balancing at individual work stations.	Job description

SAMPLE 2 (Job Title Placed at the Left)

WORK EXPERIENCE

Technician	Miller Electric Manufacturing Co., Appleton, WI	Job title, location
	Served as a technician exposed to a broad range of manufacturing processes, methods, and manufacturing engineering responsibilities. Involved in part design drawings, equipment installation drawings, office and warehouse relocation layouts, and line balancing at individual work stations.	Description
	August 1985–January 1986.	Dates

Order of Entries on the Page   In the traditional résumé, the top of any section is the most important position. Place the most important information there. Place your name, address, and career objective at the top of the page. In general, the education section is next, followed by the work section. But if you have had a relevant internship or significant full-time experience, put the work section first. Figure 19.1 shows a traditional résumé.

If your work experience is strong, you might not need a list of courses. If education is your strong point, you might want to list your courses, or even describe the contents and projects of the courses.

The Functional Résumé

The functional résumé is arranged by skills and strengths. This kind of résumé presents the applicant to the employer in the same way that the employer looks at the applicant: Are the relevant skills present? This style, in particular, allows students whose work experience is not relevant to their job area to stress skills learned in classes.
The functional résumé has these sections:

- objective
- personal data
- education
- list of employers with dates and addresses
- skills

Figure 19.2 shows how to handle each of these sections.

In the functional résumé, the following sections are handled just as in the traditional résumé:

- objective — tell your immediate occupational goal.
- personal data — include your name, address, and the address of your placement service.
- education — list your university, major, date of graduation, minors, G.P.A.

The work and skills sections are different: the work section is shorter, and skills can be presented in a *Capabilities* list or in categories.

Work Section   For the work section, give the job title, company, and dates for each position you've held. You do not, however, have to present

---

DAVID T. WEISSINGER

Address until May 10, 1986
1101 Seventh Street, Apartment 4
Menomonie, WI 54751
(715) 235–8864

Permanent Address
Route 1 Box 69–A
Durand, WI 54736
(715) 672–5084

CAREER INTERESTS	Manufacturing engineer or related position supporting product development and production operations.
EDUCATION	University of Wisconsin-Stout, Menomonie, WI 54751 Degree: Bachelor of Science, May 1986 Major: **Industrial Technology** Concentrations: **Manufacturing Engineering** 　　　　　　　　**Product Development** G.P.A.: 3.6/4.0 Magna cum laude
CAREER-RELATED EXPERIENCE	**Co-op Manufacturing Engineering** Capstone Aviation Operations, Denver, CO 80221
8/84–12/84	Involved with the Producibility Group in the design and manufacture of a special aircraft fuel pump to measure bearing film pressure. In addition, gathered data for the evaluation of material growth through nitriding, prepared linear tolerance charts, and made various process changes.
6/85–12/85	Involved with the Producibility Group in determining manufacturing capability of a helical clutch stop mechanism along with electron beam weld and gear evaluation for a constant speed drive ring gear. In addition, developed modular ECM tooling concept as a cost reduction measure for similar parts and investigated manufacturing problems, making the necessary tooling or process changes as required.
PROFESSIONAL CERTIFICATIONS	SME- certified Manufacturing Technologist. APICS-certified on Material Requirements Planning.
HONORS AND ACTIVITIES	Capstone Award for Extraordinary Achievement. Chancellor's Award for Academic Excellence (4 semesters).
CREDENTIALS	References and credentials available upon request from: Career Planning and Placement Service University of Wisconsin-Stout Menomonie, WI 54751 (715) 232–1601

FIGURE 19.1
Traditional Résumé

Jean R. Wendlandt
1427 Magnolia Boulevard
Menomonie, Wisconsin 54751
(715) 235–4176

Professional Objective	A position in materials management.
Capabilities	■ Handle administrative details under pressure ■ Motivate others based on company goals and commitments ■ Interact with the public enthusiastically so that business transactions can occur harmoniously
Experience	■ Worked with the materials management team expediting purchase orders ■ Developed the replacement parts price list, which contained over 8,000 parts ■ Responsible for weekly and monthly reports concerning repair orders, sales, and past-due customer orders ■ Communicated with plant schedulers, engineers, accountants, and nondivision personnel
	Franklin Industries, Incorporated,(March 1986–September 1986) Dayton, OH Refurbishing Center Administrator Customer Service Supervisor (September 1983–March 1986) Customer Service Coordination (March 1981–September 1983)
Educational Experience	Communication Technology Lab Supervisor (September 1987–present) ■ Developed an extensive knowledge of computers ■ Assisted communication students in lab activities Chief Executive Officer, Hi-Tech Inc., University of Wisconsin-Stout (Spring Semester 1988) ■ Managed a student-organized corporation that designed, manufactured, and sold, for 100% profit, laser-engraved card/ jewelry box.
Education	Bachelor of Science in Business Administration (May 1989) University of Wisconsin-Stout G.P.A.: 3.79/4.0
Honors	Chancellor's Award for Academic Excellence (1986–1988) Women in Management, Vice President (1986–1987) Honor Society NWTI (1978)
Credentials Available at	Career Planning and Placement Office University of Wisconsin-Stout Menomonie, Wisconsin 54751 (715) 232–1601

FIGURE 19.2
Functional Résumé

a job description. In the Sample Skills List below, notice that the writer just lists the jobs. The skills section presents the job descriptions in a different form.

Capabilities List   An effective method of presenting your skills is to list all your capabilities and then follow them with a list of experiences. The *capabilities* are skills that will help you perform the position named in your objective. You can list just one or two words, or you can write a brief description. In the following example, the writer shows clearly in a skills list that she can handle the demands of writing a manual. Notice that the sequence of the list follows the process of actually producing the manual and demonstrates that she already knows how to perform the task. Use present tense verbs in this section.

SAMPLE SKILLS LIST

CAPABILITIES

- Gather, select, and write information in a clear effective manner.

- Analyze writing projects and make decisions about content, format, organization, and style.

- Set performance objectives to clarify the purpose of a technical manual and each of its parts.

- Create understandable, step-by-step instructions for specific readers to perform specific tasks.

- Collaborate with subject-matter experts in an effective way.

- Develop plan sheets and flow charts to gather information.

- Design and lay out written material for manuals, brochures, and flyers.

Active verbs used throughout

List contains all aspects of job

The experience section then lists all the relevant projects that the author has worked on or completed at different jobs. In the following example, notice that the writer lists all the kinds of documents she has written and related skills, such as "proofread copy," that support the writing projects. Use past tense verbs in this section.

SAMPLE EXPERIENCE SECTION

EXPERIENCE

- Wrote and designed a technical operation manual for a pneumatic piston filler.

Section lists different types of projects

- Composed an educational brochure on food preservation for consumers.
- Developed a promotional brochure for a career conference.
- Gathered information and wrote articles for News and Arts section of a weekly newspaper.
- Proofread copy and created headlines for a weekly newspaper.

Verbs in past tense

1986	MRM Elgin, Menomonie, WI Student Technical Writer
1985–86	Barlow Foods, Rochester, MN Home Economist Intern
1986	Career Planning and Placement Office Menomonie, WI Student Writer
1985–86	*Stoutonia* Newspaper, Menomonie, WI News/Arts Reporter
1985–86	*Stoutonia* Newspaper, Menomonie, WI Copy Editor

For each entry: date, location, job title

Skills Categories   To arrange skills in categories, present all your capabilities and experience after a relevant topic heading. For instance, you might have subheads for management, research, evaluation, and team membership. Write a paragraph about how you obtained these skills and what level of expertise you have. Below the categories, you should list your jobs. Here is the capabilities section listed above presented as categories. The experience section would not change, and would follow the categories.

SAMPLE SKILLS CATEGORIES (Paragraph Form)

Project Management.   I have developed brochures and manuals for clients, learning how to set up schedules, review budgets, interview experts, write clear text, and design effective pages. I have proofread copy and seen the documents through all stages of the printing and production process. I have done this in industrial and free-lance situations so I feel I can use the process well.

Four basic categories
Relevant activities are listed and explained

Research.   I have gathered information by interviewing clients, observing processes, operating machines (such as disassembling the piston filler), and using the library. I have learned how to pursue a question so that I get the correct facts.

Client Interaction.   I have worked with clients to develop manuals and brochures. I understand how to interview clients, how to let them review my work, and how to interact with them so that they pick the best design to convey their meaning.

Types of Professional Documents.   I have written a technician operator manual for a pneumatic piston filler, an educational food brochure on food preservation for consumers, a promotional brochure for a career conference.

A sample functional résumé is shown in Figure 19.2.

FORMATTING THE RÉSUMÉ

The résumé must be easy to read. An employer who is reading sixty résumés in an hour will spend no more than a minute on yours. Even if there are fewer to read, most readers will not scrutinize your résumé closely on the first reading. They might later, but on the first reading they are looking for the essential information, which they must be able to find.

To make your résumé readable, use highlight strategies: heads, boldface, underlining, margins, and white space. Essentially, follow an outline format. Use these guidelines:

- Indicate the main divisions at the far left margins. Usually boldface heads announce the major sections of the résumé. Although the heads can be followed by colons, they do not have to be: the design indicates that the material that follows explains the head.
- Boldface important words such as job titles or names of majors; use underlining sparingly.
- Single-space entries; double-space above and below; the resulting white space makes the document easier to read.
- Treat items in each section the same way. All the job titles, for example, should be in the same relative space, with the same typeface and size.
- Print résumés on good-quality paper; use black ink on white paper. Shades of brown, beige, and off-white are often used. Avoid brightly colored paper which has little effect on employers, and photocopies poorly resulting in blotchy copies.
- Make your résumé easy to read by controlling the margins and type size. Be sure to leave enough margin on the left. If the page is single-spaced, and if all the sections start at the far left margin, your résumé will be very hard to read.
- Use 9- or 10-point type if you are using a word processor.

WRITING LETTERS OF APPLICATION

Your letter of application should be written with care. In this letter, you apply for a job, explain your qualifications, and ask for an interview. You also prove whether or not you can write clearly. Many employers are affected not just by what you say but also by how you say it. To make a good impression, follow these guidelines:

- Type the letter on 8½- by 11-inch paper.
- Use white, 20-pound, 100 percent cotton-rag paper.
- Use black ink.
- Use one of the letter formats explained in Chapter 18.
- Sign your name in black or blue ink.
- Proofread the letter carefully. Grammar and spelling mistakes are irritating at best and cause for instant rejection at worst.
- Mail the letter, folded twice, in a business envelope.

Parts of the Letter of Application

A letter of application has three parts: the introductory application, the explanatory body, and the request conclusion.

Introductory Application   The application paragraph should be short and to the point. Inform the reader that you are applying for a specific position. If it was advertised somewhere, mention where you saw the ad. If someone recommended that you write to the company, mention that person's name (of course, only use the name if it is someone the reader knows personally or has heard of). You do not need to give your name in this section, nor mention your year in school.

SAMPLE 1

I would like to apply for the position of Material Evaluation   Apply; tell source
Technologist, advertised in the Sunday, March 1, *Richmond
Tribune.*

SAMPLE 2

Dr. Fred Stewart has informed me that you are seeking   Apply; tell source
applicants for a quality engineer position. I would like to apply
for the position.

Body   The *explanatory body* is the heart of the letter. In it you explain, in terms relevant to the reader, why you are qualified for the job. This section should be one to three paragraphs long. Its goal is to make your strengths and skills as pertinent to the reader's needs as possible.

Choose which details to place in this section according to the employer's needs. If you are responding to an ad, show how you have the skills to meet the requirements listed in it. If the employer mentions "a knowledge of reading electrical schematics," list details that illustrate your ability to read a schematic. If the employer asks for experience in COBOL programming, point out that you had a course in it and explain what you did in the course.

If you are not responding to an ad, choose details that show that you have the qualifications normally expected of an entry-level candidate. Your research in the *Dictionary of Occupational Titles* and *Occupational Outlook Handbook* will help you here. Mention how a project on a part-time job or in a class familiarized you with the basic concerns of the company.

SAMPLE 1

I am majoring in Industrial Technology, concentrating in Product Development. I will graduate in May 1986 with a B.S. In my courses, I have concentrated heavily on all aspects of design and development from the initial idea to the finished product.

Degree data

Courses related to needs

In my mechanical design class, I made an in-depth study of shaft design. I have designed shafts with various applied loads and torques, calculating stresses, bending moments, and deflection. I have also had experience with power transmission. In a mechanical power transmission class, I selected components for drive systems. These components included electrical motors, gear reducers, couplings, shafting, clutches, and bearings. I also have a basic understanding of pneumatic and hydraulic systems.

Explain two classes

Related skill mentioned

My work experience last summer at the Grist Mill Company has made me familiar with the operation of various types of packaging machines. I was a quality control technician, responsible for the proper functioning of these machines.

Explains expertise and level of responsibility

SAMPLE 2

I have just completed a cooperative/intern program at the Dayton-Hudson Department Store where my overall responsibility entailed managing the Christmas-Trim-the-Home department. My duties included assisting with the setup and organization of the Christmas shop, and interviewing, hiring, training, and scheduling of employees. I have also gained confidence

Work experience shows expertise plus communication skills

in my professional abilities by conducting department meetings
and corresponding with upper management.

This May I will graduate from the University of Wisconsin-   Degree data
Stout with a major in Fashion Merchandising and a minor in
Business Administration. One of my major accomplishments
in college was being appointed show director for the Promotions   Project experience
Fashion Show. I was responsible for overseeing the duties of   shows managerial
the students; the skills necessary for this job included leading   capabilities
and motivating fellow students. Specific duties involved selecting
appropriate models and merchandise, arranging the location
and setup of the show, and publicizing the event.

Interview Request   In the final section, ask for an interview and explain how you can be reached. The best method is to ask a question:
"Could I meet with you to discuss this position?" Also explain when you
are available. If you need two days' notice, say so. If you can't possibly get
free on a Monday, mention that. Most employers will work around such
restrictions if they can. If no one is at the phone at your house or dorm in
the mornings, tell the reader to call in the afternoons. A busy employer
would rather know that than waste time calling in the morning. Thank
your reader for his or her time and consideration. Readers appreciate the
gesture; it is courteous and it indicates that you understand that the reader
has to make an effort to fulfill your request.

SAMPLE 1

Could I meet with you to discuss the opening? You may   Direct question
contact me during the evening at (715) 555–5555. Thank you   How to contact
for your time, and I look forward to meeting you in the near   writer
future.
                                                                Thank you

SAMPLE 2

Please consider my qualifications when you fill this position. I   More elaborate
am very interested in meeting with you to discuss these   interview request
matters. Would it be possible to arrange an interview in the
near future? If so, please call me at my home — after 5:00   How to contact
P.M. any evening. Thank you for your consideration. I am   writer
looking forward to hearing from you.   Thank you

You will find two sample application letters used as models at the end of
this chapter. The first is long, explaining how the writer's experience fits
the needs listed in the ad. The second is short, listing the writer's
qualifications.

INTERVIEWING

The employment interview is the method employers use to decide whether or not to offer a candidate a position. Typically the candidate talks to one or more people, either singly or in groups, who have the authority to offer a position — to obligate the company contractually. Since the interview is critically important, you need to perform effectively. To interview success-fully, you need to prepare well, use social tact, perform gracefully, ask questions, and understand the job offer (Stewart and Cash 188–192).

Prepare Well

To prepare well, you must investigate the company and analyze how you can contribute to it. You can investigate the company in many ways. After you are invited to an interview, ask your contact person for some company literature. Spend time in your library reference room to find information; use annual reports, descriptions in *Moody's*, items from *Facts on File*, *F&S Index*, *Wall Street Journal Index*, or *Corporate Report Fact Bank*. After you have analyzed the company, analyze what you have to offer; analyze how you can fit into the company. Answer these questions:

> What contributions can you make to the company?
> How do your specific skills and strengths fit into its activities or philosophy?
> How can you further your career goals with this company?

The more you prepare, the better you will be able to provide clear examples that illustrate how you satisfy the company's needs.

Use Social Tact

To use social tact means to act professionally and in an appropriate man-ner. Acting too lightly or too intensely are both incorrect. First impressions are extremely important; many interviewers make up their minds early in the interview. Follow a few common sense guidelines:

- Shake hands firmly.
- Dress professionally, as you would on the job.
- Arrive on time.
- Use proper grammar and enunciation.
- Watch your body language — sit appropriately, don't give the impression of lounging.
- Find out and use the interviewers' names.

Perform Well

Performing well in the interview means to answer the questions directly and clearly. Interviewers want to know about your skills, mostly from your work experience but also from your academic experience. Since this information is essential for them, be prepared to discuss it. You must be willing to talk about yourself and your achievements; if you respond honestly to questions, your answers will not seem like bragging. So-called stress interviews, in which the interviewer tries to insult you and rattle your composure are rare. For a successful interview, follow these guidelines:

- Be yourself. The worst thing is to get a job because people thought you were a kind of person that you're not. If you lose a job because of who you honestly are, you're better off.
- Answer questions freely; don't withhold facts.
- Answer the question asked.
- Be honest — if you don't know the answer, say so.
- If you don't understand a question, ask the interviewer to repeat or clarify it.
- Be sincerely interested in the company; your sincere interest can overcome other deficiencies in your candidacy.
- In your answers, include facts about your experience to show how you will fit into the company.

Ask Questions

You have the right to ask questions at an interview. After all, it's your life. If no one has explained the following items to you, you should ask about them:

- methods of on-the-job training
- your job responsibilities
- types of support available — from secretarial to facilities to pursuit of more education
- possibility and probability of promotion
- policies about relocating, including whether you get a promotion when you relocate and whether refusing to relocate will hurt chances for promotion
- salary and fringe benefits — you should know at least a salary range and whether or not you receive medical benefits

If an interviewer is reluctant to answer these questions, you might be dealing with a group that has personnel difficulties. This reluctance could be a negative factor in your decision.

The Offer   Usually a company will offer the position — with a salary and starting date — either at the end of the interview or within a few days. You have the right to request a reasonable amount of time to consider the offer. If you get another offer from a second company at a higher salary, you have the right to inform the first company and to ask if they can meet the salary. Usually you accept the offer verbally and sign a contract within a few days. It's a pleasant moment, and as Emerson told Whitman, "the beginning of a great career."

WRITING FOLLOW-UP LETTERS

After an interview with a particularly appealing firm, you can take one more step to distinguish yourself from the competition. Write a follow-up letter. It takes only a few minutes to thank the interviewer and express your continued interest in the job:

> Thank you for the interview yesterday. Our discussion of Cranston's growing
> Fluid Dynamics Division was very informative, and I am eager to contribute to
> that division.
>   I look forward to hearing from you.

◼ SUMMARY

Well-prepared résumés and letters of application help you get an interview; after an interview, the employer may offer you a job. To write these two documents effectively, good writers analyze their audience. Readers spend only about one minute on the initial review of an application, looking specifically for the applicant who will fill their needs. Job seekers must research their job area by reading books like the *Dictionary of Occupational Titles*, assess their strengths by listing skills and concepts they have mastered, investigate firms by reading in the library, and speak of other professionals in their field.

Résumés have two forms — *traditional,* which emphasizes job duties, and *functional,* which emphasizes skills. Good résumés use white space, margins, and heads to emphasize important information and to make the document easy to read. The letter of application has three parts: an introduction (for applying for the position), a body (for relating applicant skills to employer needs), and a conclusion (for requesting an interview). At the interview, applicants must use social tact, dressing and speaking well; they must also answer questions clearly and honestly, ask questions about the position, and, above all, be themselves.

■ WORKSHEET FOR PREPARING A RÉSUMÉ

☐ *Write out your career objective; use a job title.*

☐ *List all the postsecondary schools you have attended.*

☐ *List your major and any minors or submajors.*

☐ *List your G.P.A. if strong.*

☐ *List 10 or 12 important courses that gave you skills useful for performing tasks required in your job area. Include communication courses.*

☐ *List projects you carried out in those classes.*

☐ *List any other educational experiences.*
   Examples include field trips, cooperative education, internships, specialized intensive short courses, and so forth.

☐ *List extracurricular activities, including offices held and duties.*

☐ *List all your jobs, giving for each one the job title, company, job duties, and dates.*
   Phrase job duties, if possible, so that they indicate you can perform the duties expected in your job area.

☐ *List your name, phone number, current address, and permanent address if it is different.*

☐ *Choose a format — traditional or functional.*

☐ *Choose a layout design. Where will you place heads? What margins will you use? What font will you use?*

◼ WORKSHEET FOR WRITING A LETTER OF APPLICATION

☐ *State the job you are applying for.*

☐ *State where you found out about the job.*

☐ *List what the employer needs.*
In other words, the skills you feel the job requires; or list the skills the ad (if you saw one) specifically requires.

☐ *List your own skills, courses, or jobs that correspond to the employer's needs.*

☐ *Purchase good-quality paper, envelopes, and a new ribbon for your typewriter or printer if you produce the letter yourself.*

☐ *Select a format: block format is suitable.*

☐ *Write at least three paragraphs:*
-an introduction to announce that you are an applicant
-a body that matches your key skills to their key needs
-a conclusion that asks for an interview

◼ MODELS

Two sample application letters follow. The first is long, explaining how the writer's experience fits the needs listed in the ad. The second is short, listing the writer's qualifications.

1020 1/2 Ninth Avenue East
Menomonie, WI 54751

October 21, 1990

Electrotek Corp.
7745 South Tenth Street
Oak Creek, WI 53154
Attention Personnel

Subject: Production Coordinator Opening

I am responding to your ad in the October 20, 1990, issue of the Sunday *Milwaukee
Journal* for the position of production coordinator. I feel that my background in
electronics and manufacturing will fit in well at your printed circuit board manufacturing
facility.

I will graduate from the University of Wisconsin-Stout in June with a Bachelor of
Science degree in Manufacturing Engineering. I have had extensive training on
computers while at Stout. In my Production and Inventory class, I learned about
computerized production and material scheduling in a job shop environment
using a material requirements planning program. While working for the Department
of Energy and Transportation, I wrote a program to control the flow of equipment
being checked out from the electronics lab.

As shown in my résumé, I have also repaired and installed fire alarm, security,
and patient-signalling systems when I worked at Johnson Controls. Most of the
time I faced construction deadlines so I know what it is like to work under pressure.
My duties as the supervisor for the electronics lab involved assigning jobs,
coordinating schedules, training new employees, and working with the department
chairman when needed.

I am looking forward to hearing from you. Would you please call me to set up an
interview? I can be reached at my Menomonie address or by phoning (715) 235–
7980 after 4 p.m.

Sincerely,

Curt Everson

Curt Everson

enc: résumé

FIGURE 19.3
Application Letter Showing Experience

1092 South Broadway
Menomonie, Wisconson 54751

February 25, 1989

Mr. Don Swanstrom
Woodcraft Industries Inc.
525 Lincoln Avenue Southeast
St. Cloud, Minnesota 56301

Dear Mr. Swanstrom:

I would like to apply for the opening as millwright manager, advertised by your
company in the Sunday, February 23, *Milwaukee Journal.*

I am currently a senior at the University of Wisconsin-Stout and will graduate in
December. I will receive a Bachelor of Science degree in Industrial Technology
with a concentration in Wood Manufacturing Processes. I have gained experience
in the electrical, machine shop, and production woodworking areas through the
courses I have taken at Stout. I also have acquired skills necessary to be a
successful manager, including staffing, development, discipline, communication,
and motivation.

I believe I understand all the responsibilities of the job and feel I am capable of
managing your maintenance department. Is it possible to set up an interview for
this position? I am available between 2:00 and 6:00 p.m. every day and can be
reached at 235–7877.

Sincerely,

Connie Seilheimer

Connie Seilheimer

Enclosure: résumé

FIGURE 19.4
Application Letter Listing Qualifications

■ EXERCISES

1. Evaluate the wording of each of the following letter openings. Rewrite them if necessary.

   a. I was reading the Job Vacancy List published the the University I attend when I came across your request for a hotel manager. In May 19XX I will be graduating from the University of Wisconsin–Stout with a degree in Hotel Management. I would like to take this opportunity to apply for the hotel manager position that you are offering with your company.

   b. Hello, my name is Joan Foss and I am a graduating senior in college. In reply to seeing your ad in the Denver Post I wish to apply for the position of operations manager, I wish to apply for the position of operations manager. I wish to develop my knowledge in areas such as directing existing communication systems and consulting to offer clients effectiveness and adequacy of a chosen system.

2. Evaluate the following closings. Rewrite any that you feel are inappropriate. Which one do you think is most appropriate, and why?

   a. I would like to interview your company. I will be in Des Moines visiting some friends, one of whom worked for your company about six years ago. I will be there March 13–21. I would like to interview then.

   b. Thank you for your attention of this request. I would be most gratified to receive any kind of a response from you and will be more than happy to appear for a personal interview any time that it is convenient for you.

   c. I am free for an appointment at a week's notice. My number is (715) 555–5555. Thanks for your time. I anticipate your reply.

   d. May I drive to Brookfield to talk to you? I can arrange my schedule to be available for an interview on any afternoon but Thursday and Friday.

3. Which of the two following paragraphs do you think is more effective? The position applied for is an operations manager.

   a. I will be graduating in May 19XX with a Bachelor of Science degree in Industrial Technology, with a concentration in Industrial Transportation and Communication. Enclosed you will find my personal résumé, which details both my college education and my work experience. In addition to the work experience listed, I have a considerable amount of computer experience, working currently with three separate computer systems, and I feel this would be advantageous to your firm.

   b. I have worked in industry for the past four years. My current job as a Production and Machine Operator, which I work summers and breaks, has provided me with experience in inventory control, distribution, machining, assembling, shipping, payroll, record-keeping, and operating various types of machine tool equipment. Along with my work experience, my bachelor's degree in Industrial Technology has provided a broad background in the area of industrial transportation. Some of the specific classes I have taken that deal with industrial transportation and communication are Packaging, Materials Handling, Industrial Distribution, Inventory and Production Control, and Industrial Marketing Management.

4. Decide which of the following two paragraphs you think is most effective, and why.

   a. I will be graduating from the University of Wisconsin-Stout in May 19XX with a Bachelor of Science degree in Industrial Technology, including an emphasis in Packaging Engineering. While attending the university, I have learned many things that pertain to package development and consumer satisfaction.

   b. One course in particular has given me the practical experience that you are looking for. In Materials and Processes, a senior level course, I was required to design, develop, and test a package that would successfully transport the product to its destination and meet the needs of the consumers that purchase the product.

5. For each job that you have had, make a list of all the tasks you accomplished and your responsibilities. Also make a list of all the skills you needed to perform those tasks. Then make a third list of the classes you have taken, and of projects you have completed, that have developed those skills.

6. Find an ad for a position you might apply for. On the left side of a sheet of paper, list every word in the ad that suggests a skill the company is seeking. To the right, make a corresponding list of the skills and the experiences you have that fulfill those needs. Then proceed to writing assignment No. 2 below.

WRITING ASSIGNMENT

1. Write your résumé following one of the two formats described in this chapter.

2. Find an ad for a position in your field of interest. Use newspaper help wanted ads or a listing from your school's placement service. Based on the ad, decide which of your skills and experiences you should discuss to convince the firm that you are the person for the job. Then write a letter of application applying for the job.

WORKS CITED

Dictionary of Occupational Titles. 4th ed. Washington: U.S. Dept. of Labor, 1977.

Harcourt, Jules, and A. C. "Buddy" Krizar. "A Comparison of Résumé Content Preferences of Fortune 500 Personnel Administrators and Business Communication Instructors." *Journal of Business Communications* 26.2 (1989): 177–190.

Stewart, Charles J., and William B. Cash, Jr. *Interviewing Principles and Practices.* Dubuque IA: Brown, 1982.

Appendix A: Style Handbook for Technical Writers

PROBLEMS WITH SENTENCE CONSTRUCTION

PUNCTUATION

ABBREVIATIONS, CAPITALIZATION, NUMBERS, AND SYMBOLS

T his appendix presents the basic rules of grammar and punctuation that you need to write well. It contains sections on problems with sentence construction, agreement of subjects and verbs, agreement of pronouns, punctuation, abbreviations, capitalization, numbers, and symbols.

PROBLEMS WITH SENTENCE CONSTRUCTION

The following section introduces you to many common problems in writing sentences. If not avoided, all of them will make your writing less clear. Each subsection gives examples of a problem and explains how to convert the problem into a clearer sentence. Although no writer exhibits all of these errors in his or her writing, almost everyone commits several of them. Many writers show definite habitual patterns: they often write in fragments, or they use unclear pronoun references, or they repeat a word or phrase excessively. You should learn to identify your problem habits.

Identify and Eliminate Comma Splices

A *comma splice* occurs when two independent clauses are connected, or spliced, with only a comma. You can correct comma splices in four ways:

1. Replace the comma with a period and thus make two sentences.

Splice	Friction results from the movement of one material across another, uneven surfaces cause greater friction.
Correction	Friction results from the movement of one material across another. Uneven surfaces cause greater friction.

2. Replace the comma with a semicolon only if the sentences are very closely related.

Splice	A hypothesis is only an assumption, it must be tested
Correction	A hypothesis is only an assumption; it must be tested.
Splice	A hypothesis is an assumption, therefore, it must be tested.
Correction	A hypothesis is an assumption; therefore, it must be tested.

   The word *therefore* is a conjunctive adverb. When you use a conjunctive adverb to connect two sentences, always precede it with a semicolon and follow it with a comma. Other conjunctive adverbs are *however, also, besides, consequently, nevertheless,* and *furthermore.*

3. Insert a coordinating conjunction (*and, but, or, nor, for, yet,* and *so*) after the comma, making a compound sentence.

Splice	The gas-air mixture burns in internal combustion engines, it does not explode.
Correction	The gas-air mixture burns in internal combustion engines, but it does not explode.

4. Subordinate one of the independent clauses by beginning it with a subordinating conjunction or a relative pronoun. Frequently used subordinating conjunctions are *where, when, while, because, since, as, until, unless, although, if,* and *after.* Relative pronouns are *which, that, who,* and *what.*

Splice	The operation of a radar system includes three main sequences, we will study them this semester.

> **Correction**   The operation of a radar system includes three main sequences,
> which we will study this semester.

■ EXERCISES

Correct the following comma splices:

1. Enter the shipper's name in column 3, if it is the same as the manufacturer's name, write the word *same*.

2. Turn the controller power on, the controller is located to the right of the terminal.

3. The second screw is in back of the machine, it is not a removable screw.

4. The difference is that the NC machine relies upon a computer to control its movements, a manual machine depends upon an operator to control its movements.

5. They are all dissimilar in many ways, however, there are also several similarities.

6. The store managers can choose any decor they desire, there is no set interior decor.

7. A partner is personally responsible for all the debts of the partnership, these include the shares of any partner who is unable to pay.

8. All of these processes have one element in common, they all require radar contact between the pilot and the ground controller.

9. The makers of the Micro 2001 give warranties with their microwave ovens, furthermore, the magnetron is covered for seven years.

10. The cost of the Flack 2600 is $110,000 less than that of the Chryst 4230, however, both cash registers are within the price range set by management.

Identify and Eliminate Run-on Sentences

Run-on, or fused, sentences are similar to comma splices — but without the comma. The two independent clauses are run together with no punctuation between them. To eliminate run-on sentences, use one of the four methods explained above:

- Place a period between the two independent clauses.
- Place a semicolon between them.
- Place a comma and a coordinating conjunction between them.

- Make one of the clauses subordinate by using a subordinating conjunction or a relative pronoun.

■ EXERCISES

Correct the following run-on sentences.

1. This material is usually transparent it can withstand approximately 10 pounds of pressure when stretched.

2. You need to insulate the parts of the walls that are above the ground the cost figures are given below.

3. Look at the map above to see which zone you live in read the table in the column for that zone.

4. The supplier will measure for your new windows they will be installed about a week later.

5. Overcooking shrinks the egg proteins that form the gel this process causes custard to separate into curds and whey.

6. The upper rollers have a corrugated surface this rough surface feeds the rough stock into the surfacer.

7. Framing members are laid down in their approximate locations they are later moved to their permanent positions.

8. Inside partitions are put in place whenever possible the partitions act as braces.

9. Use the ruler as a guide draw chalk lines on the backing fabric to use as a guideline for stitching.

10. The Freon comes from the compressor into the condenser the condenser is the black coil on the back of the refrigerator.

Identify and Eliminate Sentence Fragments

Sentence fragments are incomplete thoughts that the writer has mistakenly punctuated as complete sentences. Subordinate clauses, prepositional phrases, and verbal phrases often appear as fragments. As the following examples show, fragments must be connected to the preceding or following sentence to gain meaning.

1. Connect subordinate clauses to independent clauses.

> **Fragment**   The company continues to lose money. *Although production has increased.*

This fragment is a subordinate clause beginning with the subordinating conjunction *although*. Other subordinating conjunctions include *where, when, while, because, since, as, until, unless, if,* and *after*.

> **Correction**   The company continues to lose money although production has increased.

> **Fragment**   The problem originates in the transformer. Which does not provide enough electric energy.

This fragment is a subordinate clause beginning with the relative pronoun *which*. Other relative pronouns are *who, that,* and *what*.

> **Correction**   The problem originates in the transformer, which does not provide enough electric energy.

2. Connect prepositional phrases to independent clauses.

> **Fragment**   The manager found the problem. *At the conveyor belt.*

This fragment is a prepositional phrase. The fragment can be converted to a subordinate clause, as in the first correction below, or made into an *appositive* — a word or phrase that means the same thing as what precedes it.

> **Correction 1**   The manager found the problem, which was the conveyor belt.

> **Correction 2**   The manager found the problem — the conveyor belt.

3. Connect verbal phrases to independent clauses.

> **Fragment**   Kinetics is a branch of dynamics. *Consisting of all aspects of motion.*

Verbal phrases often begin with *-ing* words. Such phrases must be linked to independent clauses.

> **Correction**   Kinetics is a branch of dynamics consisting of all aspects of motion.

> **Fragment**   The writer studied the statistics. *To ensure his accuracy.*

Infinitive phrases begin with *to* plus a verb. They must be linked to independent clauses.

> **Correction**   The writer studied the statistics to ensure his accuracy.

■ EXERCISES

Correct the following sentence fragments. You may need to add words to complete the thought.

1. The information needed to fill out this card is obtained from the person who performed the test. Usually a chef or manager.

2. If you have a reliable heating system.

3. We will need to buy a number of devices. To protect against physical, chemical, or atmospheric damage.

4. Some of the contacts a teacher makes with students will develop into strong lifelong friendships. Because of the understanding gained through working and learning together.

5. In any transportation mode there are two common elements: a shipper and a carrier. The shipper is the one who sends the goods. A carrier, the one who hauls whatever the shipper wants hauled.

6. Fixed costs are those costs over which the manager has no control. Such things as rent, insurance, and taxes.

7. The cycle starts with the intake stroke. The piston at the closed end, or top of the cylinder.

8. Scrap or waste can add up to a very big inventory for some companies. Depending on what they produce.

9. There are three main parts to a production run. Planning, production, and distribution.

10. MRP is a new method. Which has been found to be very effective.

Place Modifiers in the Correct Position

Sentences become confusing when modifiers do not point directly to the words they modify. Misplaced modifiers often produce absurd sentences; worse yet, they occasionally result in sentences that make sense but cause the reader to misinterpret your meaning. Modifiers must be placed in a position that clarifies their relationship to the rest of the sentence.

Misplaced modifer   Adjust the height of the mechanism head so the point of the penetrating instrument comes close to the surface of the sample *with the coarse adjustment screw.*

In this sentence, the coarse adjustment screw appears to be part of the sample.

> **Correction**   To adjust the height of the mechanism head, turn the coarse adjustment screw so the point of the penetrating instrument comes close to the surface of the sample.

> **Misplaced modifer**   Thermal pollution is worse at nuclear facilities, *which must be reduced.*

In this sentence, the modifier says that nuclear facilities must be reduced. This is not what the writer meant.

> **Correction**   Thermal pollution, which must be reduced, is worse at nuclear facilities.

◼ EXERCISES

Correct the misplaced modifiers in the following sentences.

1. The video cassette recorder is a machine connected to a television that records the sound and picture of a movie or television show.

2. Muslin is used for draping, a plain weave fabric of unfinished cotton.

3. Other simple stains can be substituted for crystal violet and safranin that behave similarly.

4. Bast fibers are fibers that grow in the stems or stalks of plants, which are often several feet in length.

5. The researcher places the mold into an autoclave, now called a charged mold because of the beads it contains.

Use Words Ending in *-ing* Properly

A word ending in *-ing* is either a present participle or a gerund. Both types, which are often introductory material in a sentence, express some kind of action. They are correct when the subject of the sentence can perform the action that the *-ing* word expresses.

> **Unclear**   *Being* a department store, our median percentage *figures* are lower than those for speciality stores or chains.

Figures cannot be a department store.

> **Clear**   Because Johnson's is a department store, our median percentage figures are lower than those for specialty stores or chains.

■ EXERCISES

In the following sentences in which an *-ing* word is used incorrectly; revise them.

1. After comparing the size and weight of the Inferno and Quickcook microwave ovens, both microwave ovens met the criteria.

2. By examining the drawing, the connection is tightened if you pull the torque limiter onto the bushing.

3. Before jumping into programming, key functions should be defined.

4. When writing a computer program, a wide number of languages are available from which to choose.

5. By measuring the weight or counting the number of items in a box, the data you collected would be quantative.

Make the Subject and Verb Agree

The subject and the verb of a sentence must both be singular or both be plural. Almost all problems with agreement are caused by the writer's failure to identify the subject correctly.

1. When the subject and verb are separated by a prepositional phrase, the inexact writer makes the verb agree with the object of the preposition.

   Faulty   The *flutes* twisting on the drill *makes* it resemble a candy cane.

   The subject *flutes* is plural; drill is the object of the preposition *on*. The verb *makes* must be made plural to agree with the subject.

   Correction   The flutes twisting on the drill make it resemble a candy cane.

   Faulty   Practically all *fields* of power engineering *involves* some use of compressed air.

   The subject *fields* is plural; *engineering* is the object of a preposition *of*.

   Correction   Practically all fields of power engineering involve some use of compressed air.

   Faulty   The *committee are* investigating the sales report.

2. When a *collective noun* refers to a group as a unit, the verb must be singular. Other collective nouns include *management, union, team, audience,* and *jury.*

Correction   The committee is investigating the sales report.

Faulty   *Each* of the steelworkers *are* highly skilled.

3. Indefinite pronouns, such as *each, everyone, either, neither, anyone,* and *everybody,* take a singular verb.

Correction   Each of the steelworkers is highly skilled.

Faulty   *Neither* management nor the *laborers* wants a strike.

4. When compound subjects are connected by *or* or *nor,* the verb must agree with the nearer noun.

Correction   Neither management nor the laborers want a strike.

■ EXERCISES

Correct the subject/verb errors in the following sentences.

1. The lower rollers carries the board slightly above the surface of the table.

2. The contents of any good program has three ingredients: documentation, a main program, and one or more subprograms.

3. Everyone know the schedule for today's meetings.

4. The types of writing Ann Beckham briefly discussed was reports, manuals, and letters.

5. Pursuing a career in research and development with major corporations have rewards too numerous to mention.

6. The qualifications for graduation in Industrial Arts Education is first to meet the entrance requirements through the A.C.T. test.

7. Screen printing is a method in which ink or a material such as paint are forced by the action of a flexible blade through a stencil mounted on a finely woven screen.

8. The cables that connect the camera to a monitor carries the video signal.

9. The attachments consists of five interchangeable whips or hooks.

10. The rotating shaft with inner threaded tracks make it move smooth and fast.

Use Pronouns Correctly

A pronoun must refer directly to its *antecedent,* the noun it stands for. Pronouns commonly used in technical writing include *it, they, who, which, this,* and *that*. Overuse of the indefinite *it* (as in "It is obvious that") leads to confusion.

As in subject-verb agreement, a pronoun and its antecedent must both be singular or both be plural. Collective nouns generally are referred to by the singular pronoun *it* rather than the plural *they*. Problems result when pronouns such as *they, this,* and *it* are used carelessly, forcing the reader to figure out their antecedents:

Problems with Number   In the following sentence, the pronoun *they* is wrong because it does not agree in number with its antecedent *machinery*.

Vague	*Machinery* for the sheeting process is steam heated. *They* are 20 feet long, 10 feet wide, and 8 feet high.
Clear	Machines for the sheeting process are steam heated. They are 20 feet long, 10 feet wide, and 8 feet high.

The antecedent *machines* is now plural. Notice that if you change *they* to *it* you will create another problem because *it* could refer either to machinery or process.

Problems with Antecedent

If a sentence has several nouns, the antecedent may not be clear.

Vague	The base and dust *collector* is the first and largest part of the lead *pointer*. It is usually round, several inches in diameter.

In this case *it* could stand for either *pointer* or *collector*. The two sentences can be combined to eliminate the pronoun:

Clear	The base and dust collector, which is the largest part of the lead pointer, is usually round, several inches in diameter.
Vague	The Atomic Energy *Commission* determines the criteria for selecting a nuclear *site*. *It* includes a specified distance from high-population zones.

It could refer to *site* or *Commission*.

Clear	The Atomic Energy Commission, which determines the criteria for selecting a nuclear site, specifies that the site be a certain distance from high-population zones.

Problems with *This*

Many inexact writers start sentences with *this* followed immediately by a verb ("this is," "this causes"), even though the antecedent of *this* is unclear. Often the writer intends to refer to a whole concept or even to a verb, but *this* should be used as a pronoun or an adjective. As a pronoun, it must refer to a noun. The writer can usually fix the problem by inserting a noun after *this* — turning it into an adjective — or by combining the two sentences into one.

> **Vague**   The *body is composed* of the lower and upper fixed members. *This* constitutes the basic physical make-up of the mechanism.

If *this* refers to *body*, the logic is odd because the sentence says the body is the physical make-up. *This* probably refers either to the whole first sentence or to *is composed*.

> **Clear**   The upper and lower members, the two main parts of the body, constitute the basic physical make-up of the mechanism.

> **Vague**   A *large number* of companies in the Twin Cities make the chemical *readily available*, even on *short notice*. Along with *this*, a local company is also using copper board so that possible purchase there is an option.

This could refer to *short notice*, a *large number*, *readily available*, or the entire first sentence.

> **Clear**   The chemicals for copper board are available from a local company and from a number of companies in the Twin Cities.

▮ EXERCISES

Revise the following sentences, making the pronoun references clear.

1. Open the switch to break the contact and to remove the path for electrons to flow. This turns off the device.

2. For example, one must be loyal, act in good faith, and in the best interest of their employer.

3. There is a polyethylene cap placed on the butter cup. It is a soft, pliable plastic.

4. These courses have prerequisites. They may be taken at the industrial college.

5. With the temperature of these materials, stated earlier, it is not recommended for toaster oven use.

6. The upper rollers have a smooth liner and a jagged surface. It feeds the stock into the surfacer.

7. The mechanism head should be adjusted so that the point of the penetrating instrument comes close to the surface of the sample. It may now be treated as outlined.

8. The previous paragraphs explain the different types of tests that are used for evaluation of retort pouches and cans. It shows that the retort can protect food for long periods of time.

9. The screw and the lever are on the back of the machine. It is not removable.

10. A hotel may not have many groups of vacationers staying in the facility but may have constant bookings from business people. This balances out the segment penetration and gives the facility a direction for which markets to pursue. This gives an idea of which segment shows the most opportunity.

PUNCTUATION

Writers must know the generally accepted standards for punctuating sentences.

Apostrophe

Use the apostrophe to indicate possession, contractions, and some plurals.

Possession   The following are basic rules for showing possession:

1. To show possession, add an 's to singular nouns:

   a corporation's profits   a business's assets   a woman's career

2. To show possession, add 's to plural nouns that do not end in -s:

   the women's caucus   the sheep's brains

3. Add only an apostrophe to plural nouns ending in -s:

   eleven corporations' profits   three businesses' assets

4. For proper names that end in -s use the same rules: For singular nouns add 's. For plural nouns, add only an apostrophe:

Ted Jones's job   the Joneses' security holdings

This point is quite controversial. For a good discussion, see the *Chicago Manual of Style*, 13th ed. (Chicago: University of Chicago Press, 1982) 160–162.

5. Do not add an apostrophe to personal pronouns in the possessive case:

theirs   ours   its

Contractions   Use the apostrophe to indicate that two or more words have been condensed into one:

I'll = I will   should've = should have   it's = it is   they're = they are.

As a general rule, do not use contractions in formal reports and business letters.

Plurals   Use apostrophes to form the plurals of letters:

A's   B's   c's

Do not use apostrophes to form the plurals of abbreviations and numbers:

BOMs   1980s

However, use the apostrophe to form the possessive of abbreviations:

IBM's decision

Brackets

Brackets are marks used to indicate that the writer has changed or added words or letters inside a quoted passage.

According to the technician, "The terminating resistors [RJ11] were not in the line."

Colon

Use colons in the following ways:

1. To separate an independent clause from a list of supporting statements or examples:

   A piston has the following strokes: intake, compression, power, and exhaust.

2. To separate two independent clauses when the second clause explains or amplifies the first:

   Just-in-time manufacturing is better than conventional methods for one major reason: it saves money.

   Do not capitalize the first word of the second independent clause.

Comma

Use commas in the following ways:

1. To separate two main clauses connected by a coordinating conjunction (*and, but, or, nor, for, yet, so*):

   Two Main Clauses   A hydraulic system replaces the electric winches and drive motor, and a new support drive and frame design reduce stress caused by load.

   Omit the comma if the clauses are very short.

2. To separate introductory subordinate clauses or phrases from the main clause:

   Clause   If the operator wants to move the telescoping tube out of the ways, she can easily do so.

   Phrases   Sensing a reduction in pressure, the regulator sends more gas into the pipeline.

3. To separate words, phrases, or clauses in a series:

   Words   The recorder contains a noise-reduction unit, a frequency-equalizer unit, and a level-control unit.

Phrases	The skill of communicating is needed in industry, in education, and in government.
Clauses	Select equipment that has durability, that requires little maintenance, and that the company can afford.

4. To set off nonrestrictive appositives, phrases, and clauses:

Appositive	Clem Stacy, the supervisor, fixed the generator.
Phrase	The engine, beginning its seventh year of service, finally needed maintenance.
Clause	The system, which is somewhat complicated, requires calibration at various intervals.

Dashes and parentheses also serve this function. Parentheses may be used frequently, but dashes should be used sparingly.

5. To separate coordinate but not cumulative adjectives:

Coordinate Adjectives	He rejected the distorted, useless recordings.

The adjectives are *coordinate* because they modify the noun independently. They could be reversed with no change in meaning: *useless, distorted recordings.*

Cumulative Adjectives	An acceptable frequency-response curve was achieved.

The adjectives are not separated by commas because the first adjective modifies the second adjective as well as the noun. *Cumulative* adjectives cannot be reversed without distorting the meaning: *frequency-response acceptable curve.*

6. To set off conjunctive adverbs and transitional phrases:

Conjunctive Adverbs	The new regulations, however, caused a morale problem.
	The crane was very expensive; however, it paid for itself in 18 months.
	Therefore, the branch plant should not be built until 1992.
Transitional Phrases	On the other hand, the maintenance crew operated efficiently.
	Performance on Mondays and Fridays, for example, is far below average.

Dashes

You can use dashes before and after material that interrupts or is an aside. Dashes give a less formal, more dramatic tone to the material they set off than commas or parentheses do.

The dash has four common uses:

1. To set off material that interrupts a sentence with a different idea:

   The fourth step — the most crucial one from management's point of view — is to ring up the folio and collect the money.

2. To emphasize a word or phrase at the end of a sentence:

   The project had one goal — cost reduction.

3. To set off a definition:

   Justifications — reports that choose between two alternatives — are written often.

4. To introduce a series less formally than with a colon:

   The figures are updated three times per week — on Tuesday, Thursday, and Friday.

Parentheses

You can also use parentheses before and after material that interrupts or is some kind of aside in a sentence or paragraph. Compared to dashes, parentheses have one of two effects: they deemphasize the material they set off, or they give a more formal, less dramatic tone to special asides.

Parentheses are used in three ways:

1. To add information about an item:

Dates	John Kennedy (1917–1963) was elected president in 1960.
Acronym for a Lengthy Title	Mr. Ford popularized the Whip Inflation Now (WIN) button.
Definition	Children less than twelve years of age should receive routine oral examinations, prophylaxis (teeth cleaning), and topical fluoride treatment once every six months.

Precise Technical   The air conditioner (Central 30,000 BTU, 2 1/2 ton) was not the
Data   model specified.

2. To add an aside to a sentence:

> Major Medical pays 100% (except as otherwise noted) of the following: hospital
> services, physician services, routine physical examinations performed and billed
> by a physician (excludes informational checkups requested by a third party).
> (Wisconsin 17–18)

3. To add an aside to a paragraph:

> Figure 4, Parts A and B, shows that management of the window and thermostat
> adjustment almost eliminate the negative thermal effect of increasing window
> area shown in Figure 3. (Thermostat adjustment reduces annual energy costs by
> about $5, regardless of window area.) For a managed south-facing window,
> energy costs remain about level even if the window has single glazing. For the
> north-facing window, double-glazing is needed to achieve the same effect.
> (Collins 9)

A Note on Parentheses, Dashes, and Commas

All three of these marks of punctuation may be used to separate interrupt-
ing material from the rest of the sentence. Each mark gives a different tone
to the material set off. Dashes tend to emphasize by giving a sense of
drama or of informality. Commas tend to be neutral. Parentheses tend to
deemphasize or to give a more formal, less dramatic tone to special asides
in a sentence.

In some instances, dashes and parentheses can avoid making an appositive seem like the second item in a series:

Commas are   The diode consists of an input, the anode, and an output, the
confusing: cathode.

Parentheses are   The diode consists of an input (the anode) and an output (the
clearer: cathode).

Commas are   The three fashion theories, trickle down, trickle up, and trickle
confusing: across, are basic to understanding fashion cycles.

Dashes or   The three fashion theories (trickle down, trickle up, and trickle
parentheses are   across) are basic to understanding fashion cycles.
clearer:

The three fashion theories — trickle down, trickle up, and trickle
across — are basic to understanding fashion cycles.

Ellipsis Points

Ellipsis points are three periods used to indicate that words have been deleted from a quoted passage.

> According to Jones (1989), "All lines . . . must have a terminating resistor"
> (p. 71).

Hyphens

Use hyphens to connect the following:

1. The parts of a compound adjective when the adjective comes before the noun:

   high-frequency system   plunger-type device   trouble-free process

   Do not hyphenate the same adjectives when they are placed after the noun:

   The system is trouble free.

2. Words that could cause confusion by being misread:

   energy-producing cell   eight-hour shifts   foreign-car buyers
   cement-like texture

3. Compound modifiers formed from a numerical quantity and a unit of measurement:

   a 3-inch beam   a 3-inch-wide beam
   an 8-mile journey   an 8-mile-long journey

   Unless the unit is expressed as a plural:

   a beam 3 inches wide   a beam 3 inches in width
   a journey of 8 miles   a journey 8 miles long

   Also use a hyphen with a number plus *-odd*:

   twenty-odd

4. A single capital letter and a noun or participle:

   A-frame   I-beam

5. Compound nouns used as units of measurement:

   kilogram-meter   kg-m

6. Adjective plus past participle (*-ed, -en*):

   red-colored table

7. Compound numbers from 21 through 99 when they are spelled out:

   Thirty-five safety violations were reported.

   Fractions when they are spelled out:

   three-fourths

8. Complex fractions if the fraction cannot be typed in small numbers:

   1-3/16 miles

   Do not hyphenate if the fraction can be typed in small numbers:

   1½ hp

9. Words in a prepositional phrase used as an adjective:

   state-of-the-art printer

10. Use suspended hyphens for a series of adjectives that you would ordinarily hyphenate:

    10-, 20-, and 30-foot beams

11. Compounds made from *half-, all-, cross-*.

    half-finished   all-encompassing   cross-country

12. Do not hyphenate:

    -ly adverb-adjective combinations

    highly complex system

    -ly adverb plus participle (*-ing, -ed*)

    highly rewarding positions   poorly motivated managers

chemical terms

hydrogen peroxide

13. Spell as one word compounds formed by the following prefixes:

anti-	co-	infra-
non-	over-	post-
pre-	pro-	pseudo-
re-	semi-	sub-
super-	supra-	ultra-
un-	under-	

Exceptions: use a hyphen

when second element is capitalized (pre-Victorian)
when second element is a number (pre-1900)
to prevent possible misreadings (re-cover)

Quotation Marks

Quotation marks are used at the beginning and at the end of a passage that contains the exact words of someone else.

According to Jones (1989), "All lines that do not connect to another unit must end in a terminating resistor" (p. 71).

If no internal documentation is present, place final quotation marks after a comma or period.

Semicolon

Use semicolons in the following ways:

1. To separate independent clauses not connected by coordinating conjunctions (*and, but, or, nor, for, yet, so*):

The storm stopped the surveyors; in fact, it stopped all work at the site.

2. To separate two independent clauses when the second one begins with a conjunctive adverb (*therefore, however, also, besides, consequently, nevertheless, furthermore*):

The machine performs better than all the others; therefore, I suggest that we purchase it.

3. To separate items in a series if the items have internal punctuation:

Plants have been proposed for Kansas City, Missouri; Seattle, Washington; and Orlando, Florida.

Underlining (Italics)

Underlining is a line typed under certain words. In books, words that you underline when typing appear in *italics*. Some laser printers can produce italics, too. Italics are used for three purposes:

1. To indicate titles of books and newspapers:

the Minneapolis *Tribune*   *The Snow Leopard*

2. To indicate words used as words or letters as letters:

You used *there are* too many times in this paper.
The capital *W* is not printing very well.

Note: You may also use quotation marks to indicate words as words:

You used "there are" too many times in this paper.

3. To emphasize a word:

Make sure that all the blanks on the contract have been filled in *before* you sign.

ABBREVIATIONS, CAPITALIZATION, NUMBERS, SYMBOLS

Abbreviations

Use abbreviations only for long words or combinations of words that must be used more than once in a report. For example, if words like *Fahrenheit, cubic inches, pounds per square inch,* or *British thermal units* must be used several times in a report, abbreviate them to save space. Below are several rules for abbreviating.

1. If an abbreviation might confuse your reader, spell it out in parentheses the first time you use it:

This paper will discuss materials planning requirements (MPR).

2. Use all capital letters (no periods, no space between letters or symbols) for acronyms:

   NASA   FTC   AT&T   FORTRAN   COBOL

3. Capitalize just the first letter, and follow with a period, abbreviations for titles and companies:

   Pres.   Co.

4. Form the plural of an acronym by adding just *s*:

   BOMs   VCRs   CRTs

5. Omit the period after abbreviations of units of measurement, except *in.* for *inch*.

6. Use periods with Latin abbreviations:

   e.g. (for example)   i.e. (that is)   etc. (and so forth)

7. Use abbreviations (and symbols) when necessary to save space on visuals, but define difficult ones in the legend, a footnote, or the text.

8. Do not capitalize abbreviations of measurements:

   10 lb   m   g   cm

9. Do not abbreviate units of measurement preceded by approximations:

   15 psi but several pounds per square inch.

10. Do not abbreviate short words such as *acre* or *ton*. In tables, abbreviate units of length, area, volume, and capacity.

Capitalization

The conventional rules of capitalization apply to technical writing. The trend in industry is away from overcapitalization.

1. Capitalize a title that immediately precedes a name:

   Senior Project Manager Jones

But not if it is generic:

The senior project manager reviewed the report.

2. Capitalize proper nouns and adjectives.

3. Capitalize trade names, but not the product:

Apple computers   Cleanall window cleaner

4. Capitalize titles of courses and departments and titles of majors that refer to a specific course:

The first statistics course I took was Statistics I.
I majored in Plant Engineering and have applied for several plant engineer positions.

5. Do not capitalize after a colon.

The boat has three parts: sail, hull, and rudder.
His explanation is simple: he claims he was gone.

6. Research shows that words set in all capital letters are hard to read. Use capital letters sparingly for first-level heads. For emphasis, underline or boldface.

Numbers

The following rules cover most situations, but when in doubt whether to use a numeral or a word, remember that the trend in report writing is toward using numerals.

1. Spell out numbers below 10; use numerals for 10 and above:

four cycles   10 machines   two-thirds of the union   1835 members

2. Spell out numbers that begin sentences:

Ninety percent of the parts were defective.

3. If a **series** contains numbers both above and below 10, use numerals for all of them:

The inspector found 2 defective parts and 12 safety violations.

4. Use numerals for numbers that accompany units of measurement:

1 gram	0.452 minute
33 ⅓%	$6.95
8 yards	5 weeks

5. In compound-number adjectives, spell out the first one or the shorter one to avoid confusion:

   75 twelve-volt batteries

6. Use numerals to record specific measurements:

   Readings of 7.0, 7.1, and 7.3 were taken.

7. Combine numerals and words for extremely large round numbers:

   2 million miles

8. For decimal fractions of less than one, place a zero before the decimal point:

   0.613

   When precision demands, place a zero after the decimal point:

   225.3 and 316.0

9. Express plurals of figures by adding just -s:

   16s   1970s

10. Place the last two letters of the ordinal after fractions used as nouns:

    ¹⁄₅₀th of a second

    But not after fractions that modify nouns:

    ¹⁄₅₀ horsepower

11. Spell out ordinals below 10:

    third mile   ninth man

12. For 10 and above, use the number and the last two letters of the ordinal:

10th day   21st year

Symbols

Like abbreviations, symbols save space but may cause confusion. Except for the symbols for *dollars* and *percent*, do not use symbols within the paragraphs of a report. To save space in tables and drawings, however, use simple symbols, such as the ones for *feet, inches,* and *degrees.* Always use symbols with numbers (7%). Do not combine symbols with spelled-out numbers (seven %). Use highly technical symbols in a report's appendix if you are certain that readers will understand them. Many fields publish books of standard symbols for their area. Find the accepted references in your field. Several sources include:

General Conference on Weights and Measures. *The International System of Units* (SI). Washington: National Bureau of Standards (Special Publication 330), 1974.

American Society for Testing and Materials. *Standard for Metric Practice,* ANSI/ASTM E380–76. Philadelphia: ASTM, 1976.

IEEE Standard for Letter Symbols for Units of Measurement (IEEE Standard 260). New York: Institute of Electrical and Electronic Engineers, 1978.

Other sources include:

Qasim, S. H. *SI Units in Engineering and Technology.* Elmsford, NY: Pergammon, 1977.

U.S. Dept. of Commerce and R. A. Hepkins. *NBS Metric Guide and Style Manual.* Tarzana, CA: American Metric, 1977.

The Chicago Manual of Style. 13th ed. Chicago: U. of Chicago, 1982.

U. S. Government Printing Office. *A Manual of Style.* New York: Gramercy, 1986.

Also helpful are the style manuals of the Council of Biology Editors, American Chemical Society, and the American Institute for Physics.

■ WORKS CITED

Collins, Belinda L., et al. *A New Look at Windows.* NBSIR 77-1388. Washington: U.S. Dept. of Commerce, National Bureau of Standards, 1978.

Wisconsin Group Insurance Board. *State of Wisconsin Group Health Insurance Benefits.* Madison: WIB, 1980.

Appendix B: Documenting Sources

HOW INTERNAL DOCUMENTATION WORKS

A COMPARISON OF METHODS OF INTERNAL CITATION

THE MLA METHOD

THE APA METHOD

NUMBERED REFERENCES

SAMPLE PAPERS

D ocumenting your sources means following a citation system to indicate whose ideas or exact words you are using. Three systems are commonly employed: the Modern Language Association (MLA) method, the American Psychological Association (APA) method, and the numbered references method. All three will be explained briefly here. For more complete details you should consult the *MLA Handbook* (3rd ed.), or *The Publication Manual of the American Psychological Association* (3rd ed.).

HOW INTERNAL DOCUMENTATION WORKS

Each method has two parts: the internal citations and the bibliography, also called Works Cited (MLA) or References (APA). The internal citation works in roughly the same manner in all three methods. The theory is this: the writer places certain important pieces of information in the text to tell the reader which entry in the bibliography is the source of the quotation or passage paraphrased. These pieces of information could be the author's last name, the date of publication, the title of an article, or the number of the item in the bibliography (1, 2, and so on).

In the MLA method, the basic reference unit is the author's last name, and sometimes a short version of the title of the work. In the APA method, the basic reference unit is the author's last name and the year of publication. In the numbered method, the reference unit is the number of the item in the bibliography.

In each method, the number of the page on which the quotation or paraphrased passage appears goes in parentheses immediately following the cited material.

	MLA	*APA*	*Numbered References*
In text	Last name (can also include first)	Last name only	number of work as it appears in References
	page numbers (without "p.")	page numbers (with "p.")	page number (with "p.")
At end	Works Cited	References	References

Each method has its strengths. The MLA method is the most flexible. The APA method indicates clearly how recent the material is. The numbered references method is unobtrusive. Since the methods vary, the rest of this chapter explains each.

A COMPARISON OF METHODS OF INTERNAL CITATION

Here is an example of how each method would internally cite the same quotation. The following passage appeared in "The Writer's Tool Chest," by Jeremy Joan Hewes, on page 49 of the July 1985 issue of *Macworld* magazine:

> Another excellent use of the computer's power as a writing tool is to check the spelling in a written work. Although word processing programs help you create and format text documents, they can't detect spelling errors.

MLA Method

The MLA method of citing this passage requires that you include the author's last name (at least) and the page number. Note that you can do this in various ways:

Jeremy Joan Hewes has noted that "another excellent use of the computer's power as a writing tool is to check the spelling in a written work" (49).

We should remember that "another excellent use of the computer's power as a writing tool is to check the spelling in a written work" (Hewes 49).

To find all the publication information for this quotation your reader would refer to "Hewes" in the Works Cited list.

APA Method

The APA method of citing this passage requires that you use just the author's last name and include the year of publication and a page number.

According to Hewes (1985), "Another excellent use of the computer's power as a writing tool is to check the spelling in a written work" (p. 40).

To find all the bibliographic information on the quotation, the reader would refer to "Hewes" in the References section.

Numbered References Method

The numbered method allows you to use a last name or not. Whether you do or not, you place in the text the number of the item in the bibliography. So if "Hewes" were the sixth item in the bibliography, you could use the number 6 to cite the source in the text. You also include a page number:

One authority feels that "Another excellent use of the computer's power as a writing tool is to check the spelling in a written work"(6,49).

THE MLA METHOD

The following section describes variations in MLA citation and explains entries in the MLA Works Cited section.

Citation Variations

Once you understand the basic theory of the method — to use names and page numbers to refer to the Works Cited — you need to be aware of the possible variations of placing the name in the text. Note that each time you refer to a quotation or paraphrase, you give the page number only; do not use "p." or "pg," and do not place a comma before the page number.

1. Author's name is part of introduction to quotation:

   Richard White, in his book for entrepreneurs, stresses flexibility in the workplace when he offers the axiom "You will realize as much from your people as you allow them to produce" (108).

2. Author is not named in introduction to quotation:

   CATS, a troubleshooting program, was designed to "improve the repair record of GE hardware in railroad companies' repair shops" (Dietz 138).

3. Author has several works in the Works Cited:

   Schendel points out that "conventional EPS molding equipment must be modified in order to mold the polyethylene foam" ("Testing Methods" 16).

4. If the work does not have an author, treat the title as the author. The title is listed first in the Work Cited list.

   Although Arcel is considered a polymer, "it may be more accurately described as an interpenetration method of PS [polystyrene] within PE [polyethylene]" ("EPS Molders" 23).

   "EPS Molders" are the first words in the Works Cited entry.

5. Paraphrases are usually handled like quotations. Give the author's last name and the appropriate page numbers.

   Swita's discussion of Techmate shows that it has superior physical properties and chemical composition (16–19).

6. Block quotations employ one minor change: place the period *before* the page parentheses. Do not place quotation marks before and after a block quotation. Indent the left margin ten spaces and double space. Do not indent the right margin.

   According to Schendel and Swita

   > in the past, cushioning materials had to be fabricated to conform to the desired inserts in the package. The process was time-consuming and required expensive fabricator techniques. New cushioning materials, however, can be molded to the desired form, thus eliminating the need for fabrication. (6)

MLA Works Cited List

The Works Cited list contains the complete bibliographic information on each source you use. The list is arranged alphabetically by the last name of the author or, if no author is named, by the first important word of the title.

Follow these guidelines:

- Present information for all entries in this order: Author's name. Title. Publication information.

> Parker, Roger C. *The Aldus Guide to Basic Design.* Seattle: Aldus, 1987.
>
> Highland, Henry A., and L. D. CLine. "Insect Resistance of Reclosable Cartons." *Packaging Technology* 16.6 (1986): 16–17.

- Double space an entry if it has more than two lines.
- Indent the second and succeeding lines 5 spaces.
- Use the author's name or initials as given in the work: I. A. Mertes; Frederick Jones.
- If an author appears in the Works Cited two or more times, type three hyphens and a period for the second and succeeding entries. Alphabetize the entries by the first word of the title.

> Muller, Arthur. "Medical Devices . . .
>
> --- *Technological Advances. . .*

Examples of common types of entry appear below. Note that on a typewriter you would underline the words that are italicized here. For more detailed instructions, use the *MLA Handbook* mentioned on p. 441.

Book with One Author

> Naisbitt, John. *Megatrends.* New York: Warner, 1982.
>
> White, Richard. *The Entrepeneur's Manual.* Radnor, PA: Chilton, 1977.

- Only the name of the publishing company needs to be used. You may drop "Co." or "Inc."

Book with Two Authors

> Cunningham, Donald H., and Gerald Cohen. *Creating Technical Manuals: A Step-by-Step Approach to Writing User-Friendly Manuals.* New York: McGraw, 1984.

- Capitalize the first word after a colon.
- A long title may be shortened, in this case to *Creating Technical Manuals* . . . Use ellipsis dots to indicate that part of the title is deleted.

Book with Editor

> Cattle, Dorothy J., and Karl H. Schwerin, eds. *Food Energy in Tropical Ecosystems.* New York: Gordon and Breach, 1985.

Article in an Anthology

Blesser, B., and J. M. Kates. "Digital Processing in Audio Signals." *Applications of Digital Signal Processing.* Ed. Alan V. Oppenheim. Englewood Cliffs: Prentice, 1978. 29–116.

- Put the inclusive pages of the article last. Do not use "Pp."

Corporate or Institutional Author

Packaging Machinery Manufacturers Institute. *Handbook for Writing Operation and Maintenance Manuals.* Washington: PMMI, 1973.

- In the text cite this entry the first time as "(Packaging Machinery Manufacturers Institute [PMMI])"; thereafter use "(PMMI)."
- This entry could also be arranged with the title first: *Handbook for Writing Operation and Maintenance Manuals.* Washington: Packaging Machinery Manfacturers Institute, 1973.
- Cite this version as (*Handbook*).

Work Without Date or Without Publisher

Jenkins, Cynthia. *Managing the Corporate Publications System.* Minneapolis: Corporate Publication Alliance, n.d.

- Use n.p. for no publisher or no place.
- If neither publisher nor place is given, write "N.p.: n.p., 1986."
- If the person's title is important, add it after the name: Schmidt, Howard, Manager of Technical Services.

Brochure or Pamphlet

ARCO Chemical Company. *Expanded Polystyrene Package Design.* Philadelphia: Atlantic Richfield, 1984. ACC-P37–841.

- If the pamphlet has an identification number, place it at the end of the entry, after the date.

Foam Packaging Systems. By Sealed Air Corporation. Darien, CT: Sealed Air, 1984.

Later Edition of a Book

American Psychological Association. *Publication Manual.* 3rd ed. Washington: APA, 1983.

Encyclopedia

"Sonar." *Encyclopedia Britannica.* 1984 ed.

Hodgson, G. F. "Die Castings." *Encyclopedia/Handbook of Materials, Parts, and Finishes.* Ed. Henry R. Clauser. Westport, CT: Technomic, 1976.

■ No page numbers appear because entries in the book are arranged alphabetically.

Article in Journal That Does Not Have Continuous Pagination

Piotrowski, John. "Eliminate Machinery Vibrations by Correcting Shaft Alignment." *Mechanical Engineering* 108.2 (1986): 80–86.

Article in Monthly or Weekly Magazine

McComb, Gordon. "A Clipboard Collage." *Macworld* Jan. 1986: 68–74.

■ If the article is printed discontinuosly, over many pages — 68–70, 72, 87–89, and 90 — give the first page only, followed by a plus sign: 68 + .

Article in Newspaper

Ward, Mark. "Power From Trash." *Milwaukee Journal* 22 Feb. 1986, states ed., sec. 2: 3.

■ Identify the edition, section, and page number. A reader should be able to find the article on the page.

■ Omit the article (*the*). If the city is not given in the title, supply it in brackets after the title.

Personal Interview

Mertes, Ione. Personal interview. 14 Feb.1986.

Telephone Interview

Mertes, Ione. Telephone interview. 14 Feb.1986.

Personal Letter

Mertes, Ione. Letter to the author. 14 Feb.1986.

THE APA METHOD

The following section describes variations in APA citations and explains entries in the APA References section.

Citation Variations

Once you understand the basic theory of the method — to use names and page numbers to refer to the References section — you need to be aware of the possible variations of placing the name in the text. Note that each time you cite a quotation or paraphrase you give the page number preceded by "p." or "pp." Do not use "pg".

1. Name as part of introduction to quotation:

   White (1979) stresses flexibility in the workplace when he offers the axiom "You will realize as much from your people as you allow them to produce" (p. 108).

2. Author is not named in introduction to quotation:

   CATS, a troubleshooting program, was designed to "improve the repair record of GE hardware in railroad companies' repair shops" (Dietz, 1986, p.32 138).

3. Author has several works listed in the References. The works in the References are differentiated by their dates, so no special treatment is necessary; if an author has two works dated the same year, differentiate them in the text and in the References with a lower case letter after each date (1985a, 1985b).

   Schendel (1985) points out that "conventional EPS molding equipment must be modified in order to mold the polyethylene foam" (p.16).

4. If no author is given for the work, treat the title as the author because the title is listed first in References:

   Although Arcel is considered a polymer, "it may be more accurately described as an interpenetration method of PS [polystyrene] within PE [polyethylene]" ("EPS Molders," 1984, p. 23).

5. Paraphrases are handled like quotations. Give the author's last name, the date, and the appropriate page numbers:

   Swita's discussion (1985) of Techmate shows that it has superior physical properties and chemical composition (pp.16–19).

6. Block quotations employ one minor citation change. The period is placed *before* the page parentheses. Do not place quotation marks before and after a block quotation. Indent the left margin five spaces and double space. Do not indent the right margin.

   According to Schendel and Swita (1986),
   > in the past, cushioning materials had to be fabricated to conform to the desired inserts in the package. The process was time-consuming and required expensive fabricator techniques. New cushioning materials, however, can be molded to the desired form, thus eliminating the need for fabrication. (6)

APA References

The reference section (entitled "References") contains the complete bibliographic information for each source you use. The list is arranged alphabetically by the last name of the author or the first important word of the title.

Follow these guidelines:

- Present information for all entries in this order:
  Author's name. Title. Publication information.

  Parker, Roger C. (1987). *The Aldus guide to basic design.* Seattle, WA: Aldus.

  Highland, Henry A., & Cline, L. D. (1986). Insect resistance of reclosable cartons. *Packaging Technology* 16 (6),16–17.

- Double space the entire list. The second line of an entry starts *three* spaces in from the left hand margin.
- Use only the initials of the author's first and middle names, with a period and a space between them.
- Capitalize only the first word and proper nouns and adjectives in the title.
- Place the date in parentheses immediately after the name.
- Place the entries in alphabetical order.
- If there are two or more works by one author, arrange them chronologically, most recent first.

  Jones, E. H. (1986).
  Jones, E. H. (1984).

Several common entries are exemplified below. Note that with a typewriter you would underline the words that are italicized here. For ease of comparison, they are the same sources as in the MLA list.

Book with One Author

Naisbitt, J. (1982). *Megatrends.* New York: Warner.
White, R. (1977). *The entepreneur's manual.* Radnor, PA: Chilton.

Book with Two Authors

Cunningham, D. H., & Cohen, G. (1984). *Creating technical manuals: A step-by-step approach to writing user-friendly manuals.* New York: McGraw.

Book with Editor

Cattle, D. J., & Schwerin, K. H. (Eds.). (1985). *Food energy in tropical ecosystems.* New York: Gordon and Breach.

Article in an Anthology

> Blesser, B., & Kates, J. M. (1978). Digital processing in audio signals. In A. V. Oppenheim (Ed.), *Applications in digital sound processing* (pp. 29–116). Englewood Cliffs, NJ: Prentice.

- Capitalize only the first word of the essay title.
- Use "pp." with inclusive page numbers.
- Use zip code abbreviations for states.

Corporate or Institutional Author

> Packaging Machinery Manufacturers Institute. (1973). *Handbook for writing operation and maintenance manuals.* Washington, DC: Author.

- When the author is also the publisher, say *Author* for the publisher.
- Do not use periods in DC.
- In text the first citation reads this way: (Packaging Machinery Manufacturers Institute [PMMI], 1973). Subsequent citations are (PMMI, 1973).
- This entry could also read *Handbook for writing operation and maintenance manuals.* (1973). Washington, DC: Packaging Machinery Manufacturers Institute. Cite it as *Handbook.*

Work Without Date or Without Publisher

> Jenkins, C. (n.d.). *Managing the corporate publication system.* Minneapolis: Corporate Publishing Alliance.

- Use "n.p." for no publisher or no place.

Brochure or Pamphlet

> ARCO Chemical Company. (1984). *Expanded polystyrene package design.* (Rep. No. ACC-P37–841). Philadelphia: Author.

- Treat the brochure like a book.
- Place any identification number after the title.

> *Foam packaging system.* (1984). Darien, CT: Sealed Air Corporation.

- Refer to this entry as (*Foam packaging*).

Later Edition of a Book

> American Psychological Association. (1983). *Publication manual* (3rd ed.). Washington, DC: Author.

Encyclopedia

> Sonar. (1984). *Encyclopedia Britannica.*

■ In the text refer to this and all works with no title this way: ("Sonar," 1984)

Hodgson, G. F. Die Castings. (1976). In H. R. Clauser (Ed.), *Encyclopedia/handbook of materials, parts, and finishes.* (pp. 121–124). Westport, CT: Technomic.

Article in Journal That Has Continuous Pagination

Wagner, J. R., & Anon, M. C. (1985). Effect of freezing rate on the denaturation of myofibrillar proteins. Journal of Food Technology, 20, 735–744.

■ In the article title, capitalize only the first word, proper nouns, proper adjectives, and the first word after a colon.

■ Underline the volume number.

Article in Journal That Does Not Have Continuous Pagination

Piotrowski, J. (1986). Eliminate machinery vibration by correcting shaft alignment. Mechanical Engineering, 108 (2): 80–86.

■ Put the issue number in parentheses after the volume.

■ You could also give the month, if that helps identify the work: (1986, February).

Article in Monthly or Weekly Magazine

McComb, G. (1986, January). A clipboard collage. *Macworld,* pp. 68–74.

Newspaper Article

Ward, M. (1986, February 22). Power from trash. *The Milwaukee Journal,* states ed., sec. 2, p. 3.

Personal Interview

(Note: The information presented for personal and telephone interviews and letters does not appear in the APA Manual, which suggests that such information not appear in the References. However, long usage of the system indicates that these entries are critical in research reports. Hence, they are included here.)

Mertes, I. (1986, February 14). [Personal interview].

■ Arrange the date so the year is first.

■ If the person's title is pertinent, place it in the brackets:

Mertes, I. (1986, February 14). [Personal interview. Manager of Technical Services, Wheeler Amalgamated, Denver, CO.]

Personal Letter

Mertes, I. (1986, February 14). [Personal Letter. Manager of Technical Services, Wheeler Amalgamated, Denver, CO.]

Telephone Interview

Mertes, I. (1986, February 14). [Telephone Interview]

NUMBERED REFERENCES

The numbered method uses an Arabic numeral, rather than a name or date, as the internal citation. The numeral refers to an entry in the bibiliography. The bibliography may use one of two organizations

- arrange the entries alphabetically
- arrange the entries by order of appearance in the text, without regard to alphabetization.

This method is commonly used in short technical reports that have only 2 or 3 references. Many technical periodicals have adopted it because it is cheaper to print one number rather than many names and dates. The difficulty with this system is that if a new source is inserted into the list, all the items in the list and references in the text need to be renumbered.

The following sample shows the same paragraph and bibliography arranged in the two different ways. Notice that the bibliography form is the APA method and that the author's name may or may not appear in the text.

EXAMPLE 1 (Entries arranged alphabetically)

SURFACE REPRESENTATIONS

A surface representation is one type of knowledge base. It is a collection of facts, rules of thumb, and educated guesses an expert has acquired through experience (*1*, p. 78). One expert, Thompson, says it is "the collection of facts used to capture all of the information in a problem area" (*3*, p. 315). The knowledge is an "empirical association" based on past experience (*2*, p. 304).

1. Edosomwan, J. (1987, August). Ten design rules for knowledge-based expert systems. *Industrial Engineering*, pp. 78–80.

2. Michaelson, R. (1985, April 4). The technology of expert systems. *Byte*, pp. 303–312.

3. Thompson, W. & B. (1985, April). Inside an expert system: From index cards to Pascal program. *Byte*, pp. 315–330.

EXAMPLE 2 (Entries arranged by position of first reference in the text)

SURFACE REPRESENTATIONS

A surface representation is one type of knowledge base. It is a collection of facts, rules of thumb, and educated guesses an expert has acquired through experience

(*1*, p. 78). One expert, Thompson, says it is "the collection of facts used to capture all of the information in a problem area" (*2*, p. 315). This knowledge is an "empirical association" based on past experience (*3*, p. 304).

1. Edosomwan, J. (1987, August). Ten design rules for knowledge-based expert systems. *Industrial Engineering*, pp. 78–80.

2. Thompson, W. & B. (1985, April). Inside an expert system: From index cards to Pascal program. *Byte*, pp. 315–330.

3. Michaelson, R. (1985, April 4). The technology of expert systems. *Byte*, pp. 303–312.

■ MODELS

The remainder of this appendix presents two sample papers. The first one is presented in numbered form, the second in APA form. Both are excerpted from much longer papers.

PHOTODEGRADABLE PLASTICS

This section examines the nature of the photodegradable plastics that are also being developed. The defintion of photodegraditon, the methods of creating photodegradability, the properties of photodegradable plastics, and the potential applications for photodegradable plastics will all be discussed here.

Definition.   Photodegradation occurs when ultraviolet light breaks down the polymer chains that plastics are comprised of. After a period of exposure, the plastic polymer becomes a fine dust that biodegrades. The ultraviolet wavelength is the only light radiation absorbed by plastic that initiates this reaction (3, p. 73). All plastics have a natural vulnerability to this degradation by ultraviolet light (4).

Methods.   This natural photodegradation in plastics can be controlled. The photodegradation rate is influenced by several factors. One significant factor in accelerating photodegradation is the use of a copolymer in the plastic's ingredient composition. A copolymer of ethylene and carbon monoxide (E/CO) is manufactured as a photodegradable PE by Dow Chemical, DuPont, and Union Carbide (2, p. 53). Another copolymer composition, called Ecolyte, adds ketone groups onto the polymer chains of either PE, PS, or PP to induce photodegradability (pp. 53–54). A second major factor is an additive to the plastic's ingredient composition. Several additive-created photodegradable plastics have also been developed, but their specific compositions are not revealed by the companies that produce them. In general, the additives increase ultraviolet light absorption and aid in the polymer chain breakdown (p. 54).

A third factor affecting the photodegradation rate is the plastic's color. A dark color, says Vic Mimeault of Ampacet Corporation, absorbs more heat and increases the rate of photodegradation (2, p. 54). Another factor is the climatic conditions that the plastic ends up in. The plastic must receive enough ultraviolet radiation to photodegrade, but the amount differs with the geographical climates. According to Ampacet's technical director, Allen Carlson, it takes normal plastic sixty times longer and photodegradable plastic 12 to 24 times longer to break down in Alaska than in Florida, which has more daylight during the year (5, p. 29).

Properties.   Photodegradable plastics have a much shorter degradation time than normal plastics, but most of their other properties are similar. For example, Ampacet's additive-induced photodegradable PE degrades 4 to 7 times faster than normal PE (5, p. 29). Canada's EcoPlastics, producer of Ecolyte, claims that their PS cup takes only sixty days to turn to dust (3, p. 73). E/CO begins to break down exponentially after only 6 hours of exposure and it disintegrates

FIGURE B.1
Sample Paper with Numbered References.

2

within one week (*2,* p. 53). The inherent properties of the base plastic can be retained, says Ampacet's Carlson, depending on the specific method used to increase the photodegradability. Carlson also notes that the photodegradation method usually does not diminish the base plastic's original processability (*1*).

Applications. Photodegradable plastics can be used in several typical applications for plastics. Application to plastic films has the greatest potential and is presently used the most. Several companies presently use photodegradable plastics for trash bags and six-pack beverage carriers (*2,* pp. 53–54). Application to the thicker material gauges of rigid plastic containers has a more limited potential, but Ampacet's Carlson says that testing is still being conducted (*1*).

REFERENCES

1. Carlson, A. (1988, February 19). [Personal Letter. Technical Director, Ampacet Corporation, Mt. Vernon, NY.]

2. Leaversuch, R. (1987, August). Industry weighs need to make polymer degradable. *Modern Plastics,* pp. 52–55.

3. Plastic dust. (1987, May 16). *The Economist,* p. 73.

4. Plastic packaging: a biodegradable breakthrough. (1987, March). Industrial Management. [Reprinted in *ECOSTAR Update,* (n.d). Mississauga, Ontario: St. Lawrence Starch Company Limited.]

5. Smock, D. (1987, June). Are degradable plastics the answer to litter? *Plastics World,* pp. 28–33.

FIGURE B.1 (continued)

ARCEL

Arcel, which is sold exclusively by ARCO Chemical Company under a licensed agreement with Japan's Sekisui Plastics, was introduced in the early part of 1984. The material is a resilient moldable copolymer. Because Arcel is extremely durable, yet resilient, it has found itself a market in protective packaging. This section will explain Arcel's composition, processing, properties, advantages, and disadvantages.

Composition

Arcel is a moldable copolymer consisting of polyethylene, polystyrene, and a blowing agent. The material consists of equal proportions of polyethylene and polystyrene. R. Chatman, product manager of ARCO, describes the blowing agent, which constitutes 12% to14% of Arcel, as a "form of pentane" (personal communication, October 27, 1985). Although Arcel is considered a copolymer, "it may be more accurately described as an interpenetrating network of PS [polystyrene] within PE [polyethylene]" ("EPS Molders," 1984, p. 23). Chatman notes that one major drawback of this material is that it must be refrigerated at 30 to 40 degrees Fahrenheit to retain enough blowing agent for preexpansion (personal communication, October 27, 1985).

Processing

The processing of Arcel is very similar to EPS with a few minor changes after the conventional steam preexpansion. The first change is in the transfer of the expanded beads; the second is in molding ("EPS Molders," 1984). Transfer of the expanded Arcel beads requires special attention. Conventional EPS equipment "airveyed" expanded beads to the storage receptacles, but because that method compacts the large and softer Arcel beads, it is unacceptable. The easiest way to avoid this problem is to place the preexpander so the beads are discharged directly into the storage receptacle (p. 24). The second change in Arcel's processing occurs during the molding cycle. Arcel needs greater steam pressure in the molding cavity over a shorter time duration. Also, the cooling cycle is said to be a bit shorter (p. 24).

Properties

Arcel, which is molded in normal densities of 2.0 pcf, commands a premium price of $2.25 per pound. Although Arcel has a high price, it does have some very desirable properties. M. Grunnet, Manager of Technical Services at Tuscarora Plastic, points out that Arcel's static compression set is only 8% at 25% compression and has a compressive strength of 17 psi at 10% deflection (personal communication, November, 1985). Arcel is extremely durable with a 43 psi tensile strength, the highest of any buoyancy and solvent resistance (ARCO, 1984, p. 6).

FIGURE B.2
Sample Paper with APA References.

2

Advantages

Arcel's major advantage is its durability. It is an excellent material choice when reusability is important. In addition, it is highly break and puncture resistant. This foam recovers well from multiple impacts that occur in the distribution cycle, allowing it to be reused many times. Arcel's strength and reusability help offset its high cost. It is, for example, reusable as automotive dunnage trays ("Moldable," 1984, p. 87).

Disadvantages

The biggest disadvantage of Arcel is its high cost, which makes it uneconomical in certain situations. Another disadvantage is its short shelf life; it must be refrigerated until it is molded.

REFERENCES

ARCO. (1984). *Arcel, Moldable Polyethylene Copolymers* (ACC-P52–849). Philadelphia: Author.

Chatman, R. (Product Manager, ARCO Chemical Company). (1985, October 27). [Telephone interview].

EPS molders: Now there's EPE. (1984, August). *Plastics Technology, 30*, 23–24.

Grunnet, M. Manager of Technical Services, Tuscarora Plastics (1985, November). Personal interview.

Moldable polyethylene cushions electronic items. (1984, September). *Packaging, 29*, 87–90.

FIGURE B.2 (continued)

■ EXERCISES

1. Edit the following sentences to correct the APA citations.
   a. In the process Golomb notes (1985) "They start with a slurry of virgin pulp fiber, and mold containers into almost any shape desired." (p. 12)
   b. Kass estimated by (1985 p. 71) that the coated aluminum trays would cost 3–4 cents more than the uncoated trays.
   c. For example (Crosby) said in 1984 (p. 23) that if dull tools are used, the machine may tend to labor harder, sound different, and form chips differently.

2. Edit the following sentences to correct the MLA citations.
   a. Parts with common processing characteristics are grouped to reduce processing and increase manufacturing productivity (Lindberg, 1983. p. 66)
   b. "However, canned products that are heat processed for longer periods showed a higher amount of TMA than pouched products," said S. S. Chia in 1983 on page 1523.
   c. Mr. B. R. Holmgren said on page 32 that "something like 90 percent of today's blisters use PVC" (p. 32).

3. Pick a paragraph from each of the two models accompanying this chapter; rewrite it, using another citation style (change the numbered to MLA or APA to the numbered references method). Also rewrite the appropriate Works cited/References entry.

4. Write a memo report in which you analyze and evaluate an index or abstracting service. Use one that is in your field of interest, or ask your instructor to assign one. Explain which periodicals and subjects the service lists. Explain how easy it is to use. For instance, does it have a cross reference system? Can a reader find key terms easily? At what level of knowledge are the abstracts aimed? Beginner? Expert? Your audience should be other class members. Your objective is to help them with their research projects.

5. Write a memo in which you analyze and evaluate a reference book in your field of interest. Explain its arrangement, sections, and intended audience. Is it aimed at a lay or technical audience? Is it introductory or advanced? Can you use it easily? Your audience should be other class members. Your intention is to help them with their research projects.

6. Examine *Ulrich's International Periodicals Directory* or *the Standard Periodical Directory* in the library. Each lists the periodicals published for all fields. Select three periodicals in your area of interest and inspect several copies of each. Then write a memo report. Analyze and evaluate the periodicals. Discuss article length, intended audience, article style. Which one is most helpful for a student research project? Why?

7. Form small groups of three or four based on your major or your interest

in a particular topic (for instance: solar heating systems, aspartame, thermoset plastics, diesel engines, industrial organization, just-in-time manufacturing). Discuss the topic. Your goal is to formulate questions about the topic. Examples might be "What kinds of process are used to make that?" "How will the new technology affect the workplace?" "What are the opinions of experts on this topic?" "Do the experts disagree? If so, how?" Once you begin to explore, you will generate many questions. Select two questions from the list. Use the library resources to find articles that contain the answers to those questions.

Write an informal report in which you explain the answers you found, based on the articles you read.

 WRITING ASSIGNMENT

Write a short research report explaining a recent innovation in your area of interest. Use at least six recent sources. Use quotations, paraphrases, and one of the three citation formats explained in this chapter. Organize your material into sections that give the reader a good sense of the dimensions of the topic. A paper like this can have many topics. Some kinds of information that you might present are

Problems and potential solutions regarding the development of the innovation

Issues debated in the topic area

Effects of the innovation on your field or industry in general

Methods of implementing the innovation

WORKS CITED

Girabaldi, Joseph and Walter S. Achtert. *MLA Handbook for Writers of Research Papers*. 3rd ed. New York: MLA. 1988.

Hewes, Jeremy Joan. "The Writer's Tool Chest." *Macworld* July 1985: 46–50.

Publication Manual of the American Psychological Association. 3rd ed. Washington, D.C.: APA, 1983.

Index